Under the Spell of Mathematics

Rik Verhulst

Under the Spell of Mathematics

The cultural phenomenon of mathematics in civilisation, art and nature

Springer

Rik Verhulst
Mathematics, Teacher Education
Karel De Grote University of applied sciences
Temse, Belgium

Translated by
Dirk Van Hemeldonck
Antwerpen, Belgium

ISBN 978-3-031-63870-1 ISBN 978-3-031-63871-8 (eBook)
https://doi.org/10.1007/978-3-031-63871-8

Mathematics Subject Classification (2020): 40-XX, 60-XX, 92-XX, 01-XX, 03-XX, 41-XX, 15-XX, 20-XX, 26-XX

Translation from the Dutch language edition: "In de ban van wiskunde" by Rik Verhulst,

Cover illustration: The two compartments of the seats count two successive numbers of the sequence of Fibonacci (21 and 34) and are therefore in a ratio 34/21 = 1,619... that approaches the golden ratio 1,618....

This Springer imprint is published by the registered company Springer Nature Switzerland AG
The registered company address is: Gewerbestrasse 11, 6330 Cham, Switzerland

Preface

Why this book?

Popularizing science is undoubtedly a trendy thing to do nowadays. Newspapers, magazines and audio-visual media are bold enough to report any recent developments in advanced research. But generally they deal more with the domain of applied science. Only rarely is mathematics, the queen of all sciences, involved. There are many reasons for this. To put it mildly and gently, we can say that a large part of the intellectual public finds no pleasure in this demon of their schooldays. Proudly and forthrightly, without fear of losing their prestige, VIPs in the world of entertainment are in the habit of telling us that they know nothing about mathematics. Furthermore, people suffer from a widespread allergy to symbols and formulas. "Explain it to me in great detail, but, please, no equations!" Most of the books on mathematics intended for popular consumption restrict themselves to specific subjects, usually of the narrative or visual kind. These include the history of mathematics, either chronologically, or by presenting various famous individuals. Or they give abundantly illustrated explanations of items such as fractals and deterministic chaos, alongside computer applications and numerical puzzles. Up till now, few authors have taken the risk of devising a broader perspective on the intrinsic nature of mathematics and presenting it to the general public. This book attempts to fill this gap. At the same time it will relate mathematics to culture in its broadest sense, as explained in the introduction below. After all, throughout the history of civilisations, mathematics has remained a beacon of culture.

Just as it's not possible to show a beautiful mountain landscape without summits and peaks, neither is it possible to reveal the true face of mathematics without looking at its actual concrete contents. But the small effort we demand from motivated climbers will be generously rewarded by the scenery that awaits them. For this reason, we have made each path accessible for everyday tourists. We wish them a very enjoyable trip, a trip which they don't need to complete all in one go, and welcome them to this small celebration of the 'Queen'.

For whom is this book intended and what can be found in it?

This book is especially intended for those who like to portray themselves publicly as mere mathematical laymen and chuckling to themselves, frequently do so with some relief. This is a target group that we assume to be much bigger than that of professional mathematicians. Nevertheless, we hope that this book will also excite the interest of those who have a more mathematical background and who are concerned with promoting their subject.

The intention is three-fold.

In the first place, we want to remove people's prejudices towards the subject, which are often the result of frustrating experiences at school and lack of understanding. By nature, the ordinary, healthy mind always seems to think rationally. Even irrationalists use logical arguments to defend their point of view and feel disappointed if they are rejected.

Next, we show that those who reject mathematics deprive themselves of a rich part of general culture. Who wouldn't like to share the ideas of all those creative minds that were indebted to mathematics and made constant improvements to our world? Or is it that the only things you know about Archimedes are his bath experiences, about Descartes his observations of a fly, and about Newton his inspiration by a falling apple?

Finally, we use some examples to illustrate the specificity of the mathematical method. At the same time, we want to demonstrate how the joy of ultimate understanding results in creative activity deriving from analogy, generalization and modification. In the construction and development of mathematical language we cannot deny some elegance and poetic freedom of expression. Was it not Paul Valéry who said: "Since I hadn't sufficient imagination to be a mathematician, I decided to become a poet"?

So, what we perhaps might achieve is that you no longer identify mathematics with the boring calculations and mindless routines you had to learn sitting at your school desk. In fact, mathematics teaching at school is generally no more than a mere run-up to applications of differential and integral calculus, those favourite instruments of the engineer and other users of applied mathematics.

It would be too pious a wish to think we could revive the interest in reading and studying mathematics of all those people who have never before noticed these other aspects of mathematics. And it would perhaps really be a utopian dream to persuade them that independent research into an intriguing problem can lead to results. But, nevertheless, who was it who said: "Mathematics is a mistress who is never tiresome and who will never deceive us"?

Looking at the enormous expansion in contemporary science and its specific language, it is a rather ungratifying task to try to popularize these subjects. This is certainly true of an abstract field like mathematics. It is nevertheless beneficial to undertake such an attempt, if only to prevent an unbridgeable gap from arising between these specialised fields of research and the conscious awareness of those who use their applications. Real mathematicians reading this book should not be irritated by the restrictions and simplifications necessitated by popularization.

An intellectual activity like mathematics inevitably has links to some aspects of philosophy, civilization and culture. These are matters where we cannot shy away from fundamental questions and historical facts. Given the limits of this book, however, these subjects can only be treated in brief, with each item merely providing material for reflection and discussion. We nevertheless hope that this succinctness will not diminish its relevance.

Often the notion of culture is interpreted in a narrow sense, restricted to the field of fine arts. Indeed, it is not too difficult to give examples of the role of mathematics in architecture, in plastic arts, in music and even in literature, where the structure, metre and use of word repetition in a text support both its substantive development and its style.

Originally, culture (colere = cultivate, produce, make...) has meant the totality of artefacts that human ingenuity has added to reality. There are many examples to be found in the field of practical convenience as well as in recreation and spirituality. Mathematics has arisen precisely out of the practical problems of livestock farming, land surveying, irrigation, astronomy, navigation, barter, etc. Over the course of time it also proved to be an instrument for the engineer creating new applications. But few people nowadays realize how much daily use they make of maths when, for instance, they use their bank card or listen to a CD.

In a broader sense, however, the notion of culture also means the complete history of the civilization of peoples with their perception of reality in diverse places and times. Thanks to the emancipation of the mind, the vision of life of successive generations has evolved enormously, especially in western civilization. The development of mathematics has also provided a substantial contribution to the waves of history. The transition from nomadic culture to agrarian, then from agrarian to industrial, and finally from industrial to modern information technology culture, has coincided each time with specific changes in the mathematical world. Those developments are linked with the names of scientists and philosophers who were mainly also mathematicians.

This book also gives some attention to these aspects, however briefly.

We try to carry out this project in four chapters, the order of which follows its own logic. However, the chapters can be read independently of each other. Readers who are initially reluctant to exert themselves had better start with chapters three and four. There they can get an idea of the role and meaning of the other chapters. However, for a popularizing text on science that wants to avoid trivialization, there is no other alternative than to show concrete results from time to time. It would hardly be satisfying to read a sentence of the kind: `Just trust that what is stated here is important and significant'.

In the first chapter we show how mathematics is put together and how it grows.

In the second chapter we give some examples of the specific way in which a mathematician acquires understanding in several problem areas and how that understanding becomes fertile for application and creative renewal. In the third chapter we give a brief overview of the relations between the most momentous waves of civilization and the characteristic developments within mathematics that promoted them. In the fourth and final chapter we illustrate the presence of mathematical aspects in natural phenomena and in the various fields of art.

Word of thanks

I would like to thank the following persons for their support and contributions to the realization of this book:

Dr. Habiba Kalantarova at Springer Nature for her unconditional belief in the project and for the encouraging cooperation with her team.

My colleague and friend Dirk Van Hemeldonck for the accurate English translation and the necessary adaptations to the lay-out as well as to the beautiful drawings by Ludo Vereecken, my other colleague and friend.

Glyn Hughes, as a native speaker, for checking the final English text.

Of course, those who contributed to the original version in Dutch still deserve my enormous thanks.

Rik Verhulst

Paul Valéry (1871–1945)

French author and poet. After law studies he published some works of literature, including a study on Leonardo Da Vinci. After an acute existential crisis in 1892, in which he abandoned his idolatry for literature, he continued to dedicate himself solely to what he himself called ' The Life of the Spirit '. Under the influence of André Gide he returned to poetry with poems including *La jeune Parque, Le cimetière marin* and *Charmes* (carmina). Besides scientific and mathematical deliberations his diary *Cahiers* also includes numerous famous aphorisms with a sceptical and cynical undertone.

Contents

Chapter 1

How is mathematics put together?

All human knowledge starts with intuitions, strolls on along concepts and ends in ideas.
Immanuel Kant

In this chapter we run through the different components that give contemporary mathematics its specific shape and appearance.

Firstly, mathematical thinking manifests itself as an innate property of the human intellect. The logic of this intellect is very probably nothing more than a refined form of adaptation to the reality in which it struggles to survive, more by ingenuity than instinct.

Secondly, mathematics makes use of its own artificial language. Moreover, through that language, science communicates with the phenomena of reality in an apparently successful way. Like any other language, mathematics uses concepts and statements about these concepts. How do these abstract concepts make their entrance? From where do we get the inspiration to make 'true' statements about them?

The relevance of mathematical statements is not their 'truth' according to any kind of reality, but it is their consistency in a deductive network, knitted together by the needles of the rules of logical reasoning. Theorems and proofs are thus the essence of mathematics.

How do we keep from making blunders, i.e., preventing contradictions from creeping in? In order to guarantee the consistency of a theory we have to pair it to a model which embodies the theory. In this respect absolute consistency, in the sense that a theory can guarantee its own freedom from contradictions, has proved to be an illusion.

What is ultimately essential in the structure of an axiomatic-deductive theory? How can we navigate safely through the archipelago dotted abundantly with islands that are independent mathematical structures?

Finally: what do we learn by looking down from the meta-theoretical ivory tower onto the fieldwork of our mathematical methods? How do we deal with questions like: 'Does this true theorem nevertheless turn out to be unprovable?' or 'Can we prove that we cannot prove everything?'

R. Verhulst, *Under the Spell of Mathematics*, https://doi.org/10.1007/978-3-031-63871-8_1

1 Mathematics and the human mind.

What is the origin of human intelligence?
In what does the human brain differ from that of animals?
How does it work and what are its features?
Are we alone capable of language and conscious reflection?

These are the questions that have occupied, and continue to occupy, many eminent thinkers in various branches of science. Definitive answers to these questions have still not been found. Not everyone worries about these problems, yet everybody makes daily use of their brain without understanding precisely what happens when so doing.

When we speak, do we first summon up images in our mind? Do we then check some list to find the sound symbols corresponding to the concepts that match these images? Do we then use the rules of an agreed grammar in order to construct meaningful sentences? Finally, do we issue motoric orders to our vocal cords to enunciate these sentences? Do we add nuances by including a touch of mime and emotion that we retrieve from a hidden arsenal? We have no idea how it all takes place. Nevertheless, the fact is that whenever people are together, there is a lot of babble and chatter.

When we organize information and make conclusions that give a new appearance to that piece of information, are we then consciously using the laws of logic, such as syllogism, modus ponens or the law of excluded middle? Obviously not! However, on hearing that a recent childbirth was not a girl, we conclude that it must have been a boy. A reductio ad absurdum proof works in precisely the same way. For the ordinary sound mind this is more like a reflex than the conscious use of a formal scheme.

So fish swim and birds fly without pondering why the former manage better with fins and the latter with wings. Certainly, fins and wings are the result of a

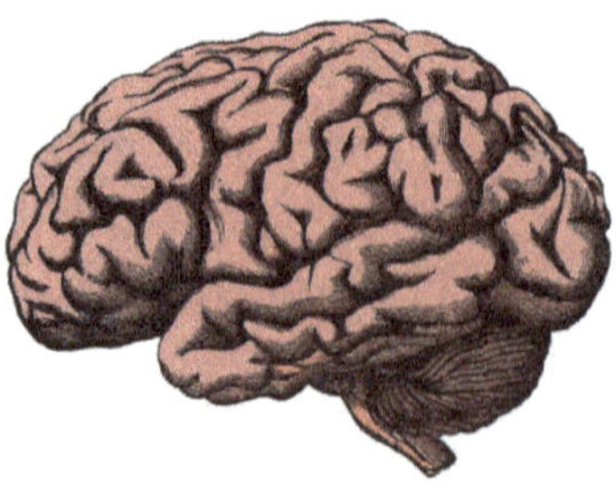

The structure of the **human brain** differs from that of other animals in several important aspects. This is seen, for instance, in the neo-cortex where the centres for advanced cognitive skills are situated. Also the prefrontal cortex is larger in humans than in any other species..

long process of evolution, in which adaptation to the specific environment of water or air was optimized. But we refrain from stating that this process has yet reached a definitive end point. Essentially, fins tell us something about water and wings something about air, and vice versa.

The human brain is also the product of an amazingly long evolutionary process. Certain steps in this process can be traced by examining its volume and structure. The complexity of the brain corresponds to successes in adaptation and survival. Probably the human brain carries within it the specific features of the world in which it lives. Reason refers to the universe just as the universe refers to reason. Descartes still needed the artificial construction of occasionalism to explain this harmony (which was not evident at the time, since the theory of evolution had yet to be discussed). According to this doctrine, on all occasions it was Divine Goodness that brought about the miraculous coincidence of reality and thinking. Even now many scientists and philosophers wonder how the operations of the mind have such a successful hold on the reality of phenomena. Why does the universe seem to conform willingly to the formal pattern knitted by the logical laws of reasoning?

For any given volume, the sphere is the wrapping with the minimal surface, accordingly to three-dimensional Euclidian geometry. We can deduce this theoretically using the technique of extremes in differential calculus without making observations or experiments. But why does a drop of water, falling freely, actually take on a spherical form in order to minimise its surface tension and hence its potential energy?

Many phenomena in nature can be simulated using such optimization principles, for example, aiming for minimum use of energy or maximum output. We all know the phenomenon of a stick that appears to be bent when partially immersed in water. This is one aspect of Snell's law of refraction referring to light changing from one medium to another. Based on the theory it can be precisely predicted how this will happen: the sines of the angles of incidence and refraction, respective to the normal line *n*, relate to each other in the same way as the respective speeds

v_1 and v_2 of light in both media. This is premised on the minimization of displacement time and not of distance.

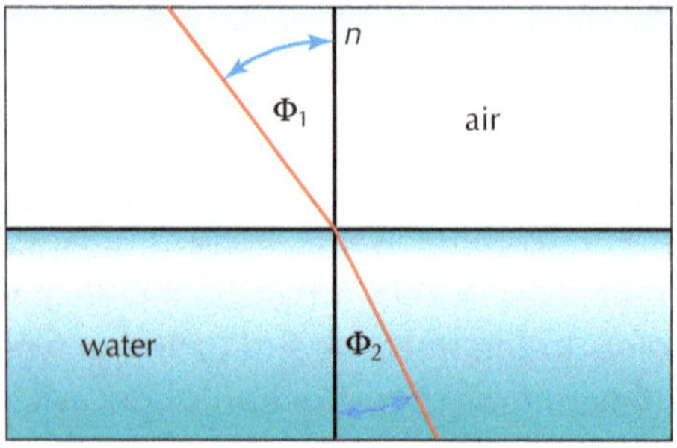

$$\frac{\sin \Phi_1}{\sin \Phi_2} = \frac{v_1}{v_2}$$

But why does light prefer fast displacement over short and, moreover, why is it so shrewd as to use sines? Even Aristotle would not have known.

It is clear that we have brains that are especially suited for describing 'how' things work, but which are less adept at discovering 'why' things are the way they are. These question words typify one of the differences in attitude between the ancient Greeks and the scientists of the seventeenth century. But the rationality used was essentially the same.

As far as western rationality in general and western science in particular are concerned, there has never been a lack of scepticism and sense of relativism. Important evidence of this is easily found among the Greeks, who were themselves the pioneers of rationality, logic, philosophy and mathematics. And this evidence is not exclusively limited to the so-called Sophists.

Is objective conceptive knowledge attainable?

What can we know and what not?

What is the efficiency of this knowledge?

Over the centuries all great minds have adopted changing and refining points of view on this. Realists, nominalists and other Abélards, convinced of the irreconcilability of their premises, never achieved a synthesis in their dispute over the universals (general concepts). We return to this in section 3.6 of this chapter.

It is striking that scepticism and relativism are not practised to the same fanatical degree in everyday life as in the intellectual sport of non-commital philosophizing. We survive because we rely on our senses, even though we know from experience that they do not always function flawlessly. When we notice a tree in our path, when we hear the wind in its leaves, touch its bark and smell its fruit, then we presume that our senses agree and that the tree really is there. We step aside and walk round it, which is very much to our advantage and safety.

In this same way we also cherish the convenience of the occasionally mistrusted sciences. When a sceptical postmodernist turns the tap on during his morning shave and gets no water, he will presume that the water has been shut off or that there is a leak somewhere, rather than screaming: "Didn't I tell you that this law of communicating vessels is just a fairy story!"

If, in order to survive, we have to rely on our senses and the functioning of our organs – breathing, heartbeat, digestion and glandular secretions – which we cannot deliberately control, and if we freely let them go their own way, why can we not show the same trust in and submission to the operation of our brains? It would not be consistent to put the mind in a different category. Both life and conscious life came into being by the same process of increasing complexity and optimization.

Here our intention is not to come down firmly on either side of the problem of materialism or dualism, nor that of realism or nominalism. These dilemmas have nothing to do with the question of trust. Both Plato and Hilbert were part of the real world and so was their thinking. No single philosopher can put himself outside the universe to see through it. The process of his thought structures is inevitably a process in and from the same universe.

We also make no exclusive choice in the dilemma between determinism and non-determinism. Both seem to play their part in the great adventure which, to our own joyfous satisfaction, we can experience in this cosmos.

Neither is it our intention to once again boost our anthropological arrogance at being the champions of evolution. Our restrictions do not have to spoil our enthusiasm. We cannot imagine what a may bug discerns of the world with its antennae. In order to succeed it may even combine this with some alternative logic. But it turns out that antennae and this alternative logic are suitable means for survival.

Our plea in this section can be summarized as: "Do not just feel well in your body, but also in your mind."

Abélard and Héloïse. Pierre Abélard (1079–1142)

French theologian and philosopher. He became involved in the scholastic discussion between realists and nominalists, trying to reconcile their points of view. The dramatic story of his love for his pupil Héloïse is known from their famous correspondence.

2 Mathematics and language

Where does the capacity for language come from?
What is language?
What is the function of language?

These are not easy questions. Even a razor-sharp analyst like Wittgenstein had to make subtle distinctions in his preliminary simple answers and admit their inadequacy.

When somebody tells another person "John is off his head again", then the meaning of this statement does not only depend on the words, conventional concepts and structure, but equally on the social context, the previous shared history and the experiences of 'somebody', the 'other person' and 'John'. And we have not even mentioned the stress, gestures and mood accompanying 'head' and 'again'.

Language is not just a means of unambiguously representing realities in sound or written symbols. Nevertheless, the sender has a certain message in mind and uses the historically cultivated conventions of his social group to package the message as well as possible. By so doing he expects the receiver to perform the inverse operation in as adequate a way as possible. A definitive test for the result of this process does not seem to exist. Yet we succeed in sharing information and teaching, not least language itself.

It is remarkable that the capacity for language is not specified by the structure of the language itself, but in the principle of connecting abstract symbols with meanings. Imagine a Chinese baby girl brought to another country right after birth - say Denmark or Portugal - and that her education starts there. Then she will not learn her original mother tongue but Danish or Portuguese. Her genetic origin does not determine her for acquiring Chinese. She will, however, have the capacity for language.

This makes it possible for people to create an artificial language. Its construction can be geared to specific needs or purposes. Both Esperanto and mathematics are such languages. Moreover, mathematics seems to be a language that enables man to 'communicate' with nature. The successes in various domains of science cannot be denied. Of course, these do not derive from the symbols chosen by the mathematicians, but from the intrinsic qualities of the concepts and from the universality of the methods that appeared in the development of mathematics in all cultures.

Egyptians, Babylonians, Chinese, Hindus, Greeks, Arabs, Aztecs and Incas have very different mother tongues. A simple translation would not make them match.

Yet what is written about numbers and shapes, as tools for agrarian culture as well as for mystical or recreational purposes, is essentially the same in all of them. One plus one is always two, and the sides and angles of a square are always equal.

The result of this universality of mathematics is that all over the world, present-day mathematicians can effortlessly use the same symbols for the same mathematical items. Thus they understand each other perfectly. And however much the growth of mathematical knowledge is also determined by history, most of them (within the same field of specialization) are aware of the results already obtained and are working on the same conjectures, i.e., searching for their proofs.

This unity of thinking and language between different peoples has yet to occur in other aspects of civilization and culture. Wouldn't we be acting in bad faith if we failed to recognise the particular place and role of mathematical language?

Ludwig Wittgenstein (1889–1951)

Austrian philosopher. After studies in engineering he went to England where he came into contact with Bertrand Russell. He became a professor at Cambridge, succeeding G.E. Moore.
In his *Tractatus logico-philosophicus* he analyzed the role of language in philosophy. Famous is its final sentence: "Whereof one cannot speak, thereof one must be silent". (1918)

3 Where do mathematical concepts come from?

This sounds like someone asking for sex education.

Do you know what a 'triangle' is? Just describe it yourself. The number-theoretic concept 'three' will likely appear in your description. But will it also contain the fundamentally difficult geometrical concept 'angle'?

Maybe you used 'points' and 'lines'? So, do you have the line figure in mind or the area it encompasses? Are the lines part of the area or not?

However easy it is to recognise a triangle among a set of figures, it is not so easy to give a precise definition of it. Just as it is not so easy to describe the colour 'green' precisely. Note: this is not meant to be a political joke.

According to Aristotle, concepts can only come about through the senses and by the process of observation, analysis, abstraction and definition. All subsequent empiricists agreed completely with this. And for many concepts this seems to work. But do we see in nature forms that suggest the abstraction of a triangle? Clearly we also have to consider other methods. We will try to analyse the usual methods using some examples.

3.1 Defined concepts

'What is a prime number?' This is a far easier task. 'A prime number is a natural number with precisely two different dividers (one and itself)'. The number 'one' does not correspond to this definition (only one divider) and hence is not a prime number. But 'two', 'three', 'five', 'seven', etc. are.

Notice that this definition was produced by the so-called 'specification' principle. From the set of natural numbers, we pick those with the specific characteristic of having precisely two different dividers. The concept 'prime number' hinges on two other concepts, i.e. 'natural number' (and especially the number 'two') and 'divider', which must precede in the conceptual hierarchy.

What is the meaning of the relationship 'is a divider of' in the set of natural numbers? Also this we can define by yet another preceding concept, i.e. the multiplication of natural numbers. The number a is a divider of the number b if and only if a third number c exists which, multiplied by a, makes product b.

Examples

'Three' is a divider of 'six' because a natural number 'two' exists for which the product of 'two' and 'three' equals 'six'.

'Three' is, however, not a divider of 'seven' because there is no natural number c that multiplied by 'three' would equal 'seven'.

But then again we have to define what 'multiplying' is. It's clear that we can't proceed eternally in developing a backwards running sequence of concepts consecutively depending on each other. If we are ever to understand something, then we have to let this sequence start with concepts we do not define in this manner. So we have to start with so-called 'basic concepts' ('undefined terms').

The procedure for introducing geometrical concepts is similar. So, going backwards and appealing to ever more elementary concepts, we end up at 'points' and 'lines' as basic concepts.

3.2 Basic concepts

How can we give meaning to the basic concepts 'point' and 'line' without using the specification principle to define them? If these basic concepts lack substance, so too do all their derivatives.

In the game of chess, the bishop is one of the elemental pieces. Just try to describe what a bishop is to someone who has never heard about chess. Obviously the description should not be based upon the appearance of the piece. In chess sets of various ethnic origins, such as Asian or African, both material and shape can differ considerably. But what determines all these different looking 'bishops' implicitly (indirectly)? Well, it's the rules of position and movement that apply to all of them as well as their relationship to the other pieces in the game. So 'bishop' is strictly speaking a compendium of these rules and relations.

English chessmen

German chessmen

This we also apply to 'points' and 'lines'. We will determine their mutual relationships.

Examples

'A line contains an infinite amount of points.'

'Two different points belong to exactly one line.'

Note that by this we do not explicitly state what a 'point' or a 'line' looks like, but rather that how you have to deal with them. This does not only open the possibility of starting the game of geometry. It also gives the freedom, should it arouse our interest, to change the appearance of 'points' and 'lines' in, let's say, 'Asian' or 'African' versions, i.e. to model them in different ways. We can link 'point' and 'line' in a very sensible way to representations quite different from the usual dots and stripes. We will come back to this later.

So, besides explicit definitions using the specification principle, we also have at our disposal implicit relational rules to determine mathematical concepts. These are the axioms, of which Poincaré claimed that they are nothing more than veiled definitions.

How will we now describe these relational rules and how can we understand them? And how will we make abstractions?

3.3 Classification of terms

To replace the vague intuitive ground, on which historically all mathematical plants had to grow, by a common solid base, a theory was constructed at the end of the nineteenth century. This theory that became the mother of all mathematical theories is the so-called set theory of Georg Cantor (1845–1918). Its development did not proceed without difficulties and controversies, but finally mathematics firmly overcame this growing pain. This theory is built on two basic concepts, i.e. 'set' and the relation 'is an element of'.

It is worth noting that in the relation rules of 'points' and 'lines' it was precisely this set-theoretical terminology that we used. In this we considered 'lines' as sets and 'points' as elements. This gives us the advantage that all 'rules of the game' that we agreed upon for the set theory also apply to the so-called 'incidence-axioms' of geometry. We will not describe the whole construction of this theory of sets, but restrict ourselves to those aspects that are relevant to the creation of concepts.

Imagine that an infant was given a box of beautiful coloured blocks that are varying in colour, size and (simple) shapes. Maybe the child has no knowledge of the concepts 'colour', 'big', 'small' and 'shape'. How can this fascinating set of blocks help the infant to discover these concepts by itself? Well, instead of building architectural masterpieces, it is more inventive to arrange the blocks in distinct subsets. In this the infant can decide on a common feature of the blocks to be put in the same subset. We give some examples using a very small box of 'building blocks', containing only six of them:

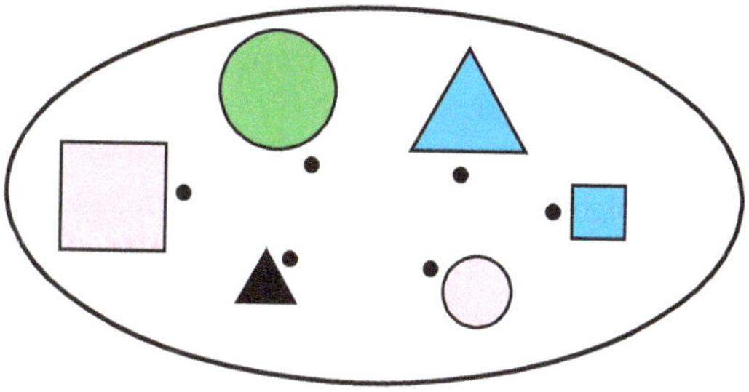

All these blocks are individually distinct. But the human mind enables us, through the registrations of our senses, to notice that certain blocks are related by having 'something' in common although they are not fully identical. And, using our fanciful imaginative powers, this can be done in various ways. Here are three different examples:

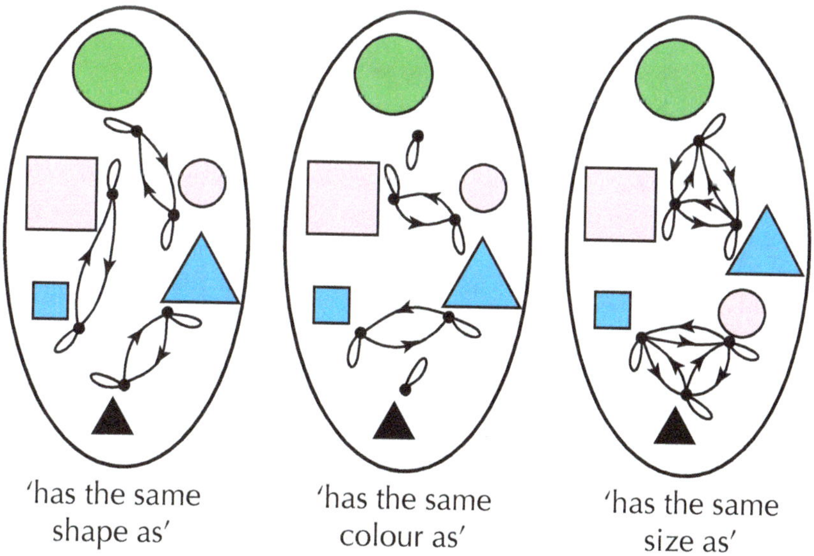

By making relationships between the objects in one of these ways, we declare them to be equivalent with regard to that 'something' feature that they represent. Next we classify them into subsets according to that feature.

For these three examples this results respectively in:

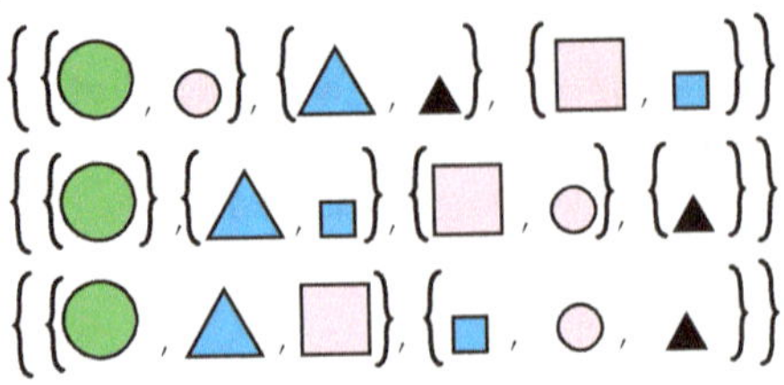

Now, following the nominalists, we add a third step and stick a name tag onto each class, i.e. on each subset we obtained with the above mentioned procedure. Of course, mum may give a helping hand. This results in the diagram below showing the creation of the concepts 'circle', 'triangle' and 'square' following the three-step method: relating, classifying, naming. Of course, we have to take into account the restrictions of the available set of blocks.

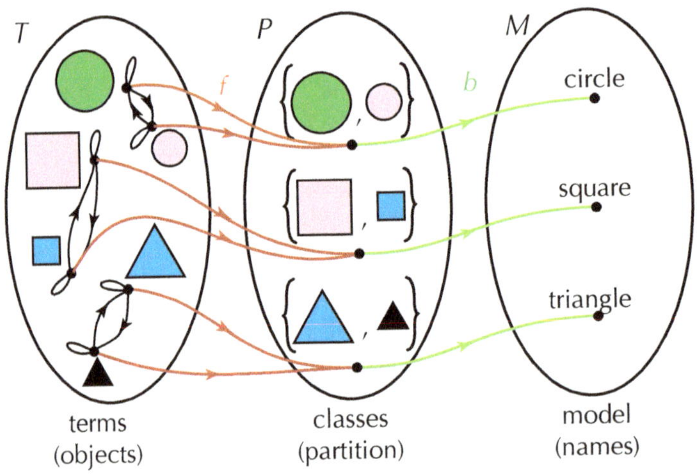

As an exercise (Oh dear!), you can make the corresponding representations for the concepts 'green', 'blue', 'red' and 'black' or for the concepts 'big' and 'small'.

This three-step method is somehow the cradle of most mathematical concepts. The set of blocks 'T' is in this the set of terms. The relationship 'R' between the terms is an equivalence relation in T. The set P of the obtained subsets is a partition of T. The mapping f is the so-called 'canonical mapping' of T onto P. The set M of the associated concepts is a representation (model) of P in names or in symbols. The one-to-one mapping b of P onto M is a bijection.
Note that no two terms are identical and that each term unambiguously determines precisely one class in the partition constructed. Both the expression 'the class of ○ ' as well as the name 'circle' determine the same class in the classification of shapes,

i.e. 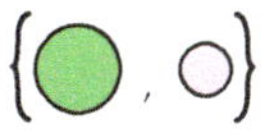 .

Different terms, however, can determine the same class. So in the given example:

'the class of ' and 'the class of ' are equal.

Now we will illustrate the birth of some important mathematical concepts using these instruments.

Example 1 Natural numbers

A mother of two children asks her three-year-old infant to lay the table for the family meal. How can that little one lay the correct number of plates on the table without counting, because it hasn't learned to do that yet? Remarkably enough, the infant copes better with the theoretical solution than with its physical aspects. It can indeed distinguish the relationship 'is as many as' by making a one-to-one match between the members of the family and the plates needed. It thus says for itself: 'a plate for mummy', 'a plate for daddy', 'a plate for sister' and 'a plate for me', simple as that! For the notion 'natural number', only the recognition of the relationship 'is as many as' is needed.

We use such a one-to-one relationship (bijection) in a set of sets. We will here describe these sets with character symbols rather than with the usual 'apples' and 'pears'. Among them we must now recognise the sets with the same amount of elements.

That way the sets with only one element are all classified in the same class and are provided with the label 'one', the pairs in another class with the label 'two', etc.

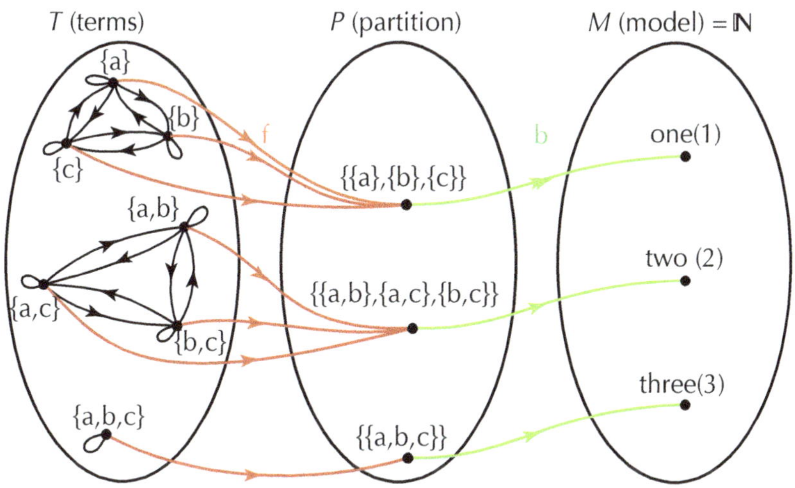

Example 2 Rational numbers

Fractions are taught worldwide by means of the principle of splitting up some magnitude into a number of equal shares (denominator), from which are then taken a particular number of shares (nominator). In this way, using two natural numbers, a fractional magnitude of the original magnitude is constructed. The ultimate motivating didactic aid thereby seems to be its application to cakes. For our convenience, here we will just take the example of a rectangle, which may also represent something delicious.

The rectangle shown below has firstly been divided in six equal parts, from which then two parts have been taken. The coloured rectangle represents the fraction 2/6, relative to the original rectangle.

$\frac{2}{6}$ 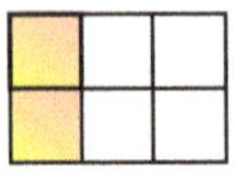 $\frac{1}{3}$ 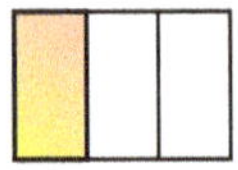

The shaded rectangle is thus the magnitude associated with the pair of natural numbers (6, 2), on which we based the realization of the construction. Because the pair of numbers (3, 1) generates the same partial magnitude, we call the terms (3, 1) and (6, 2) equivalent and the fractions 1/3 and 2/6 equal. We also state that the fractions 1/3 and 2/6 determine one and the same rational number. Note that for the equivalent terms (3, 1) and (6, 2) it applies that:

$$3 \times 2 = 1 \times 6 \text{ (crosswise products } \begin{matrix} 3 & \times & 1 \\ 6 & & 2 \end{matrix}\text{).}$$

This offers the possibility to extend this equivalence relation to all pairs of integer numbers (including negative numbers) and to base this relation on the multiplication of integers rather than on equal partial magnitudes. But to avoid a zero denominator the first component of the pairs must not be zero.

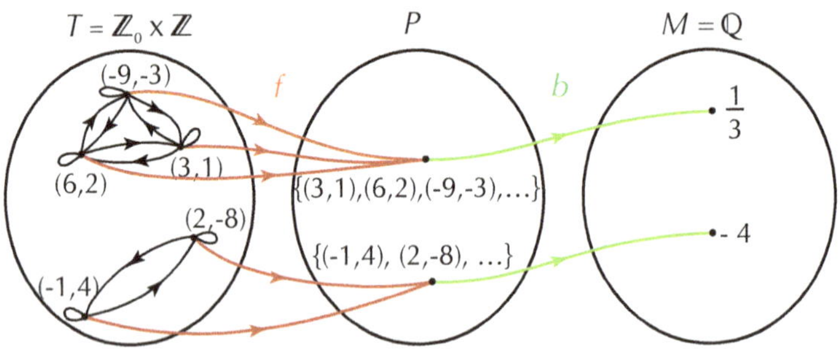

$\mathbb{Z}$ is the set of all integer numbers (positive and negative).

$\mathbb{Z}_0$ is the set of all integer numbers except 0.

$\mathbb{Z}_0 \times \mathbb{Z}$ is the set of all pairs of integer numbers of which the first element is not 0. That is the set of terms T.

P is the set of all classes of equivalent terms (pairs with the same crosswise products).

These classes establish the set of the rational numbers $\mathbb{Q}$ that can be represented in the form of fractions. So fractions constitute a model M for $\mathbb{Q}$.

Example 3 Real numbers

We can make the decimal expansion of the rational number 1/3 by means of continuing divisions and we will get 0.33333... . The three dots show that an infinite number of digits 3 is still to be expected. We call this a 'repetitive decimal form' with period 3.

In this way any rational number can be transformed into a repetitive decimal form. The sequential remainders of the division of the nominator by the denominator indeed all have to be smaller than this denominator, so there is a finite number of different remainders in the sequence. If ever a zero appears then all the following digits in the decimal form will be zero. Usually these are not written.

Example $\frac{7}{20} = 0.35000... = 0.35 = 0.350 = 0.3500$ etc.

However, if none of the remainders in the sequence is 0, then sooner or later a remainder in the sequence has to appear a second time, because of finite possibilities. But starting from there the whole sequence of quotients and remainders will repeat. So the decimal form is repetitive.

Example $\frac{10}{7} = 1.42857\ 42857\ 42857\ ...$.

Reciprocally, we can always transform a repetitive decimal form into a fraction.

Example

If $a = 4.257\ 57\ 57...,$

then $1000a = 4257.\ 57...$ and $10a = 42.57...,$

so $1000a - 10a = 4215$ which results in: $a = \frac{4215}{990} = \frac{281}{66}$

And also $a = 0.999...$ equals $1.000...,$

because $10a - a = 9.99... - 0.99... = 9$

results in $a = \frac{9}{9} = 1 = 1.000...$.

So, a repetitive decimal form with period 9 can always be transformed into a repetitive decimal form with period 0 by adding 1 to the digit before the infinite sequence of nines and then changing all the nines to zeros. Those zeros may however be omitted.

Example 34.723999... = 34.724000... = 34.724

On the other hand, there are also non-repetitive decimal forms.

Example 0.1010010001000001...

The three dots here have the meaning that the principle of construction of the sequence of decimal digits (each time one more 0 digit comes between two 1 digits) is infinitely repeated.

The numbers defined by such a non-repetitive decimal form are called irrational numbers. Together the rational and the irrational numbers make the set of real numbers.

We can link each real number to a point on a line. To do this we have to gauge (or calibrate) the line, which means marking two different points on it as 0 and 1.

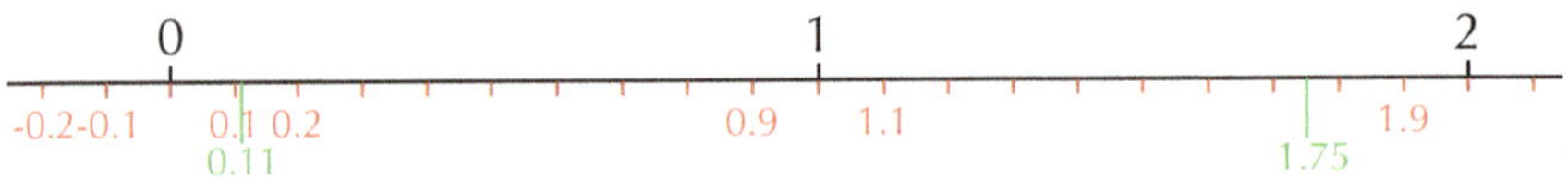

Some decimal forms, however differing in notation, can determine the same point of the calibrated line and are hence representations of the same real number. This can be the case for decimal forms with period 0 or 9, or with a finite number of decimals. So, for instance, the forms 1.799..., 1.800... and 1.8 determine the same point. Two non-repetitive decimal forms of different notation will however always determine two different points of this calibrated line and hence are representations of two different real numbers.

We can therefore define the concept 'real number' using decimal forms as terms that can be classified by the equivalence relation 'determine the same point on a calibrated line', which will arrange them in classes that correspond precisely with the real numbers. A calibrated line is therefore a perfect model for the set of real numbers ℝ.

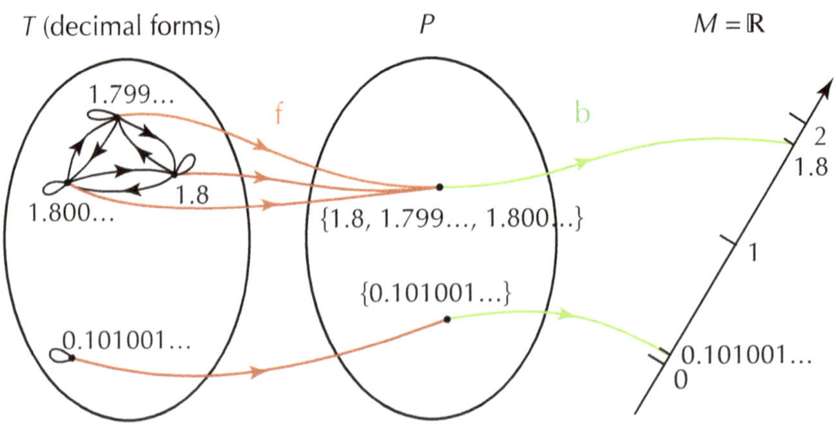

Example 4 The Real numbers modulo 1

In this example we illustrate a double particularity. First the fact that we can use the concepts, which we obtain by the three-step method as classes of terms, again as new terms in a new context. Then we can repeat this three-step method once more, by defining some other new equivalence relation. In this way we create piles of classes and of classes of classes, each time producing more complex concepts.

Secondly we notice that also geometrical concepts can arise from number-theoretic concepts and vice versa. The 'real numbers modulo 1' apparently seem to correspond to 'oriented angles'.

By reflecting on its own mental patterns, mathematics offers instruments that open unexpected creative possibilities. We will illustrate this on several occasions further on in this book.

We take the real numbers as terms and declare two of these terms as being equivalent if and only if their difference is an integer number. That means that their decimal parts have to match, i.e. they can only differ in the integer part before the point. All real numbers having the same tail (mantissa) and a different head (index) are therefore categorized in the same class.

Example $a = 7.3495$ and $b = 2.3495$ are equivalent since $a - b = 5$ and 5 is an integer number.

Essentially we are now only interested in 'decapitated real numbers' such as 0.3495. A good model for the set of the thus obtained new classes is the half open interval [0,1[. This interval we can curl up to a circle with perimeter 1. Each real number of the form 0. ... determines a point on this circle, and respective to an origin marked as 0, also an arc, alias an angle.

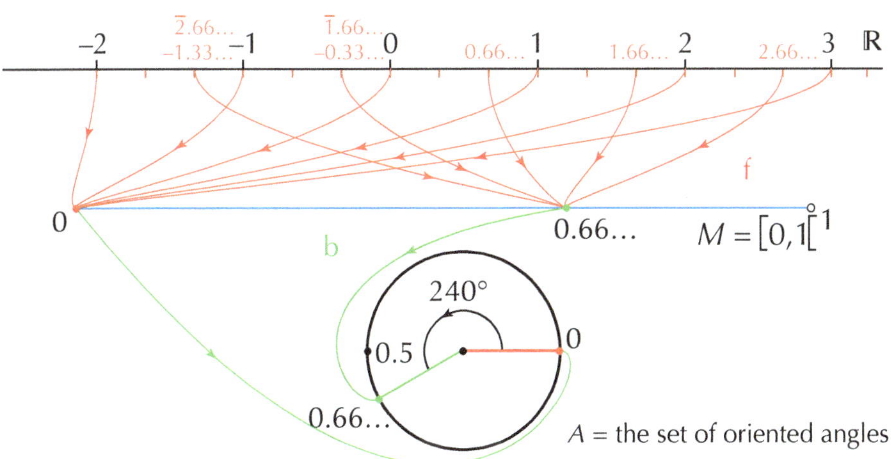

$\overline{2}.66...$ is the logarithmic notation of $-1.33...$ Where $\overline{2} = -2$ is the index and where $.66... = +0.66...$ is the mantissa. Indeed, we have: $-2 + 0.66... = -1.33...$

By considering real numbers as terms and by decapitating them to make them equivalent, the (oriented) angles are born as new concepts. Very remarkable: a guillotine that creates new life!

It seems astonishing that we can obtain a geometrical concept from number-theoretic terms. But then again, this only concerns superficial appearances such as, let's say 'Asian' and 'African' versions, which have nothing to do with the substantial structure of the created concepts.

Later we will see that a mathematician can implement this structure, if necessary by using an isomorphism, in a multitude of models depending on how much his poetical inspiration and the context inspire him. So later it will become clear that we can calculate in an identical way whether it be with oriented angles, with decapitated real numbers, with the complex numbers having modulus 1, with matrices of a particular form, with rotations around one point, etc. We will indeed always calculate in the same group structure. Those different representations with the same structure (isomorphic structures) do not only offer a semantic richness, but will also make it possible to interpret all the results obtained in one model immediately in all other models.

At first sight this may seem a commonplace, but from the next simple example it will become clear that in this way we nevertheless come to surprising results. An oriented angle clearly has two different halves which differ 180° (i.e. half a circle) from each other on the circle. This becomes clear by their construction by means of the bisector.

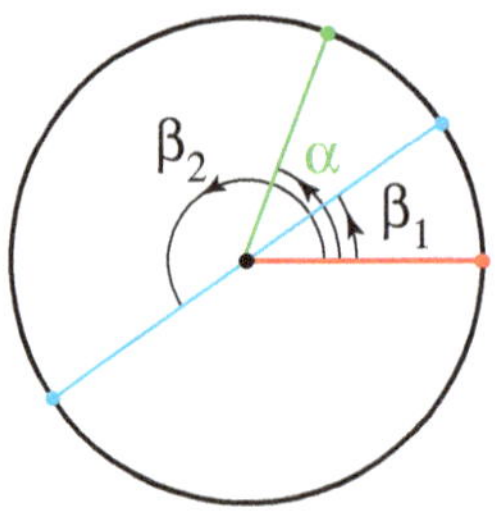

$$2\beta_1 = \alpha$$
$$2\beta_2 = \alpha$$
$$\beta_1 \neq \beta_2$$

So a decapitated real number has to have two different halves that differ from each other by 0.5 (half the length of the interval). For instance, the two halves of 0.3 are 0.15 and 0.65 because both $2 \cdot 0.15$ and $2 \cdot 0.65$ are (after decapitation) equal to 0.3.

So already now you can conclude, without knowing the concepts and without any effort, that a complex number having modulus 1, or such a special matrix we mentioned before, or such a rotation, must also have two different square roots (in their context multiplicative notation is used instead of additive). In section 6 this will be explained extensively.

3.4 Abstract creation of concepts

The Holy Spirit, according to a beautiful Bible story, had the capacity to create life. Now we will show that even a simple spirit, if mathematically educated, can do this at will. Although unfortunately it is not virgins that have to be fertilized, but terms.

Example 5 The bidecimals' light of day

As a set of terms, we take decimal expressions provided with two points. Just call them bidecimals.

Examples

$$\ldots 431.25.8666\ldots$$

$$\ldots 999.\overline{135}.4700\ldots \quad \text{where } \overline{135} = -135$$

In themselves these are just superficial appearances without any meaning. To link them to concepts, we have to define an equivalence relation. Depending on our inspiration, we can even do this in different ways. Thus by using the same terms, we can nevertheless create several different concepts. We shall restrict ourselves to two different versions.

Definition 1

We look at the bidecimal term as being composed of two parts: the part before the first point we consider as a decapitated real number, i.e. as an element of the interval [0,1[, and the part remaining behind the first point as an ordinary real number, i.e. an element of $\mathbb{R}$.

Thus

$$\ldots b_2 b_1.z.a_1 a_2\ldots \text{ is equivalent to } \ldots b_2{'}b_1{'}.z{'}.a_1{'}a_2{'}\ldots$$

if and only if

$$0.b_1 b_2\ldots = 0.b_1{'}b_2{'}\ldots \text{ (in } [0,1[\text{)}$$

and $$z.a_1 a_2\ldots = z{'}.a_1{'}a_2{'}\ldots \text{ (in } \mathbb{R}\text{)}$$

Hence equivalent bidecimal terms of this first kind will determine equal objects of the first kind.

Examples

$$13.\,27.99\ldots9\ldots = 13.\,28.00\ldots0\ldots$$

$$\ldots9\ldots99.\,15.34 = \ldots0\ldots00.\,15.34$$

$$\ldots9\ldots9936.\,\overline{13}.99\ldots9\ldots = 46.\,\overline{12}.00\ldots0\ldots$$

We seek a useful representation for these objects. A line segment, closed at the left edge and open at the right, is a good model for the interval [0,1[and a calibrated line is a good model for ℝ. So in the first instance we can represent such an object as a point on an infinite strip, closed at the left and open at the right, that represents the set-theoretical product of [0,1[with ℝ,i.e. [0,1[x ℝ. This results in the first model (left figure).

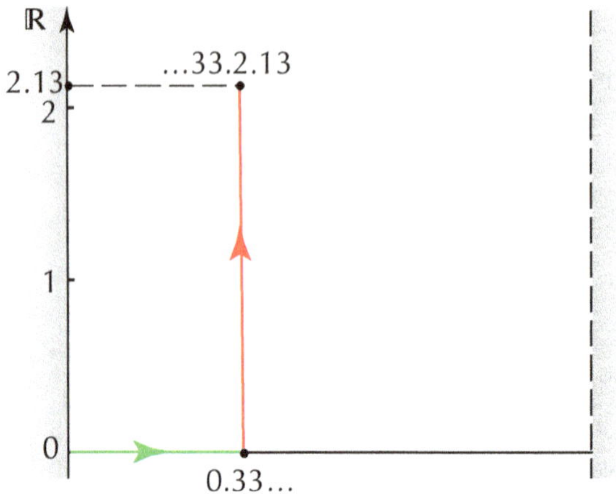

first model

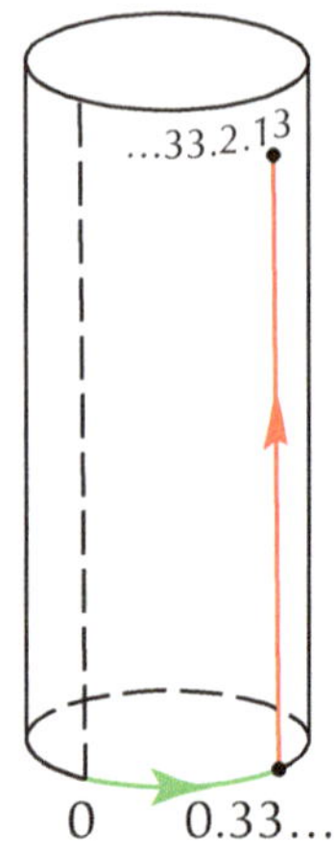

second model

We can, however, also coil that strip and make the open edge match the closed one so that we get a cylinder shaped surface. That gives a second representation (right figure). Thereby the first component, i.e. a decapitated real number, determines a rotation along a parallel circle. The second component, i.e. a real number, determines a shift parallel to the axis of rotation. The composition of that rotation and that shift we call a screw motion, which is just an elementary transformation of that cylinder (mapping between points of the cylinder). Each point on the cylinder then determines, respective to the fixed origin, a screwing which in itself represents a bidecimal of the first kind.

To be able to do calculations with these new concepts we need an addition of the bidecimals of the first kind that corresponds to the composition of screwings. For this addition we add up the first components like we did with the decapitated real numbers in the interval [0,1[, and the second components we add up like real numbers in ℝ.

Example

231.14.564 + 88.3.27 = 210.17.834.

since 0.132 + 0.88 = 1.012 = 0.012 (in [0,1[)

and $\qquad 14.564 + 3.27 = 17.834$ (in $\mathbb{R}$)

When calculating in this group of bidecimals of the first kind, one result is that we get two different halves for each bidecimal.

Example

$$2a = \ldots 33.2.13 \Leftrightarrow a = \ldots 661.1.065 \text{ or } a = \ldots 666.1.065$$

since $\qquad 2 \cdot 0.166\ldots = 0.33\ldots$ and $2 \cdot 0.666\ldots = 1.33\ldots = 0.33\ldots$ (in $[0,1[$)

and $\qquad 2 \cdot 1.065 = 2.13$ (in $\mathbb{R}$)

This also implies that each screwing has two different square roots:

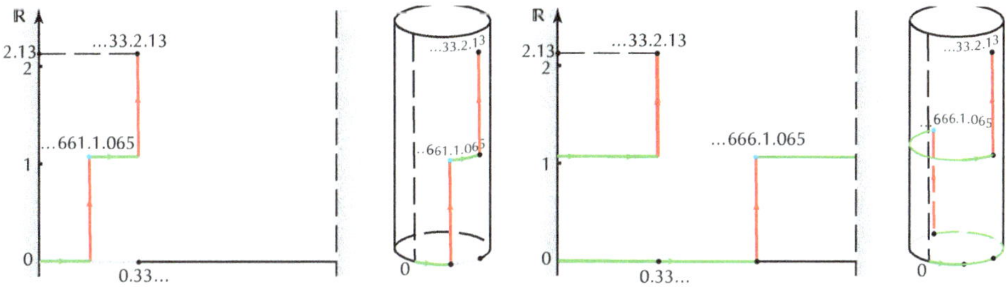

Definition 2

This time we consider the bidecimal terms as consisting of three different components: the part before the first point as a decapitated real number, the part between both points as an integer, and the part behind the second point again as a decapitated real number.

Thus: $\qquad \ldots b_2b_1.z.a_1a_2\ldots$ is equivalent to $\ldots b_2'b_1'.z'.a_1'a_2'\ldots$

if and only if

$$0.b_1b_2\ldots = 0.b_1'b_2'\ldots \text{ (in } [0,1[\text{)}$$

and $\qquad z = z'$ (in $\mathbb{Z}$)

and $\qquad 0.a_1a_2\ldots = 0.a_1'a_2'\ldots$ (in $[0,1[$)

Equivalent terms of this second kind determine equal objects of this second kind.

Examples

$$13.\,27.\,99\ldots9\ldots = 13.\,27.\,00\ldots0\ldots$$

$$\ldots9\ldots99.\,\overline{13}.\,99\ldots9\ldots = 0.\,\overline{13}.\,00\ldots0\ldots$$

We seek a representation for these objects of the second kind. We can represent both decapitated real numbers on the set-theoretical product of two line segments open at one side, which is a rectangle open at two of its four edges. The integer component we can represent as a point on the calibrated line. This results in a first representation.

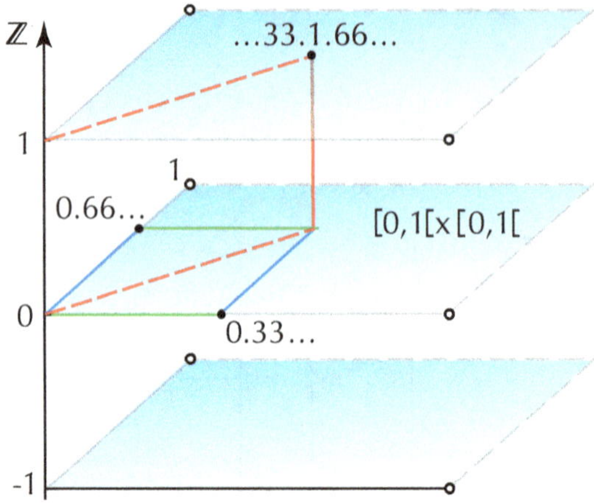

If we coil each rectangle so that the open edge matches the opposite closed edge to make it a cylinder and if we then coil this cylinder and fit together its remaining edges, we get a torus (like a doughnut). Then the bidecimals of the second kind are represented by a pile of tori. On this pile the first component (decapitated real number) determines a rotation along a parallel circle. The second component (integer number) determines a shift to a level in the pile. And the third component (decapitated real number) determines a rotation along a meridian circle. The composition of these three mappings is an elementary transformation of the pile of tori (mapping between points of the pile). The order of the mappings in this composition is arbitrary. So we get a second representation:

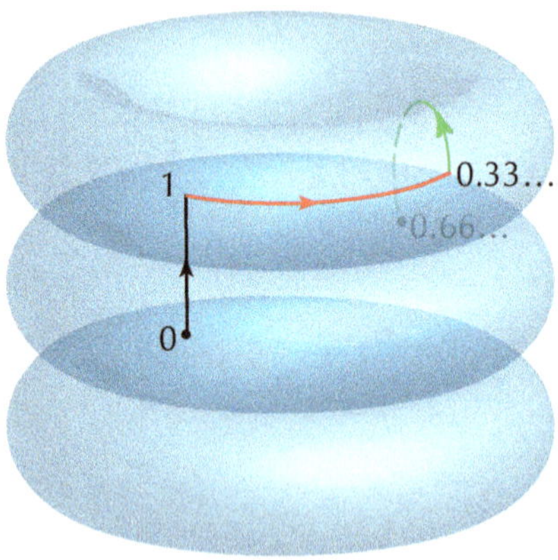

By defining, in an analogous way to what we did for the bidecimals of the first kind, an addition of the bidecimals of the second kind that corresponds to the composition of these elementary transformations on the pile of toruses, we end up with a construction showing results that are surprising.

So a bidecimal of the second kind with on odd integer component has no halves, while a bidecimal with an even integer component has four different halves.

Examples

$$2x = 723.4.183 \quad \Leftrightarrow \quad x = 5361.2.0915$$
$$\text{or} \quad x = 5366.2.0915$$
$$\text{or} \quad x = 5361.2.5915$$
$$\text{or} \quad x = 5366.2.5915$$

This shows that by the technique of classification of equivalent terms we can create concepts at will in an abstract way.

Our imagination can implement the artificial creatures, which have appeared this way, into meaningful contexts in which they can be usefully integrated. In the course of history lots of mathematical concepts have been created in this abstract way, and often only much later did they get a useful interpretation and application. We will give more examples of this further on.

3.5 Number of concepts associated with a finite set

By acquiring insight into the way in which classification generates concepts the mathematician can even calculate how many different concepts can be obtained from a finite set of terms.

For example, in this area, what is the maximum we can expect from the infant with a construction box containing six blocks? It means that we count the total number of possible partitions into non-empty subsets of the set of blocks. The reader who does not like to calculate can read the following 'as the crow flies' (i.e. cutting corners) and in the end proceed to an act of faith. But anyway, the infant is entitled to a correct explanation!

a) With one class

There is just one, all blocks go into the same class. The corresponding concept can be called 'construction box'.

b) With two classes

For our convenience we label those classes '1' and '2'. Then we decide for each of the six blocks whether we put it in class 1 or in class 2. This gives $2 \times 2 \times 2 \times 2 \times 2 \times 2 = 2^6$ different mappings of the set of blocks onto the set $\{1,2\}$.

Because absolutely no class can be empty, we must map blocks on both 1 and on 2. Therefore we ignore two mappings, i.e. those which map all blocks exclusively on 1 or exclusively on 2. The remaining $2^6 - 2 = 62$ mappings we call surjective mappings.

Because the order of the classes in partition plays no role, there are twice as many surjective mappings as there are partitions (we can always change the order in which we consider the two classes). Therefore the number of partitions with two classes is 62 : 2 = 31.

For example, one partition is linked to the concepts 'big' and 'small'. Observe that this corresponds indeed to two surjective mappings, i.e. the one that maps those blocks of the 'large' class on 1 and those of the 'small' class on 2, and the other mapping that swaps the meaning of 1 and 2.

c) With three classes

First we label the classes '1', '2' and '3', then we calculate the number of all mappings of the blocks onto the set {1, 2, 3 }, i. e. 3^6.

The number of non-surjective mappings (those that do not use at least one of the target numbers and hence create an empty class in the set of blocks) is $3 + 3 \times 62$, because three mappings send all blocks to only one target number (so two numbers are not used) and moreover there are mappings that use exactly two target numbers and leave one untouched. Above, in b, we calculated that there are 62 of them.

Thus the number of surjective mappings of the blocks on {1, 2, 3 } is $3^6 - 3 - 3 \times 62 = 540$. Not all of them determine different partitions because the three classes can be taken in arbitrary order. They can be arranged in $3 \times 2 \times 1 = 6$ different ways, since we have three possibilities for choosing the first class, then two choices left for the one we pick as second and only one possibility left for the third one. So the number of partitions is 540 : 6 = 90.

One of them is associated with the concepts 'circle', 'square' and 'triangle'.

d) With four classes

We label the classes '1', '2', '3' and '4'. The number of mappings of the blocks onto {1, 2, 3, 4} is 4^6.

The number of non-surjective mappings among them a is $4 + 4 \times 540 + 6 \times 62$: four mappings only use one target number, we have four choices of omitting one target number and for each choice we have 540 mappings on exactly the three remaining target numbers (see c). Finally we have 6 different choices to choose pairs of target numbers we could omit, i.e. {1,2}, {1,3}, {1,4}, {2,3}, {2,4} and {3,4}, and for each of these six pairs 62 mappings on exactly the two remaining target numbers (see b).

This adds up to $4^6 - 4 - 4 \times 540 - 6 \times 62 = 1560$ surjective mappings on the four target numbers. Since we can arrange the four classes in $4 \times 3 \times 2 \times 1 = 24$ different ways, the number of partitions in four classes is only $1560 : 24 = 65$.

One of them is related to the concepts 'green', 'blue, 'red' and 'black'.

e) With five classes

We, or rather you, can 'easily' calculate this in the same way. But now there is a shorter way with the so called 'pigeon-house-principle'. Since we have to arrange six blocks into five classes, and no class is to stay empty, we will end up with four of the classes containing one block and one class containing two.

We only have to count the number of choices for that class containing two blocks, so the number of pairs out of five i.e. $5 + 4 + 3 + 2 + 1 = 15$ (see method in d). So there are fifteen partitions in five classes.

One of them is related to the concepts 'triangle', 'big circle', 'small circle', 'big square' and 'small square'.

f) With six classes

There is precisely one such partition, i.e. each block in its own class.

Phew! So all together exactly $1 + 31 + 90 + 65 + 15 + 1 = 203$ concepts can be related to the set of only six blocks (terms), no more and no less. This infant should not abandon his construction box too soon. Ten little blocks generate already 115975 partitions (concepts). Fortunately you don't have to calculate that in the previous way. We will now explain an easier method based on the recursive method. Using this method these larger numbers can also be found

We introduce a symbol for the number of all partitions possible of a set of n elements into i classes, be it $P(i, n)$. Then we deduce a formula that expresses $P(i, n+1)$ respective to $P(i-1, n)$ and $P(i, n)$ and this is:

$$P(i, n+1) = P(i-1, n) + i \cdot P(i, n) \qquad (*).$$

We call this a recursive formula since it helps us to calculate a term using the previous terms. We will now give a proof of this formula.

If we add one element to the set of n elements, then we can change each partition into $i-1$ classes of the original set of n elements into a new partition of the set of $n+1$ elements into i classes just by arranging the new element into a separate class of its own. And each partition of n elements into i classes can be changed in i different ways into a partition of $n+1$ elements into i classes by adding the new element in turn to each of the i classes.

Example

Let $n = 3$ and $i = 2$ where $V = \{a,b,c\}$ and $W = \{a,b,c,d\}$ then $P(i-1, n) = P(1, 3) = 1$ is the number of partitions of the set V into one class, i.e. $\{\{a,b,c\}\}$, and $P(i, n) = P(2, 3) = 3$ is the number of partitions of V into two classes, being $\{\{a\},\{b,c\}\}$, $\{\{b\},\{a,c\}\}$ and $\{\{c\},\{a,b\}\}$.

Starting from these partitions we now construct the partitions of W into i (=2) classes:

first type: $\{\{a,b,c\},\{d\}\}$ number: as many as $P(i-1, n) = P(1, 3) = 1$

second type: $\{\{a,d\},\{b,c\}\}$, $\{\{a\},\{b,c,d\}\}$
$\{\{b,d\},\{a,c\}\}$, $\{\{b\},\{a,c,d\}\}$
$\{\{c,d\},\{a,b\}\}$, $\{\{c\},\{a,b,d\}\}$

number: $i \cdot P(i, n) = 2 \cdot P(2, 3) = 2 \cdot 3 = 6$

The number of partitions of W into i (=2) classes is thus:

$$P(2, 4) = P(i, n+1) = P(i-1, n) + i \cdot P(i, n) = P(1, 3) + 2 \cdot P(2, 3) = 1 + 2 \cdot 3 = 7$$

Using the recursive formula (*) and knowing that for any n

$$P(1, n) = 1 \quad \text{and} \quad P(n, n) = 1$$

we can set up a triangular diagram in which each one of the different $P(i, n)$ can easily be calculated from the term on top of it and its left neighbour. Just like a knight's move in chess.

Example

$$P(4, 7) = P(3, 6) + 4 \cdot P(4, 6) = 90 + 4 \cdot 65 = 350$$

n \ i	1	2	3	4	5	6	7	8	9	10
1	1									
2	1	1								
3	1	3	1							
4	1	7	6	1						
5	1	15	25	10	1					
6	1	31	90	65	15	1				
7	1	63	301	350	140	21	1			
8	1	127	966	1701	1050	266	28	1		
9	1	255	3025	7770	6951	2646	462	36	1	
10	1	511	9330	34105	42525	22827	5880	750	45	1

The sum of the numbers on the sixth row is the number of all partitions of a set of six elements (let's say the six blocks in the construction box):

$$1 + 31 + 90 + 65 + 15 + 1 = 203 \text{ (Bingo!)}$$

The number of all partitions of a set of 10 elements is:

$$1 + 511 + 9330 + 34105 + 42525 + 22827 + 5880 + 750 + 45 + 1 = 115975$$

The reader who really did fly like a crow, was probably already gulping during this section, but the only thing he should understand is that it is worthwhile to reflect on the way people do think. This pattern of thinking will then become an instrument for the creation of new concepts. Such a process is specific to mathematics.

3.6 The struggle for the universals

Also in every day life outside mathematics we find this three-step-scheme (equivalence relation R on the terms; partition P of those terms; a mapping f from the classes onto a model M) as a method in the creation of concepts.

In the livestock in a farmyard we can, for instance, distinguish cows, sheep, pigs, horses, etc., either by noticing some equivalence in their anatomical characteristics, or by keeping them kind by kind, or by hanging name tags around their neck. It is none of our business how the farmer keeps order in his affairs.

However, in scholastic times the realists, the nominalists and the other related spirits were frothing at the mouth to postulate the ontological priority (order of being) of R, P, f or M.

For the supporters of Plato, who according to Thomas Mann was the philosopher of the definite and indefinite article, the universal concepts (ideas), such as THE horse, are transcendent with regard to the individual real objects, such as A particular horse. So the 'ideas' already existed even before the first object of their species was created (universalia ante res). For those realists M has therefore priority in the order of being. By the mapping f, the objects, arranged in the classes of the partition P, refer to the 'ideas' in M and only in the second instance do they take part in that 'being'. Exactly for this reason we can distinguish the relation R between those things.

For the supporters of Aristotle, more moderate realists, the universal concepts are immanent to the material objects. Exactly by analysing the common features of the individual horses and by arranging them in one categorical class, the abstract concept 'horse' arises (universalia in res). It is therefore the active intellect that, starting from the preceding potential reality of the objects (terms), and through relations and classifications, establishes the secondary reality of the concepts. By that immanency we are thus predestined to the discovery of the relation R.

The rigorous nominalists, however, state that concepts only arise by the linguistic activity of the human mind that does no more than sticking names on the reality of the material objects (universalia post res). For them the priority therefore lies in the mapping f, which consequently shapes the classification P in the classes of the objects that got the same name.

Nowadays it seems incredible that a person was banned by the church for taking a certain point of view with regard to this matter, applied for example to the concept of 'the Holy Trinity'. In our time, mathematicians do not bother any more about all those priorities because they realize that all these components R, P, f and M coincide inextricably and mutually define each other.

The role of these elements is different however: the relation R has an analysing function, the partition P has an abstracting role, the mapping f has a terminological task and the model M has a semantic character.

Duns Scotus (1266–1308)

Franciscan, theologian and philosopher. With his critical comments on the Sententia of Petrus Lombardus, the most important dogmatic handbook of that time, and on the medieval Aristotle-doctrine he acquired fame and prestige. He recognized the existence of the universalia but acknowledged them beside a 'what-ness' (quiditas) also a 'this-ness' (haecceïtas), which he found more important. He also put free active will above passive receptive thinking. Together with William of Ockham (approx. 1288–1347) he became the major pioneer of nominalism and later empiricism.

4 Truth and consistency

How to reason with mathematical concepts?

What is the truth of propositions about these concepts based upon? And how do logical deduction laws extend the truth to their consequences?

4.1 True or false

Of objects we do not say that they are true or false, at most that they are 'real' or 'fictitious' (whatever we may mean by that). 'True' or 'false' are qualities that we assign to certain linguistic sentences that we call propositions (statements or judgements).

Examples

'John is Mary's brother',

'This table is made of wood',

'It is raining'.

These are sentences that are either true or false, but not both, and for which a third possibility is excluded. Such propositions are the domain of binary logic (bivalent logic).

Not all sentences are propositions.

Examples

'Give me that knife, will you?',

'Who is running there?',

'Get lost!'

These are not propositions, for the simple reason that we cannot decide whether they are true or false. For these sentences it makes no sense to discuss their truth.

Also in mathematics, propositions about created concepts can be true or false according to the actual model they are interpreted within (structural context).

Example

'18 plus 9 equals 27' is true in the model of the addition of natural numbers, but false in the model of the addition of hours (clock calculations modulo 24), because in this last model 18 plus 9 equals 3 ($18 + 9 = 27$ and $27 - 24 = 3$).

The truth of propositions thus depends on the chosen context in which they would be verified. If that context is not available, unlike in the given examples, the truth of simple synthetic propositions can therefore not always be decided. This is however not of any substantial importance in logic or in mathematics. These sciences only deal with the question of how the 'truth-value' of those simple propositions is reproduced in compound propositions that are constructed out of the simple propositions by the use of words such as 'not', 'and', 'or', 'if... then', etc.

Those words that connect propositions are called connectives or logical operators. It is not the content or meaning of the propositions that matters, only how the connective, by its pure formal definition, transforms the truth-value of the simple propositions into the compound proposition. So, the logical meaning of the connective 'and' (conjunction) is defined by the condition that the simple propositions which are linked by 'and' have both to be true in order for the compound proposition to be true. In mathematical logic this is represented as follows: let p and q be simple propositions, and let us note 1 if the proposition is 'true' and 0 if it is 'false', then the definition of 'and' is set by the diagram:

p	q	p and q
1	1	1
1	0	0
0	1	0
0	0	0

or shorter:

p	and	q
1	1	1
1	0	0
0	0	1
0	0	0

This truth table is easy to remember: just multiply the truth-values of the propositions p and q to obtain the truth-values of 'p and q'. That is why we also call the conjunction ('and') the logical product. How convenient these truth tables can be will become clear later.

At first sight all this could seem rather obvious. Nevertheless it took 23 more centuries since the foundation of logic by Aristotle (384–322 BCE) before mathematicians in the second half of the 19th century (George Boole, August de Morgan, Charles S. Peirce, Gottlob Frege, Guiseppe Peano and others) finally, by treating it in a mathematical way, gave proposition logic its definite appearance. The calculus of that mathematical logic, which is now used in the algebra of digital electronics, made the break-through of the computer possible in the twentieth century.

The reason why precisely mathematicians also find solutions for linguistic problems – whereas obviously linguistic intuition seems to fail – is to be found in the typical methodical view by which they examine those problems from an abstract angle. Just as Goethe had already mentioned: "Mathematicians are like the French, whatever you tell them, they translate it into their own language and consequently it has become something totally different".

Just try for yourself to answer the following questions:

1 Which of the next propositions contain the same information?

 a 'If it is raining, I do not go for a bicycle ride'.
 b 'If it isn't raining, I go for a bicycle ride'.
 c 'It isn't raining or I go for a bicycle ride'.
 d 'If I go for a bicycle ride, it isn't raining'.
 e 'I go for a bicycle ride or it is raining'.
 f 'It is not possible that I do not go for a bicycle ride and that at the same time it is raining'.

2 What is the negation of the proposition:

 'If I am late, I will be told off'?

3 What is the maximum number of words like 'and', 'or', 'if ... then', etc., by which two sentences can be connected to a compound sentence, giving it different logical nuances?

4 If we allow repetitions and combinations, how many such connectives are at least needed for describing all these nuances?

To make the answers to these questions comprehensible for everybody, we feel obliged to give a short initiation in proposition logic.

George Boole (1815–1864)

English mathematician and logician. In 1854 he published his *An investigation of the laws of thought*, in which he shows how the logical operators can be taken as arithmetic operations. This finally fulfilled the aspiration of Leibniz for a *characteristica universalis*. Boolean algebra forms the theoretical basis for the modern computer. However, it still took almost a full century before the technological conditions for its realization were fulfilled.

4.2 Connectives

a *Unary connectives*

There are connectives that operate on the truth-value of only one proposition. The connective 'not' is certainly the most important among them. From the intuitive use of language we know that negating a proposition switches over its truth-value. The negation of a true proposition is false and that of a false one is true. This leads to the next truth-table for the negation ('not') as a unary connective.

p	not p	or shorter	not	p
1	0		0	1
0	1		1	0

Do you know any other unary connectives than 'not' in language? Let's try and think how many different unary connectives can exist. How can we find the answer to such a question? Do we have to search dictionaries or just fantasise in our imagination? Of course not, we just put our mathematical spectacles on and at once the problem looks trivial.

The possible truth-values of one simple proposition are 1 and 0. A unary connective transforms these values into 1 or into 0. So, we simply have to count in how many ways that transformation can take place.

1 onto 0 and 0 onto 1 was one of these possibilities (the negation).

1 onto 1 and 0 onto 0 is a second possibility (the confirmation).

1 onto 1 and 0 onto 1 is a third possibility (the unary tautology)

1 onto 0 and 0 onto 0 is a fourth possibility (the unary contradiction).

And that's all there is for unary operators. We only had to reflect on the number of ways that, for the truth-table of ' * p', we could fill in a column of two symbols (1 or 0) under the operation symbol '*'. And that is exactly as many as the possibilities of choosing twice from two symbols, and that is $2 \times 2 = 2^2 = 4$.

*	p		...	p		...	p		...	p
0	1		0	1		1	1		1	1
0	0		1	0		0	0		1	0

b Binary connectives

By means of truth-tables we define some more interesting binary connectives in a similar way as was done in 4.1 for the conjunction ('and').

The *disjunction*, i.e. the inclusive 'or ' (in Latin: 'vel')

p	q	p or q	or shorter	p	or	q
1	1	1		1	1	1
1	0	1		1	1	0
0	1	1		0	1	1
0	0	0		0	0	0

So the disjunction is true if at least one of both simple propositions is true, and only false if both are false.

The *exclusive or* (either ... or ...) 'xor' (in Latin: 'aut')

p	q	p xor q	or shorter	p	xor	q
1	1	0		1	0	1
1	0	1		1	1	0
0	1	1		0	1	1
0	0	0		0	0	0

Thus the exclusive or 'xor' is true if precisely one of the simple propositions is true and the other false.

The *implication* 'if ... then' (symbol: $\rightarrow$)

p	q	$p \rightarrow q$	or shorter	p	$\rightarrow$	q
1	1	1		1	1	1
1	0	0		1	0	0
0	1	1		0	1	1
0	0	1		0	1	0

In an implication '$p \rightarrow q$' we call the proposition p the *antecedent* and the proposition q the *consequence*. Thus the implication is only false if the antecedent is true and the consequence false. The truth-values in the last two rows are often puzzling, but this shouldn't be so if one just observes that in this nothing is to be understood: it only deals with a definition of the particular connective. However, this matches historical logical intuition. In Latin this is expressed by the rule *Ex falso sequitur quod libet* (From the false follows whatever). Thus also truth.

The *equivalency* 'if and only if ... then' (symbol: $\leftrightarrow$)

p	q	$p \leftrightarrow q$	or shorter	p	$\leftrightarrow$	q
1	1	1		1	1	1
1	0	0		1	0	0
0	1	0		0	0	1
0	0	1		0	1	0

Thus the equivalency is only true if both simple propositions have the same truth-value, in other words: if they are equivalent.

Of course, we also get binary connectives by taking exactly the negation of the connectives already introduced. That means by switching the truth-values in the defining columns to the other value. For example, by negating 'and' we get 'not-and', shortened as 'nand' (the Sheffer's stroke) and by negating 'or' we get 'not-or', shortened as 'nor' (the Peirce's stroke). Observe that the negation of an equivalency gives the exclusive or 'xor' and vice versa.

Charles Sanders Peirce (1839–1914)

American mathematician, chemist and science philosopher. He showed already in 1880 that all logical operators could be expressed by means of the operator 'nor' (Peirce's law). Because his results were only published in 1933, Henry Maurice Sheffer (1882–1964), who published the same findings in 1913, took the honours. In 1886, Peirce also came to the perception that it is possible to translate the logical operators in electronic circuits, thus laying the basis for the modern computer.

4.3 Solutions to logical problems

Now we have the tools available for providing, in an indisputable way, correct answers to the questions put in 4.1.

1 Which of these sentences are logically equivalent?

If we represent the proposition 'It is raining' by p and the proposition 'I go for a bicycle ride' by q, then we can formalise the sentences concerned as:

a	p → (not q)	'If it is raining, I do not go for a bicycle ride'.
b	(not p) → q	'If it isn't raining, I go for a bicycle ride'.
c	(not p) or q	'It isn't raining or I go for a bicycle ride'.
d	q → (not p)	'If I go for a bicycle ride, it isn't raining'.
e	q or p	'I go for a bicycle ride or it is raining'.
f	not((not q) and p)	'It is not possible that I do not go for a bicycle ride and that at the same time it is raining'.

We need no more than to calculate the truth-table of each of these compound propositions, each time starting from the same possible truth-values of the simple propositions p and q. We can then check which match completely, and thus are equivalent. In this, we have to process firstly the connectives in the innermost parentheses, and then we have to apply the next connectives to the resulting truth-tables. This results in:

a

p	→	(not	q)
1	0	0	1
1	1	1	0
0	1	0	1
0	1	1	0

b

(not	p)	→	q
0	1	1	1
0	1	1	0
1	0	1	1
1	0	0	0

c

(not	p)	or	q
0	1	1	1
0	1	0	0
1	0	1	1
1	0	1	0

d

q	→	(not	p)
1	0	0	1
0	1	0	1
1	1	1	0
0	1	1	0

e

q	or	p
1	1	1
0	1	1
1	1	0
0	0	0

f

not	((not	q)	and	p)
1	0	1	0	1
0	1	0	1	1
1	0	1	0	0
1	1	0	0	0

This shows that a and d contain the same information, also b and e, as well as c and f. Simple as can be if only you know how to proceed. Using intuition does not really work out too well.

2 *What is the correct negation of the proposition: 'If I am late, I will be told off'?*

That is: 'I am late and I am not told off'.

The negation of the implication 'p → q' is thus equal to the conjunction 'p and (not q)'. Again, this can easily be checked by truth-tables:

not	(p	→	q)
0	1	1	1
1	1	0	0
0	0	1	1
0	0	1	0

p	and	(not	q)
1	0	0	1
1	1	1	0
0	0	0	1
0	0	1	0

If you would have guessed something else for that negation, just formalise your answer and calculate its truth-table to convince yourself whether your version is true or false. Many people think that they should form a conditional sentence again or that 'being late' should somehow be negated. But what must be negated is the conditional link between 'being late' and 'being told off'. All mathematicians know this: one counter example is enough for denying a theorem.

3 *What is the maximum number of connectives a language can produce?*

The possible distributions of truth-values of two simple propositions are respectively (1, 1), (1, 0), (0, 1) and (0, 0). They number $2 \times 2 = 2^2 = 4$. A binary connective maps these four values onto 1 or 0. So this can be done in $2 \times 2 \times 2 \times 2 = 2^4 = 16$ different ways. So there exist exactly sixteen binary connectives.

It is useless to look for them in a dictionary, nor in a foreign language dictionary. There is not a single language that produced this completeness by intuition, for this an abstract mathematical approach was needed first.

Now that we have put on these abstract spectacles, nothing will keep us down any more. We are even free to create connectives that operate on three or even on *n* simple propositions. This is not known in any language, except in a formal language or in a computer language. We can even calculate how many connectives of each of these 'arities' exist.

For ternary connectives (operating on three simple propositions at the same time) there are $2 \times 2 \times 2 = 2^3 = 8$ possible distributions of truth-tables. The ternary connective will map these on 1 or 0. So there are $2^8 = 256$ different ternary connectives! The general formula for the number of *n*-ary connectives is $2^{(2^n)}$.

This shows once more that insight is not only understanding but also creativity.

4 *What is the least number of connectives needed at least if we want to express really all possible logical nuances?*

As a first step we show that 'not', 'and' and 'or' are sufficient to describe all unary and binary connectives. The following equivalencies can easily be checked by calculating their truth-tables.

'not (not p)'	is equivalent to	'p' itself.
'p or (not p)'	is equivalent to	the unary tautology
'p and (not p)'	is equivalent to	the unary contradiction
'(not p) or q'	is equivalent to	'p $\rightarrow$ q'

And since

'p $\leftrightarrow$ q'	is equivalent to	'(p $\rightarrow$ q) and (q $\rightarrow$ p)'
'p xor q'	is equivalent to	'not (p $\leftrightarrow$ q)'
'p nor q'	is equivalent to	'not (p or q)'
'p nand q'	is equivalent to	'not (p and q)'

These last four connectives can also be changed into compound propositions containing only the connectives 'not', 'and' and 'or'. This also applies to the nine still missing binary connectives that have not been introduced yet. We however leave this as an exercise for the 'diligent' reader.

In a second step we show that 'not', 'and' and 'or' can be described by only using the connective 'nor' (or using only the connective 'nand'). You can calculate to check that:

'p nor p'	is equivalent to	'not p'
'(p nor p) nor (q nor q)'	is equivalent to	'p and q'
'(p nor q) nor (p nor q)'	is equivalent to	'p or q'

In everyday speech one uses 'neither ... nor ...' instead of 'nor'. The whole treasure of connectives a language owns can be expressed by only using the connective 'neither ... nor ...', however with appropriate repetition. What a cluster of 'neither ... nor ...' you would have to produce to express, for instance, a simple equivalency...

Our mother tongue is not so dull as not to provide sufficient variety in the choice of connectives. This makes it easier to produce all logical nuances. However, the answer to question 4 proved to be very important in computer science. All logic gates can apparently be constructed as a network of 'nor' gates (or 'nand' gates). So one can build all the complicated functions needed in information processing by using only one 'basic-chip'.

The story of the connectives does not end here. They can in fact be considered as operations on propositions. Note that operations like + and · on numbers do have some well defined properties.

Examples

$a + b = b + a$ (commutativity)

$(a \cdot b) \cdot c = a \cdot (b \cdot c)$ (associativity).

Also connectives can own such properties.

Example

'p or q' is equivalent to 'q or p'

'(p and q) and r' is equivalent to 'p and (q and r)'.

But they also have some more specific properties.

Example

'p and p' is equivalent to 'p' (idempotency).

Given the scope of this book, however, it is not relevant to introduce all these properties.

The first integrated circuit, also called chip or IC, was presented on September 12th, 1958 by Jack Kilby of Texas Instruments. In these chips, the electronic circuits are printed in a miniaturized form on a silicon chip and then wrapped in a ceramic or plastic sheath. Since then, the number of transistors, which are used as switches in them, has grown to a phenomenal number and already reached the millions by now. The invention of the chip is undoubtedly one of the most important milestones in the history of computers.

4.4 Tautologies

A tautology is a compound proposition that is true in all cases, irrespective of the initial truth-values of the simple propositions of which it has been built. Thus they are propositions that are true purely on the basis of their structure. So they are pre-eminently suited for deriving new true propositions from other true propositions. For this reason we also call them argument forms.

Of course, the equivalency of propositions with the same truth-table, i.e. of equivalent propositions, is a tautology. Tautological equivalency we mark by a left right double shafted arrow, i.e. $\Leftrightarrow$. Therefore, the tautological equivalency determines a relation between propositions, particularly the equivalency. The ordinary equivalency, marked by the single shafted arrow $\leftrightarrow$, is however an operation on propositions and is not true in all cases.

Like all rationalists, mathematicians are of course keen on tautologies, the pre-eminent instruments for correct reasoning and writing correct proofs.

We give some examples of frequently used tautologies:

not(not p)	$\Leftrightarrow$	p	(double negation)
$p \to q$	$\Leftrightarrow$	(not p) or q	(factorization of an implication)
$p \to q$	$\Leftrightarrow$	(not q) $\to$ (not p)	(contraposition)
not(p and q)	$\Leftrightarrow$	(not p) or (not q)	(De Morgan's laws)
not(p or q)	$\Leftrightarrow$	(not p) and (not q)	(De Morgan's laws)

These tautologies can easily be verified by truth-tables. From previous tautologies one can, however, directly derive new tautologies without the need for calculating their truth-tables.

Example

The rule for the negation of an implication can also be proved in the following way:

not $(p \to q)$	$\Leftrightarrow$	not ((not p) or q)	(fact. of an impl.)
	$\Leftrightarrow$	(not (not p)) and (not q)	(De Morgan's law)
	$\Leftrightarrow$	p and (not q)	(double negation)

There also exist tautologies that do not have the form of an equivalency, but for example the form of an implication. A tautological implication we also mark by a rightward double shafted arrow, i.e. $\Rightarrow$.

Example

$p \Rightarrow p$	referring to the factorizing of an implication this can also be written as: '(not p) or p is a tautology' (this is the law of the excluded third possibility)
p and q $\Rightarrow$ p	(elimination of the conjunction)
p and q $\Rightarrow$ q	(ditto)
p and (p $\rightarrow$ q) $\Rightarrow$ q	(modus ponens or detachment rule)
p $\Rightarrow$ ((not p) $\rightarrow$ q)	which can also be written as
(p and (not p)) $\Rightarrow$ q	(inconsistency law)

This last argument form expresses the fact that, if in a theory both a proposition and its negation occur, one can conclude whatever. In other words, one single contradiction makes the whole theory absurd. So this argument form does not tolerate Hegelian situations. We come back to this in section 4.7.

4.5 Reasoning and proofs

With the argument forms we can get correct new information from given (non contradictory) information. As an illustration of this we give following 'suitable' example:

Given:	'if this text is not clear then it is not good'	(1)
and	'this text is too difficult or it is not bad'	(2)
and	'the text is easy'	(3)

We first formalise the propositions starting from the following simple propositions:

'the text is clear' = c

'the text is good' = g

'the text is easy' = e

The given propositions are then:

	(not c) $\rightarrow$ (not g)	(1)
and	(not e) or (not (not g))	(2)
and	e	(3)

We first simplify proposition (2) by the law of double negation and the factorization of an implication:

$$\text{(not e) or (not (not g))} \Leftrightarrow \text{(not e) or g}$$

$$\Leftrightarrow e \rightarrow g \qquad (4)$$

Then we use the conjunction of (3) and this implication (4) in the modus ponens:

$$\text{e and } (e \rightarrow g) \Rightarrow g \qquad (5)$$

We rewrite proposition (1) with the contraposition:

$$\text{(not c)} \rightarrow \text{(not g)} \Leftrightarrow g \rightarrow c \qquad (6)$$

By applying the modus ponens on these last two propositions (5) and (6) we finally get:

$$\text{g and } (g \rightarrow c) \Rightarrow c$$

So 'the text is clear'! Or what would you have expected?

The validity of an argumentation can also be shown by a truth-table. The implication ((1) and (2) and (3)) →c proves to be tautological. But therefore the conclusion c should be known before this implication was written down, which is not evident.

In mathematical theorems, always supplementary to the 'given' (antecedent), the 'to prove' (consequence) is formulated. But no mathematician is keen on calculating complicated truth tables, neither do they like to formalise the argument forms used. Generally only reasoning in the 'naked common mind' is done, with the corresponding argument forms only vaguely at the back of the mind. Logic is in itself not a source of inspiration for mathematical creativity, but it is in a certain sense its conscience.

4.6 The substitution principle

The tautological character of an argument form is insensible to the initial truth-values of the propositions out of which it is built. That is why in a tautology we can change each simple proposition by whatever other proposition, even a compound one. This is called the substitution principle.

Example

You could check by calculating the truth-table that

$$(p \rightarrow q) \text{ and } (q \rightarrow r) \Rightarrow (p \rightarrow r)$$

is indeed a tautology, irrespective of whatever propositions you take for p, q and r.

If we now for instance replace p by 's or t', q by 'v and not w', and r by 'u nor s' then we get the following new tautology:

$$((\text{s or t}) \rightarrow (\text{v and not w})) \text{ and } ((\text{v and not w}) \rightarrow (\text{u nor s})) \Rightarrow$$

$$((\text{s or t}) \rightarrow (\text{u nor s}))$$

This substitution principle makes the logical tautologies universally applicable.

4.7 Contradiction and inconsistency

If a proposition is true, then its negation is false and vice versa. That is precisely the definition of negation. It is obvious that in binary logic not both a proposition and its negation can be true (or false). The proposition 'p and not p' is thus a contradiction:

p	and	(not	p)
1	0	0	1
0	0	1	0

But what is worse: fencing in such a 'disease germ' as a local disaster is never possible. Because it would show then that in the theory, by the law of inconsistency and the modus ponens, any arbitrary proposition q as well as its negation 'not q' are provable:

$$(\text{p and not p}) \text{ and } ((\text{p and not p}) \rightarrow \text{q}) \Rightarrow \text{q} \quad (\text{see } 4.4)$$

Since q is arbitrary here, according to the substitution principle it can be replaced as well by 'not q'.

If we encounter only one contradiction in a logical-deductive theory, then in the wink of an eye all propositions become contradictions, which means: the whole theory gets absurd. It is evident that mathematicians, and with them all convinced rationalists, are scared to death of contradictions. A kind of Hegelian compromise such as the dialectic dynamics of thesis, antithesis and synthesis is not to be expected in their midst.

It is however remarkable that already at the birth of mathematics and also later on during the important phases of its development, paradoxes have always been formulated that shook its very foundations.

One remembers the bewilderment of the Pythagoreans when they, due to nothing less than their own 'royal theorem', discovered that the proportion of the diagonal of a square to its side could not be rational. That was in stark contradiction to the axiom they tacitly accepted, which states that all proportions of magnitudes could be expressed using integer numbers. Their mystic belief that everything in the universe was based on integer numbers was given a fatal jolt by this. It was intellectually not fair that they initially tried to keep this a secret. However, a contemporary of theirs by the name of Eudoxos, had already thought of a solution for giving the irrational proportions a place within Greek geometry. Because of the problems with number theory it was, however, kept safely separated from the splendid construction work that geometry was. For a long time this kept the development of Greek mathematics within the very narrow spectre of the visual three-dimensional space.

Also the paradoxes of Zeno were a nuisance to the Greeks: Achilles, who seemingly could never catch up with the tortoise; the arrow which stands still at all places in its flight and hence seemingly does not fly; etc. One must already be very acquainted with the continuum and the subtleties of infinitesimal calculus to be able to refute them.

Also in the 17th and 18th centuries controversies arose in the field of the newly devised differential- and integral calculus. The ill-considered use of infinite sums (series) gave rise to apparent contradictions.

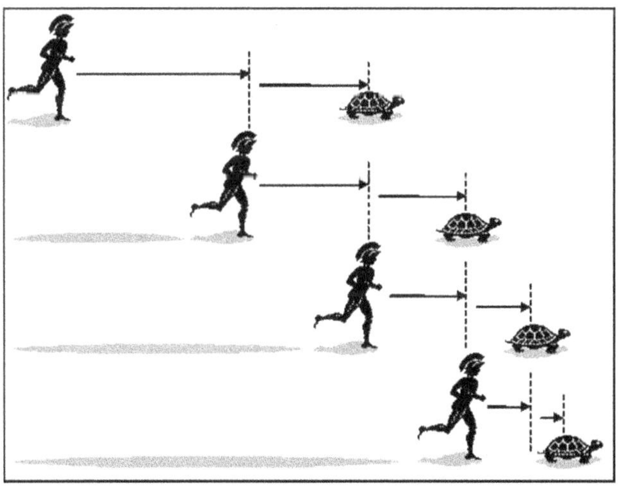

Achilles chases after the tortoise. According to Zeno from Elea he can, however, never catch up with it because in the time that he bridges his original delay, the tortoise itself also has gone a bit ahead, and that fact repeats itself continuously forever. In the figure above three of those phases are shown. Zeno's contempories were unable to refute his arguments because they hadn't acquired the necessary mathematical concepts yet.

Example

Does the series $1 - 1 + 1 - 1 + 1 - 1 + \ldots$ really represent a number? If so, which number? By adding parentheses (in an improper way) this series can firstly be rewritten as $(1 - 1) + (1 - 1) + (1 - 1) + \ldots$ and so the sum seems to be 0. But if this series is written as $1 - (1 - 1) - (1 - 1) - \ldots$ it seems to add up to 1.

It took two more centuries before people realized how they could manipulate such a series and how not.

A lot worse were those paradoxes that were not only of a mathematical but also of a logical nature. These led, at the end of the nineteenth century, to a real crisis about the fundamentals. This has discouraged a lot of mathematicians. A famous example, which is known as 'Russell's paradox', resembles the logical paradox of the Cretan who says 'I am lying'. Of such paradoxes many alternatives are given: 'this statement is not true'; 'the barber of the village shaves all villagers who do not shave themselves'; etc.

The mathematical world had hardly welcomed the set theory of Georg Cantor as the ultimate paradise in which all mathematical theories could find their nourishment, when suddenly paradoxes in this theory seemed inevitable. Bertrand Russell defined, for example, a set V of all elements that have the property as not to contain themselves.

In symbols: $V = \{x \mid x \notin x\}$.

If one reflects on the question whether V also contains itself,

i.e. $V \in V$,

then this can only be under the condition that V does not contain itself,

i.e. $V \notin V$.

So they found to their own bewilderment that $V \in V$ if and only if $V \notin V$. This is a pure contradiction.

A lot of mathematicians were persuaded that in this way the complete structure of mathematics, and hence that of all its applications, was built on a minefield. Once more ingenuity and subtle thinking was needed to avert this morbid misfortune.

Since then various schools of mathematicians have also chosen different directions. The logicians and the intuitionists appeared not to be as successful in this matter as the formalists. However, also the axiomatic formalization of a theory proved to have its own fundamental restrictions.

The sudden emerging of paradoxes in a young theory, however, also has positive aspects. Namely, they have always paved the way to a deepening of that theory. According to the words of Sylvester: "paradoxes are the slaughterers of naive presuppositions".

4.8 The indirect proof

The indirect proof is frequently applied in mathematics. Outsiders have always had their doubts regarding that apparent box of tricks. How can one prove a theorem by starting from the disavowal of what one has to prove, to end up at the disavowal of exactly that which has been given? Not without reason this form of argumentation got the name 'demonstratio ex absurdo'. That technique of proving, nevertheless, is justified by the tautologies of contraposition and of the excluded third possibility: $(p \rightarrow q) \Leftrightarrow (\text{not } q \rightarrow \text{not } p)$ and 'p or (not p) is a tautology'.

Example (in the set of natural numbers)

'If the number d is not a divider of a multiple b of the number a, then d is not a divider of a'.

Proof:

d is either a divider of a or is not a divider of a.

Let d be a divider of a, then there exists a number p such that $d \cdot p = a$.

Since b is a multiple of a, there also exists a number q such that $a \cdot q = b$ and thus $(d \cdot p) \cdot q = b$ which implies $d \cdot (p \cdot q) = b$. From this last equation we conclude that d is a divider of b, which is a contradiction of what has been given.

So we have to abandon the supposition that d would be a divider of a. So d is not a divider of a.

The so-called intuitionists, lead by the Dutch mathematical L.E.J. Brouwer, reject the law of the excluded third possibility. They do not want and are not able to use the indirect way of proving. But in that way they also have to throw away, together with the bath water, so many of the dear babies of mathematics. And that is really a bridge too far for most mathematicians, who therefore prefer other, more fertile ways of avoiding imminent paradoxes.

Luitzen E.J. Brouwer (1881–1966)

Dutch mathematician, founder of intuitionism. He did innovative work in topology, for example by his inductive definition of the concept dimension and with his fixed-point theorem. According to that theorem an entirely haired skull must have a whirl somewhere after each combing treatment. According to Brouwer's point of view mathematics is a free creation of the mind apart from each experience and independent of language and logic. In his doctoral thesis *The fundamentals of mathematics* from 1907 he explained his wayward interpretation of the mathematical constructivist intuitive method.

4.9 From logic to the computer

Until now, it has not been clear what mental processes come into action in our brain when we set up a proof or solve a problem.

The German mathematician G.W. Leibniz (1646-1716) hoped that someday a universal language for reasoning would be developed that could be converted into a simple calculation technique. The principle of the wish for convenience of lazy people! But in this way, he cherished the idea of the computer more than three centuries before its first realization.

How mathematicians have succeeded in moulding the laws of logic into the form of (Boolean) algebra, and made it suitable for electronic circuits, is a poetic tale full of imagination and ingenuity. We shall give a brief sketch of the steps that were taken.

Step 1: From connectives to (Boolean) polynomials

In binary logic the truth-value of a proposition p can be 'true' or 'false', respectively represented by 1 or 0. We write this truth-value symbolically as v(p). So v(p) = 1 or v(p) = 0.

Example 1 the conjunction 'and' as the 'logical product'

from

p	and	q
1	1	1
1	0	0
0	0	1
0	0	0

it can be seen that

$$v(p \text{ and } q) = v(p) \cdot v(q) \qquad (1)$$

In the formula (1) '·' can be interpreted as the ordinary multiplication of natural numbers, but also as the so-called 'logical multiplication' of the values 0 and 1 in the two-element Boolean algebra.

Example 2 the negation 'not' as the 'logical complement'

from

not	p
0	1
1	0

it can be seen that

$$v(\text{not } p) = 1 - v(p) \qquad (2)$$

In the formula (2) '−' means the usual arithmetic subtraction. Because this operation is not defined in Boolean algebra, we will write here v(not p) as $\overline{v(p)}$, which we then call the *'logical complement'* of v(p). This is thus the operation that changes 0 into 1, and vice versa.

Example 3 the disjunction 'or' as the 'logical sum'

At first sight one of the truth-values, namely the one for v(p) = 1 and v(q) = 1, poses a problem here.

in	p	or	q
	1	1	1
	1	1	0
	0	1	1
	0	0	0

The last three lines comply with:

$$v(p \text{ or } q) = v(p) + v(q) \qquad (3)$$

If we interpret the operation '+' in formula (3) as the ordinary addition of natural numbers then this formula would not be valid for the first line.

To have this sum formula functioning also for the first line we have to consider an addition for which 1 + 1 = 1. This is true in the two-element Boolean algebra. This algebra is completely defined by the next tables for its operations:

+	0	1
0	0	1
1	1	1

.	0	1
0	0	0
1	0	1

and by the properties of these operations, corresponding to those of the logical operators 'or' and 'and'.

It is, however, also possible to set up a general formula for the disjunction using the ordinary arithmetical operations +, − and · .

Which is why we rewrite 'p or q' as 'not (not p and not q)'.

With the formulas from example 1 and 2, this leads to:

$$\begin{aligned} v(p \text{ or } q) &= v(\text{not}(\text{not } p \text{ and not } q)) \\ &= 1 - v(\text{not } p \text{ and not } q) \\ &= 1 - v(\text{not } p) \cdot v(\text{not } q) \\ &= 1 - (1 - v(p)) \cdot (1 - v(q)) \\ &= 1 - 1 + v(p) + v(q) - v(p) \cdot (v(q) \\ &= v(p) + v(q) - v(p) \cdot v(q) \end{aligned}$$

You can check, just by using ordinary arithmetic, that all lines of the truth-table of the disjunction comply with this formula.

Example

For v(p) = 1 and v(q) = 1 it shows v(p or q) = 1 + 1 − 1 · 1 = 1 .

Since all sixteen binary operators can be expressed by the operators 'and, 'or' and 'not', all these can be changed into Boolean polynomials, as they are just ordinary algebraic polynomials of two variables v(p) and v(q). If we simplify by substituting v(p) = x and v(q) = y, then the given examples show that:

	in Boolean algebra	in ordinary algebra
not p	$\overline{x}$	$1 - x$
p and q	$x \cdot y$	$x \cdot y$
p or q	$x + y$	$x + y - x \cdot y$
p nand q	$\overline{x \cdot y}\ (= \overline{x} + \overline{y})$	$1 - x \cdot y$
p nor q	$\overline{x + y}\ (= \overline{x} \cdot \overline{y})$	$1 - x - y + x \cdot y$

So the binary logic has been transformed into the algebra of polynomials.

Step 2: From Boolean polynomials to electric circuits.

Electric circuits are constructed by electric conductors and switches.

If a circuit is complete, i.e. if electricity runs through the conductor, we mark this by the value 1. If the circuit is not complete, i.e. if a switch breaks the loop and electricity doesn't run, we mark this by the value 0. We can represent a simple proposition p by a switch in a circuit.

If v(p) = $x = 1$ then the switch is closed (turned 'on').

If v(p) = $x = 0$ then the switch is open (turned 'off').

Compound propositions are thus represented as circuits containing switches.

Examples

The 'and' operator corresponds to a series connection because the electricity only runs through if both switches are closed.

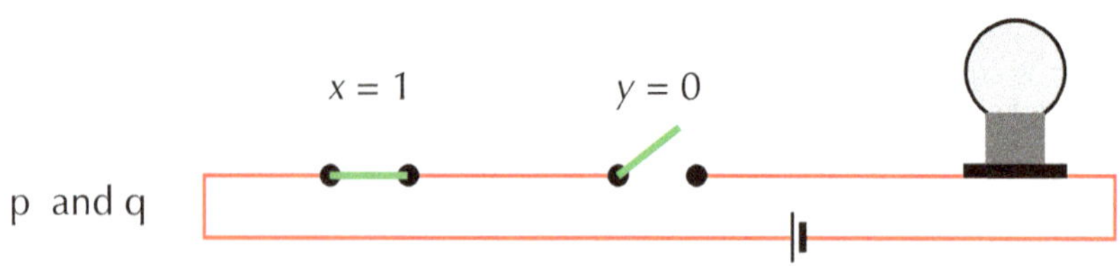

if v(p) = $x = 1$ and v(q) = $y = 0$ the lamp does not light up

The 'or' operator corresponds to a parallel connection because the electricity flows as soon as one of the switches is closed.

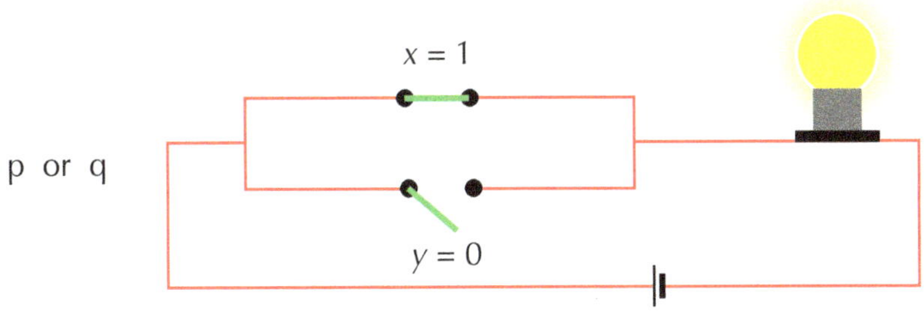

if v(p) = x = 1 and v(q) = y = 0 the lamp lights up

The 'not' operator corresponds to a double-throw switch (2-way switch) that puts the circuit into the other position.

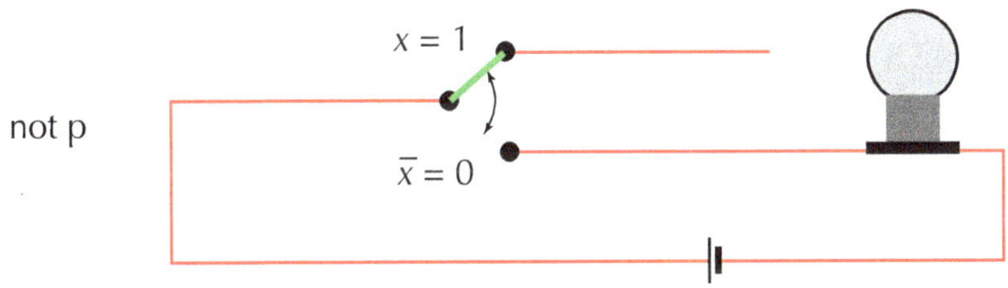

if v(p) = x = 1 the lamp does not light up
(so v(not p) = $\overline{x}$ = 0)

All other operators can be defined by 'and', 'or' and 'not'. They can all be represented by a combined circuit containing series and parallel connections of double and single throw switches.

Step 3: From Boolean Polynomials to integrated circuits

Remember that all logical operators, and especially 'and', 'or' and 'not', can be constructed by using only one single operator, namely 'nor' (as well as 'nand'). Suppose that we have of a basic chip available that works as a 'nor' operator,

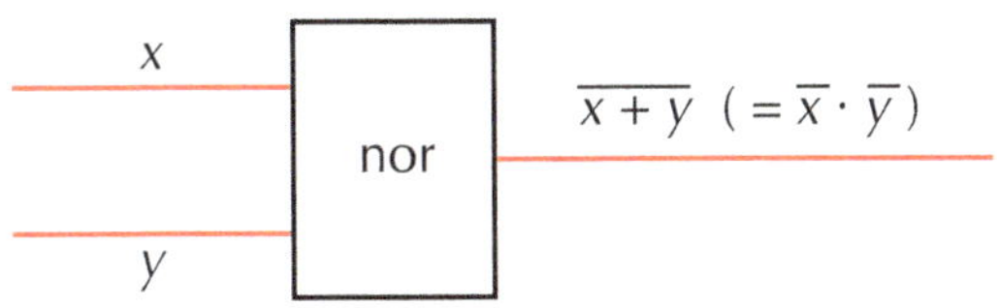

then we can translate all logical operators into integrated circuits by multiple use of this basic chip.

Example

Since 'not p' ⇔ 'p nor p' where v(p) = x:

Since 'p or q' ⇔ '(p nor q) nor (p nor q)' where v(p) = x and v(q) = y:

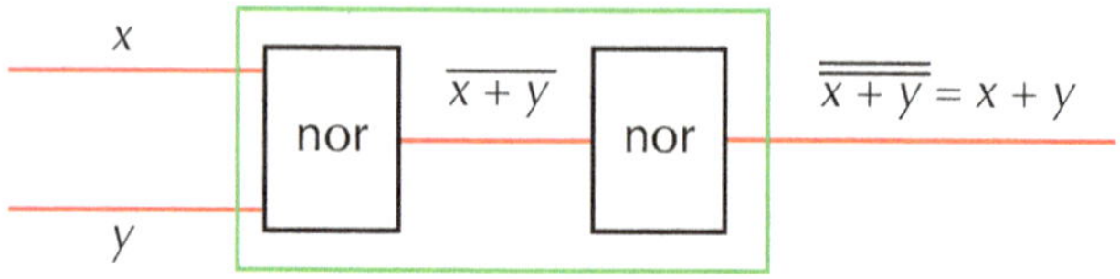

or

Since 'p and q' ⇔ '(p nor p) nor(q nor q)' where v(p) = x and v(q) = y:

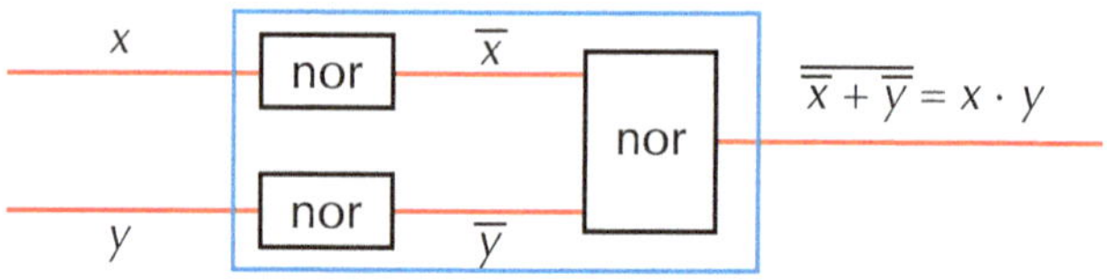

and

But how can we process information with these tools? We'll give a simple example of this.

In a council consisting of three people, a proposal is decided on by a simple majority of votes. A member votes by pushing or not pushing a button under the desk. If the button is pushed, then its value is 1 (yes). If not pushed then its value is 0 (no).

Which Boolean function expresses this situation and how can it be materialized in an electric circuit?

Let x, y and z be the value (0 or 1) of the three switches. If at least two buttons are pushed, then the result of the Boolean function, $f(x,y,z)$, has to be equal to 1. The circuit has to close and allow the electric current to power the green light in favour of the proposal. So:

if $(x,y,z) = (1,1,1)$ then $f(x,y,z) = 1$ and $xyz = 1$

if $(x,y,z) = (1,1,0)$ then $f(x,y,z) = 1$ and $xy\overline{z} = 1$

if $(x,y,z) = (1,0,1)$ then $f(x,y,z) = 1$ and $x\overline{y}z = 1$

if $(x,y,z) = (0,1,1)$ then $f(x,y,z) = 1$ and $\overline{x}yz = 1$

And $f(x,y,z)$ has to be equal to 0 for all other values of the variables x, y and z. The Boolean function concerned is thus:

$$f(x,y,z) = xyz + xy\bar{z} + x\bar{y}z + \bar{x}yz$$

By means of the properties we can simplify it to:

$$\begin{aligned} f(x,y,z) &= (xyz + xyz + xyz) + xy\bar{z} + x\bar{y}z + \bar{x}yz \text{ (idempotency)} \\ &= (xyz + xy\bar{z}) + (xyz + x\bar{y}z) + (xyz + \bar{x}yz) \text{ (commutativity)} \\ &= xy(z + \bar{z}) + xz(y + \bar{y}) + yz(x + \bar{x}) \text{ (distributivity)} \\ &= xy \cdot 1 + xz \cdot 1 + yz \cdot 1 \text{ (law of complements)} \\ &= xy + xz + yz \text{ (neutral element)} \end{aligned}$$

This function is rendered by the following combination of gates:

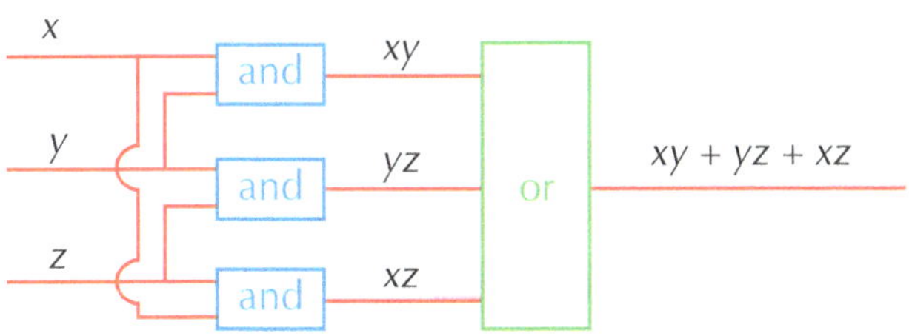

By transforming the logic connectives in polynomials the language of logic can be expressed by (integrated) circuits using the available technology.

The ENIAC (Electronic Numerical Integrator And Computer) was the first electronic and full computer. It was devised at the University of Pennsylvania between 1943 and 1946, mainly for the purpose of processing ballistic tables. The gates (switches) were still made by means of radio tubes instead of transistors. It was an enormous construction that weighed 30 tons and whose 18000 tubes filled a room of 10 by 16 meters.

4.10 Elegance in language

We can also represent the symbolism of language in a more compact and efficient way. In order to simplify the complexity of colloquial spoken and written language, mathematicians introduce proper symbols for the concepts and relations they want to use. In logic the following symbols are used, for instance, for the most frequent connectives:

$\neg$ (not), $\wedge$ (and), $\vee$ (or), $\rightarrow$ (if...then),
$\leftrightarrow$ (if and only if... then), $\uparrow$ (nand), $\downarrow$ (nor).

So only one of the four unary connectives, and only six of the sixteen possible binary connectives are provided with a symbol. Moreover, at first sight there is little uniformity and system in these choices. In such a situation the poetical passion of the mathematician could deliver better and more elegant results. Let's give it a try.

A unary connective maps the two possible truth-values 1 and 0 of a proposition onto 1 or 0. Let us represent, for instance, the initial truth-values 1 and 0 by two halves of a disc, separated by a (horizontal) diameter. In this, the upper half represents the value 1 and the lower half value 0. Like this:

Next we agree that a connective maps a coloured sector onto 1 and a colourless sector onto 0.

Example

By this agreement the diagram 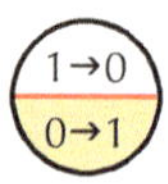

defines the mapping '1 → 0, 0 → 1' completely. This diagram is thus a symbolic representation of the connective 'not':

not	p
0	1
1	0

is represented by 

negation confirmation tautology contradiction

We can continue this system for the binary connectives. The four possible truth-values for two propositions are (1, 1), (1, 0), (0, 1) and (0, 0). We represent them by the four quarter sectors of a disc, separated by two (oblique) perpendicular diameters. Like this:

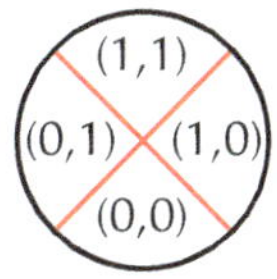

We map the coloured quarters onto 1, the colourless onto 0.

Example

By these agreements the diagram 

represents the mapping '(1,1) → 1, (1,0) → 0, (0,1) → 0, (0,0) → 0'.

So this diagram is a symbol for the connective 'and':

p	and	q
1	1	1
1	0	0
0	0	1
0	0	0

is represented by

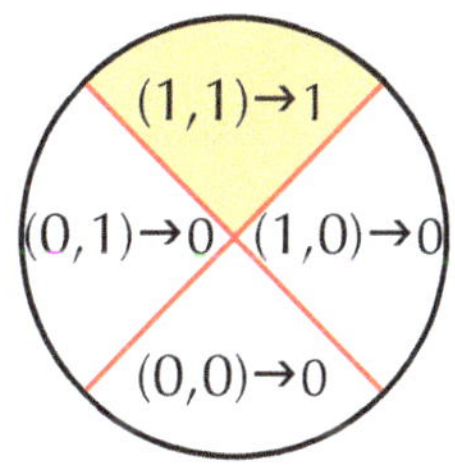

The analogous diagrams for the sixteen possible binary connectives are thus:

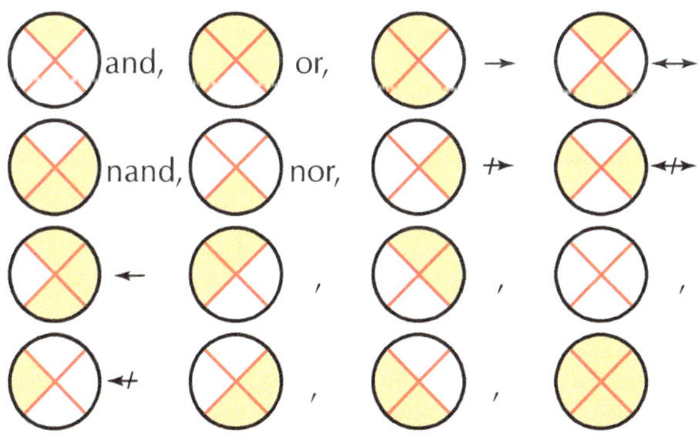

If we did this for blind people, we could make small 'coins' on which the coloured part would be elevated and the top would be marked by a notch.

The above may show that playful and efficient expression can also go together in mathematics. In this regard, most of the mathematical symbols that are familiar to us, like +, =, √, etc. have their own interesting but not always obvious history.

5 Theory and model

In section 3 we showed how mathematical concepts came into being. We start with basic concepts that have no meaning or particular content by themselves. And then we construct derived concepts, by means of the specification principle or by the classification of terms.

To be able to work with those basic and derived concepts we must formulate basic rules (axioms). These describe relations between the basic concepts, which they implicitly define this way. Just remember the rules for moving the bishop in chess.

The truth of these rules of the game is irrefutable. Just as it has no sense to dispute the validity of the rules in the chess game, it would be ridiculous to reject the axioms of the mathematics game.

In section 4 we gave examples of logical argumentation forms by which we can derive 'guaranteed' true new propositions from true propositions. This way we prove new theorems, by means of the argumentation forms and starting from the axioms.

As we showed in section 4.7, from one contradiction there follows the contradiction of all propositions and therefore the absurdity of the whole theory. It is thus necessary to ensure that the axioms themselves do not show any contradictions. How this can be obtained is one of the primordial problems of mathematics (and in fact of each rational game). A lot of sophisticated, and not always very encouraging results have been published on this matter.

In mathematics, the consistency of a set of axioms is tested by means of model or representation theory. Thereby mathematicians link the basic concepts and the axioms of a theory, of which they want to show that it is free of contradictions, to a corresponding representation in a model. That model they generally select from another theory of which the consistency is postulated. In fact, they try to hang the consistency of a new 'coat' onto a 'solid coat rack' in a trusted consistent 'vestibule'.

From the results of Kurt Gödel (see section 7) it has become clear that unfortunately there exists no possibility to show the consistency of a game, which has some degree of complication, starting from its own rules. So we always try to refer to something else in order to legitimise the things we undertake, just like lawyers in lawsuits or pedagogues in their publications. To make the above a bit more understandable we will first give one historical example from mathematics and then one that is related to a new game.

In the first half of the nineteenth century K.F. Gauss, J. Bolyai and N.I. Lobatchevsky elaborated the first designs for a non-Euclidean geometry. This was done by denying the parallel axiom (postulate) of Euclidean geometry. That axiom states that through a point not on a given line, there exists exactly one line parallel to the given line. In the so-called 'hyperbolic geometry' this parallel axiom is replaced by the proposition that through a point not on a line there exist an infinite number of lines 'parallel' to the given line.

The faith in the exclusively correct truth of Euclidean geometry was by then still so strong that any attempt to develop a non-Euclidean geometry was considered nonsense. Moreover it would soon prove to be inconsistent, this was at least the opinion of the antagonists. But then the convinced addicts thought of the trick of illustrating the consistency of the new geometry by a model that was borrowed from Euclidean geometry itself. If one believed that at least Euclidean geometry was consistent, then by this representation model one could also no longer doubt the validity of hyperbolic geometry. Such a model is for example the following:

Take as a representation for the 'plane' an open disk (i.e. a disk without the points on its edge) considered as an infinite collection of points. Define a 'line' in that 'plane' to be any chord (again without the edge points) in this disk. Consider 'parallel lines' to be the chords that do not have a single point in common. Therefore:

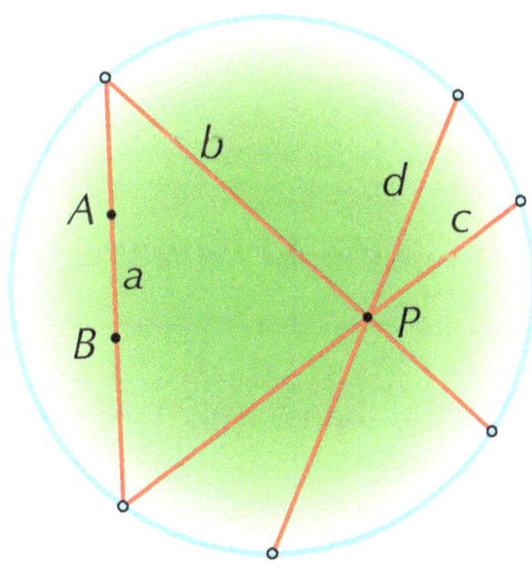

In this model it is obvious that:

- two given points of this 'plane' are on exactly one 'line'.
- an infinite number of lines through one point are 'parallel' to a given 'line'.

For example, the 'lines' *b*,*c* and *d*, through the point *P* are all 'parallel' to *a*.

This model kills two birds with one stone. In the first place it shows that hyperbolic geometry is consistent under the condition that Euclidean is consistent.

In the second place it shows that the parallel axiom of Euclid is independent of the other so-called incidence axioms of Euclidean geometry. Those last axioms are all valid in that model, but not the parallel axiom.

Until that moment a lot of mathematicians had already unsuccessfully attempted to deduce the parallel axiom from the other incidence axioms. The presented model shows clearly that this is impossible. The representation theory is thus also suitable to prove meta-theoretical propositions like 'the proposition p is not provable'!

The problem was then transferred to the question: 'is Euclidean geometry consistent?' For that geometry it is possible to construct a model as well, based on the model of the real numbers, by replacing points by their coordinates and by replacing lines by equations of the first degree. This can be realized by introducing a cartesian coordinate system.

Example

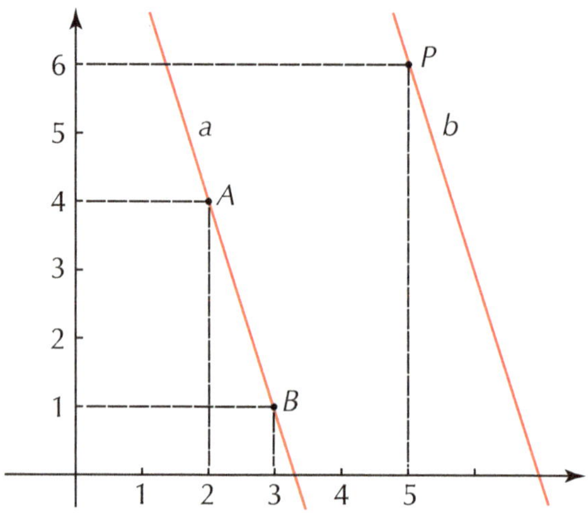

The point A corresponds in this model with the coordinate (2, 4), the point B with (3, 1) and the point P with (5, 6).

The line a through A and B is determined by the equation:

$$(3 - 2)(y - 4) = (1 - 4)(x - 2)$$

That can also be written as: $y = -3x + 10$.

You can easily check that the coordinates of A and B comply with the equation.

The line b through P, parallel to the line a, has equation:

$$y - 6 = -3(x - 5)$$

Also written as: $y = -3x + 21$.

Observe that this equation contains the same direction coefficient as the equation of line *a*, namely -3, and that the coordinate of *P* complies with this equation. So that last equation corresponds, in this model, with the unique line *b* through *P* and parallel to *a*.

Now the consistency of the real numbers comes into question. It was found that this can be linked to the consistency of the natural numbers. Because, starting from natural numbers, one can define rational numbers and then again from the latter the real numbers. But like this, won't we end up in an endless street?

Still in 1900, on the occasion of his presentation to the international mathematical community in Paris, the German mathematician Hilbert pronounced the expectation that someone would succeed in proving the consistency of the theory of natural numbers by the resources of that theory itself, i.e. that an absolute consistency proof for it would be possible. In 1931, Kurt Gödel showed that this will remain an idle hope for ever. For both the mathematical as well as the philosophical world this was a baffling result (see section 7).

To give you a better feel of the situation you would encounter if you set up a theory (or a game!), by the introduction of basic concepts and axioms (rules of the game), we will give a simple example of a new game: the 'trip-trap-game'.

In this game the following basic concepts appear: trips, traps and bingos. We also give the rules of the game (axioms):

'three trips make exactly one trap'

'two traps make exactly one bingo'

'altogether there are 36 trips, 12 traps and 4 bingos'.

Being given only this information, you can probably not get a clear view of this game at all.

Moreover, are these rules of the game consistent? From merely this formulation, you can not decide this. For this purpose we will give a representation model further on. Although you still don't know what this is all about and whether the rules of the game are indeed consistent, you can nevertheless already derive some other rules (you can prove some theorems).

Examples

'six trips make a bingo'

'three trips and one trap make a bingo'

'a maximum of 16 bingos can be constructed', etc.

Principally, we are exactly in the same situation when we build geometry: points, lines and planes are basic concepts that have no content by themselves. By the introduction of axioms like 'two different points determine exactly one line', etc.

we can however reason and prove new propositions without necessarily having a need for representations or concrete 'things'. It is this situation that Bertrand Russell referred to when he said that "Mathematics may be defined as the subject in which we never know what we are talking about, nor whether what we are saying is true." Not to mention other pathological aspects...

As we know from the practice of geometry, we can represent points, lines and planes by images on which the propositions about them can be verified. Whether we represent points by 'dots' and lines by 'stripes', or points by 'coordinates' and lines by 'equations', both representations are good enough to inspire us about theorems and to support our reasoning.

That we can do as well for the trip-trap-game.

We take for example a deck of 52 playing cards. Let us agree that the four aces represent the 'bingos', the twelve face cards the 'traps' and the remaining 36 numeral cards the trips. Three numeral cards can thus be exchanged for one face card and two face cards for one ace. In this way the axioms are however still not intuitively recognizable.

A better model would be a deck of 52 cards with the following pictures on them:

36 cards like 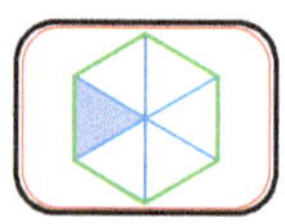 'a trip'

12 cards like 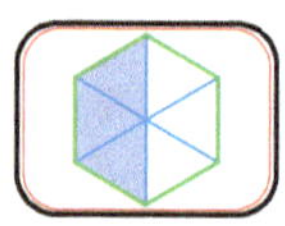 'a trap'

4 cards like 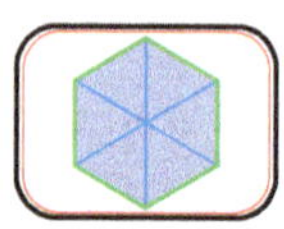 'a bingo'

Both the basic rules (axioms) as well as the derived rules (theorems) are now graphically supported and consequently they are easier to remember.

To get everybody reasoning, you can play this game for example as follows:

- deal out the deck of 52 cards evenly among four players
- the dealer can start by picking a card from the player on his right-hand side
- then he may lay down all the bingos he has as well as all the bingos he can make by combining some or all of his other cards
- next, the player on his left-hand side picks one of his cards, etc.

- the player who ends up with no cards first, wins the game.

To help the mathematics teacher, you may think of a way of slightly boosting motivation for the game of geometry. The students will appreciate it, though, if you arrange for the game to stop sooner or later!

So this is the countenance of the mathematical theories: basic concepts, axioms, defined concepts (definitions) and derived propositions (theorems and their proofs). Thus there exists a variety of theories (games) that we call structures, from which each builds its own independent rational universe.

Felix Klein (1849–1925)

German mathematician who concentrated especially on the study of non-Euclidean geometries and group theory. He was professor in Erlangen and Göttingen. His name has been linked to the so-called 'Erlanger-programm', in which a geometry is defined as the theory of the invariants under a specific group of transformations. Later, after Beltrami had provided, by means of the pseudo-sphere, a model showing that the consistency of two-dimensional non-Euclidean geometry was linked to the consistency of three-dimensional Euclidean geometry, Klein developed some more models.

6 Structures

We warn those unsuspecting readers who have an allergy to formulas, that in this section where mathematics shows off its most beautiful side, they will have to browse through a barrage of schooldays' knowledge from time to time.

In section 3 we constructed some number sets: natural numbers, rational numbers, real numbers, 'decapitated real numbers', etc. In such a set of numbers we can define operations. We define binary operations in a number set by assigning to two numbers of that set another number of that set. So, addition of natural numbers is an operation that assigns to the pair of natural numbers (3,5) the sum 8 . We say that, in the set of natural numbers $\mathbb{N}$, the operation + maps the pair (3, 5) onto 8.

Generalizing, the addition in $\mathbb{N}$ is a mapping of the form:

$$\mathbb{N} \times \mathbb{N} \rightarrow \mathbb{N}: (a, b) \rightarrow a + b$$

Such a mapping, which we call an operation, can have certain properties. For example:

for all a, b:	$a + b = b + a$	(commutativity)
for all a, b, c:	$(a + b) + c = a + (b + c)$	(associativity)
for all a:	$a + 0 = a$	(0 is neutral element)

The set of these basic properties makes a set of axioms for $\mathbb{N}$ provided with the operation +. $\mathbb{N}$ provided with the operation + is more briefly written as $\mathbb{N}$,+ .

Thus we obtain a structure that is completely characterized by the basic properties and in which reasoning is purely based on the form of those axioms and not on the nature of the elements. This offers the advantage that each time the mathematician meets an analogous structure, of which the elements at first sight have another appearance or meaning, he knows that the same reasoning and results apply in that structure.

We illustrate this with the example of the 'decapitated real numbers' (see section 3.3 example 4). The set of those numbers we represent by the interval [0,1 [. In that set we defined an addition, derived from the ordinary addition of real numbers: first we add up the numbers as usual and then we set the part before the decimal point at zero. This we call: adding up modulo 1.

Examples

$$0.3452 + 0.89 = 1.2352 = 0.2352 \text{ (in } [0,1[\text{)}$$
$$0.3452 + 0.6548 = 1.0000 = 0 \quad \text{(in } [0,1[\text{)}$$

The following properties apply, among others, to this addition in the set [0,1[:

for all a, b:	$a + b = b + a$	(commutativity)
for all a, b, c:	$(a + b) + c = a + (b + c)$	(associativity)
for all a:	$a + 0 = a$	(0 is neutral element)
for all a there exists a number b such that:		
	$a + b = 0$	(symmetrical element)

Due to this set of properties we say that [0,1[,+ (i. e. [0,1[provided with the operation +) is a commutative group. This is one of the most important and most fertile structures in mathematics.

We will now take a closer look at a particular result in the group of decapitated real numbers. Besides the neutral element 0 there still exists another self-symmetrical element. That means: an element a, such that $a \neq 0$, for which $a + a = 0$ applies.

That element a is 0.5 , since $0.5 + 0.5 = 1.0 = 0$ (in [0,1[,+).

From this follows that any equation of the form $2x = b$ in [0,1[,+ has two solutions that differ by 0.5.

Example

$$2x = 0.32 \quad \Leftrightarrow \quad x = 0.16 \text{ or } x = 0.66$$

since $\quad 2 \cdot 0.16 = 0.32$ and $2 \cdot 0.66 = 1.32 = 0.32$ (in [0,1[, +).

Let's look at the set of oriented angles, represented by the points on a circle with perimeter 1 and provided with a reference point O, which in itself represents the zero angle.

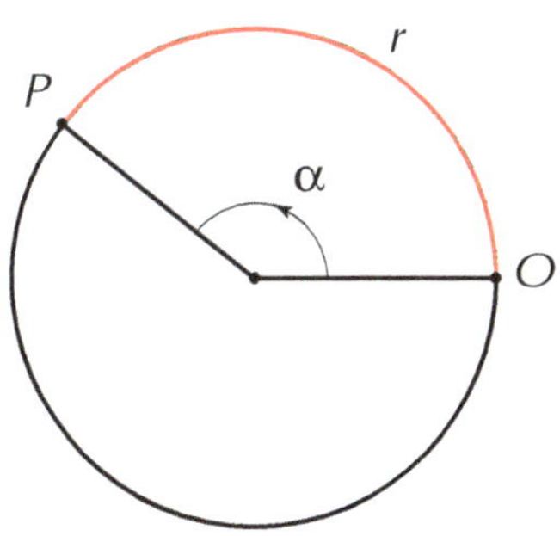

With each point P of this circle there corresponds a 'decapitated real number', since the length of the circle-sector OP is expressed as a number between 0 and 1. This sets up a one-to-one correspondence between the set of oriented angles H

and the elements of [0,1[. We add up oriented angles in the same way as the correspondent 'decapitated real numbers'. Both additions hence have precisely the same algebraic properties. We say that $H,+$ and $[0,1[\,,+$ are isomorphic groups.

Graphically represented as:

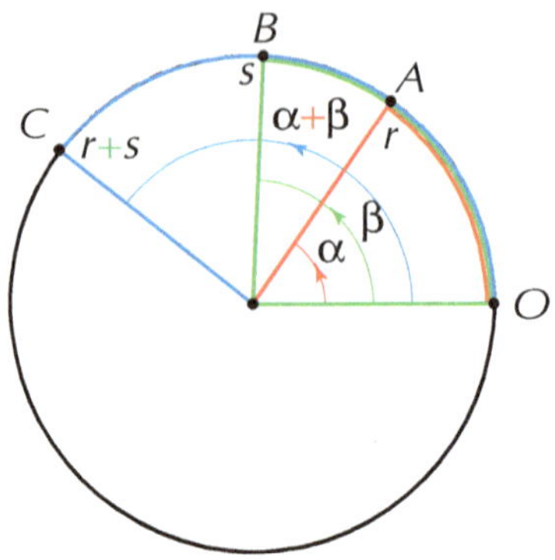

$\alpha+\beta$ in $H, +$ → $r+s$ in $[0, 1[\,, +$

In the figure below we project the image points A and B of the angles α and β on two perpendicular diameters, which we consider as cartesian axes. The coordinates of A and B relative to these axes are respectively (cosine α, sine α) and (cosine β, sine β).

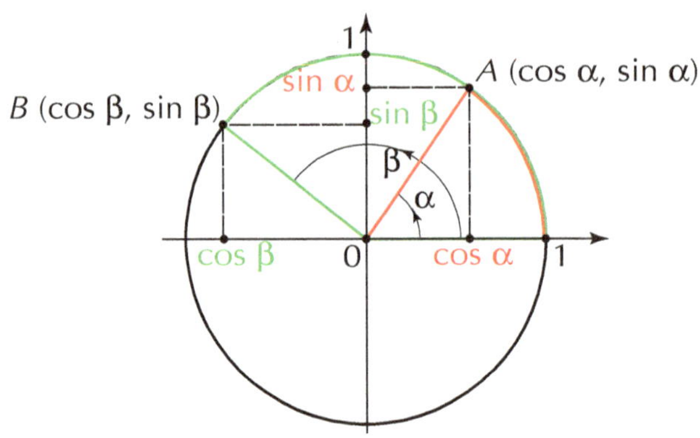

Let $\cos\alpha = p$, $\sin\alpha = q$, $\cos\beta = r$ and $\sin\beta = s$ then, remembering our dear secondary school mathematics, the following formulas apply:

$$p^2 + q^2 = \cos^2\alpha + \sin^2\alpha = 1$$

$$r^2 + s^2 = \cos^2\beta + \sin^2\beta = 1$$

$$\cos(\alpha+\beta) = \cos\alpha \cdot \cos\beta - \sin\alpha \cdot \sin\beta = p \cdot r - q \cdot s$$

$$\sin(\alpha+\beta) = \cos\alpha \cdot \sin\beta + \sin\alpha \cdot \cos\beta = p \cdot s + q \cdot r$$

Let C be the image point of the angle $\alpha + \beta$ then the coordinates of this point C relative to the same axes are $(p \cdot r - q \cdot s, p \cdot s + q \cdot r)$.

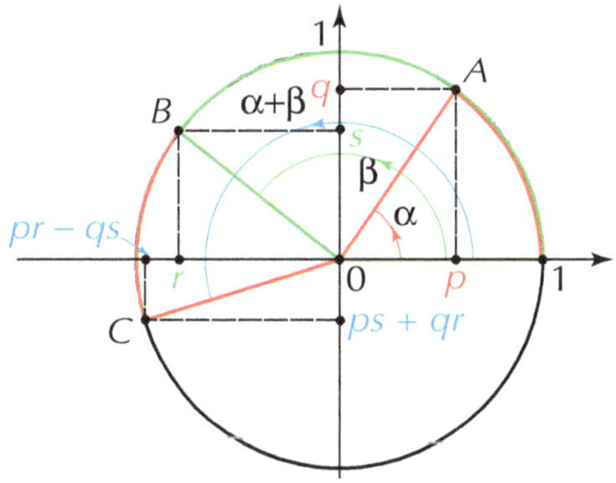

In mathematics an ordered pair of real numbers (a, b) can also be called a complex number. Usually it is then written $a + bi$ and is represented by a point in the so-called Gauss plane, being the plane provided with two cartesian axes of which the second determines the multiples of the imaginary unit i.

Under these conditions it is clear that:

$1 + 0i$ (– 1) has image point O, i.e. it represents the zero angle.

$0 + 1i$ (= i) has image point R, i.e. it represents the right angle.

$-1 + 0i$ $(= -1)$ has image point S, i.e. it represents the straight angle.

$p + qi$ has image point A, i.e. it represents the angle α .

$r + si$ has image point B, i.e. it represents the angle β.

$(p \cdot r - q \cdot s) + (p \cdot s + q \cdot r)\, i$ has image point C, i.e. it represents $\alpha + \beta$.

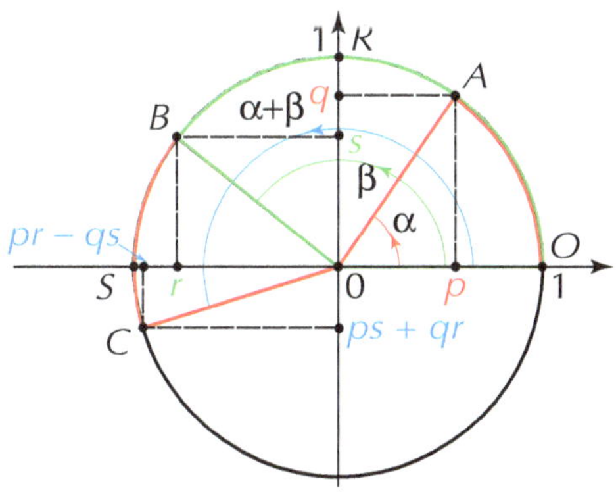

Allowing the addition of angles to correspond to the multiplication of those complex numbers that are situated on the circle, we get then, for example, the following results:

$i \cdot i = i^2 = -1$ since $90° + 90° = 180°$,

which corresponds to the fact that S is the image point of 180° and thus also of i^2.

$$(p + qi) \cdot (r + si) = (p \cdot r - q \cdot s) + (p \cdot s + q \cdot r)i$$

corresponding to the fact that C is the image point of $\alpha + \beta$ and thus also of $(p + qi) \cdot (r + si)$.

It is convenient that we can deduce the multiplication rule for the complex numbers from the usual multiplication rule for real numbers if we add the statement that $i^2 = -1$.

Examples

$$\begin{aligned}(2 + 3i) \cdot (1 - 4i) &= 2 \cdot 1 + 2 \cdot (-4i) + 3i \cdot 1 + 3i \cdot (-4i)\\ &= 2 - 8i + 3i - 12i^2\\ &= 2 - 8i + 3i - 12 \cdot (-1)\\ &= 14 - 5i\end{aligned}$$

$$\begin{aligned}(p+qi) \cdot (r+si) &= p \cdot r + p \cdot (si) + (qi) \cdot r + (qi) \cdot (si)\\ &= pr + (ps)i + (qr)i + (qs)i^2\\ &= pr + (ps)i + (qr)i + (qs) \cdot (-1)\\ &= (pr - qs) + (ps + qr)i\end{aligned}$$

The set $\mathbb{C}_1$ of all complex numbers $a + bi$ for which $a^2 + b^2 = 1$ provided with this multiplication, is completely isomorphic to the groups $H,+$ (oriented angles) and $[0,1[,+$ (decapitated real numbers) mentioned earlier. The group $\mathbb{C}_1, \cdot$ is however of the multiplicative kind (written as a multiplication) instead of the additive nature of the other two (which we wrote as additions).

The double of an oriented angle (or of a decapitated real number) thus corresponds to the square of such a complex number. And the two different halves of an oriented angle (or of a decapitated real number) correspond to the two different square roots of the complex number.

All these isomorphic groups are no more than just different representations of one and the same intrinsic structure.

So each time, the same mathematical individual is involved, however in different clothes. That wardrobe can of course still be extended, depending on the needs of the actual contexts. In this respect, we don't have to put budgetary restrictions on imagination or on romanticism. As an illustration, we give two more representations of this group.

A rotation in the plane is completely defined by the centre (fixed point) which is turned about and by the (oriented) angle about which is turned. The composition of rotations with the same centre corresponds to the addition of the rotation angles concerned. The set R_o of the rotations with the same centre O, provided with this composition law ∘, is therefore also a multiplicative version of a group isomorphic to H, +.

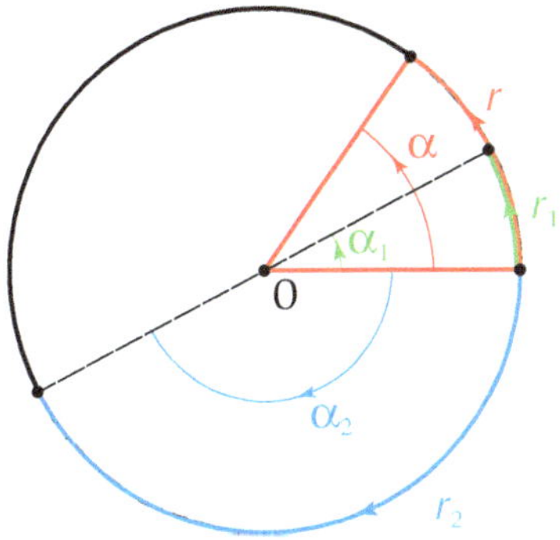

These rotations will then also have two different square roots, this means: for each rotation r, there exist two different rotations r_1 and r_2 such that

$$r_1 \circ r_1 = r \quad \text{and} \quad r_2 \circ r_2 = r.$$

Rotations about O belong to the so-called 'linear transformations' of the plane. These are transformations that are completely defined by the images of two basic elements E_1 and E_2 on the axes (see figure below). This is clearly the case for a rotation about the centre O.

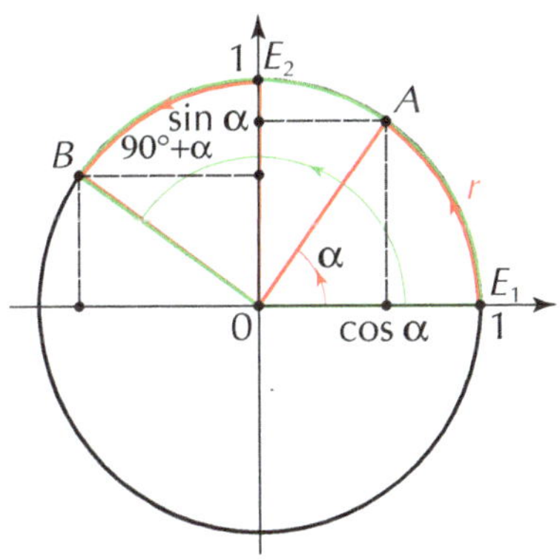

If α is the rotation angle of the rotation r, then r maps the point E_1 with coordinate (1, 0) onto the point A with coordinate (cos α, sin α), and the point E_2 with coordinate (0, 1) onto the point B with coordinate (cos(90° + α), sin(90° + α)).

By means of the formulas for cos(α+ β) and sin(α + β) mentioned earlier and because of the meaning of 'cos' as the projection onto the first axis and 'sin' as the projection onto the second axis we can calculate:

$$\cos(90^\circ + \alpha) = \cos 90^\circ \cdot \cos\alpha - \sin 90^\circ \cdot \sin\alpha$$
$$= 0 \cdot \cos\alpha - 1 \cdot \sin\alpha$$
$$= -\sin\alpha$$
$$\sin(90^\circ + \alpha) = \cos 90^\circ \cdot \sin\alpha + \sin 90^\circ \cdot \cos\alpha$$
$$= 0 \cdot \sin\alpha + 1 \cdot \cos\alpha$$
$$= \cos\alpha$$

The coordinates of the images of the basic elements, that define the rotation unambiguously, are usually written in matrix form, i.e.

$$\begin{bmatrix} \cos\alpha & \sin\alpha \\ -\sin\alpha & \cos\alpha \end{bmatrix} \quad \text{and here is } \cos^2\alpha + \sin^2\alpha = 1$$

If we compose two rotations with rotation angles α and β respectively, then we know that the rotation angle of the composition is $\alpha + \beta$. If we consider the matrix of the composition of two rotations to be the product of the matrices of those two rotations, then we get the following result:

$$\begin{bmatrix} \cos\alpha & \sin\alpha \\ -\sin\alpha & \cos\alpha \end{bmatrix} \cdot \begin{bmatrix} \cos\beta & \sin\beta \\ -\sin\beta & \cos\beta \end{bmatrix} = \begin{bmatrix} \cos(\alpha+\beta) & \sin(\alpha+\beta) \\ -\sin(\alpha+\beta) & \cos(\alpha+\beta) \end{bmatrix}$$

$$= \begin{bmatrix} \cos\alpha\cdot\cos\beta - \sin\alpha\cdot\sin\beta & \cos\alpha\cdot\sin\beta + \sin\alpha\cdot\cos\beta \\ -\cos\alpha\cdot\sin\beta - \sin\alpha\cdot\cos\beta & \cos\alpha\cdot\cos\beta - \sin\alpha\cdot\sin\beta \end{bmatrix}$$

In this diagram we recognise the specific rule of the row-column multiplication of matrices:

$$\begin{bmatrix} a & b \\ c & d \end{bmatrix} \cdot \begin{bmatrix} p & q \\ r & s \end{bmatrix} = \begin{bmatrix} ap+br & aq+bs \\ cp+dr & cq+ds \end{bmatrix}$$

The sum of the products of the elements of the i-th row of the first matrix with the respective elements of the j-th column of the second matrix makes the element on the i-th row and the j-th column of the product matrix.

This shows that the set of the matrices M of the form

$$\begin{bmatrix} a & b \\ -b & a \end{bmatrix} \quad \text{with } a^2 + b^2 = 1,$$

provided with the matrix multiplication, is also a multiplicative group that is isomorphic to $\mathbb{C}_1, \cdot$ and $R_o, \circ$ as well as with the additive groups $H, +$ and $[0,1[, +$.

It is an enormous saving of mental energy and a convenience for the applications if one takes notice of the isomorphism of these structures. And this is only possible by bringing to the fore the intrinsic properties (axioms) of such structures rather than the particular appearance of their elements.

Although it might possibly not be simple for a layman in mathematics to have a deep understanding of all this, it should nevertheless become clear how the different phases of the mathematics game proceed:

1 How are concepts introduced?
2 How are true propositions about them brought together in an axiomatic system?
3 How is supervision kept, with the help of logic, over the correct transition of those propositions into new theorems?
4 How is the theory moulded into a structure?
5 How are such structures implemented in isomorphic models?
6 How does the transfer of properties and results between the isomorphic models work out?

The structures, which mathematics creates by this method, are consequently available to all branches of science that attempt to give their research field a rational basis, in order to boost its transparency and efficiency. The success tale of that cooperation still continues.

The French mathematician **Evariste Galois** (1811–1832), who died too young, is considered the founder of group theory and of the theory of equations, also called the Galois theory. On the occasion of a prize contest he had sent in a paper *Concerning the conditions for the solvability of equations by radicals*, with Fourier as his intermediary, to the 'Académie des Sciences'. When Fourier died shortly afterwards, the paper got lost. On the eve of a duel, fearing a dramatic outcome, he wrote a letter to a friend containing the sketch of his most important discoveries in the field of the symmetry groups of the solutions of equations. These writings became the basis of one of the deepest and most fertile structures in mathematics.

7 Meta-theory and the theorem of Gödel

Each mathematical structure forms a rational mini-universe that completely stands by itself, apart from all the other varieties that claim legitimacy as well. Sometimes it is possible to travel from one universe to another or to incorporate some of them in more general entities. For others this is, however, impossible because of the contradictions that would appear. Think, for example, of the incompatibility of Euclidean and non-Euclidean geometry.

The mathematician finds himself in this matter in an extremely comfortable position. Without any hesitation he can move into each of those structures and conform to the rules of the game, like a player who is playing chess, draughts and cards without getting confused by the different rules of these games. He can also examine those structures from the outside and, from a higher level, make statements and raise questions on them.

Examples

'Is this set of axioms consistent?'

'Does this proposition really depend on those others?

'Can this proposition indeed be proved in this structure?"

'Can the theory be completed?' i.e. 'Are all true propositions also provable?'

These are meta-theoretical questions about the theory concerned. To solve those problems, usually other means are called upon than only those of the theory itself. Just think of the examples from the model theory in section 5. This way, however, the meta-theory lives a life apart, without any link to the theory itself. One of the 23 fundamental problems, which the German mathematician David Hilbert presented in 1900 to the international mathematicians community in Paris, was exactly the meta-meta-theoretical question whether it is possible, by means of the theory itself, to solve the problem of its consistency, i.e. whether an absolute consistency proof is possible.

Already in 1931, aged only 25, Kurt Gödel published a remarkable paper concerning his research on this subject. Because the results were so profound and bewildering, not only for mathematics but also for philosophy, we will make an attempt here to briefly outline its main ideas.

The theory of the natural numbers is the mother of a multitude of derived theories. "The integer numbers have been created by God, everything else is man's work", L. Kronecker postulated earlier. For the system of the natural numbers an elegant axiom system was created by G. Peano (1858–1932). So it was obvious that people attempted to resolve the question that Hilbert had suggested in the first place for the theory of the natural numbers.

In the axioms of Peano the successor structure of the natural numbers had been systematized. The concept successor of a natural number, together with the principle of the inheritance of properties from the predecessor (induction principle), thus makes the essence of the formal construction of that theory.

In their *Principia Mathematica* the English mathematicians and philosophers Bertrand Russell and Alfred North Whitehead tried to show that the fundamentals and the methods of mathematics and logic are essentially the same and that arithmetic hence could be reduced to logic. While they were so diligently busy relating more and more theorems concerning the natural numbers in a formal piece of knitwear, in the hope of being able to complete the theory sooner or later, this expectation was suddenly floored by the already mentioned paper by K. Gödel, with the title: *Über formal unentscheidbare Sätze der Principia Mathematica und verwandter Systeme* (On Formally Undecidable Propositions Of Principia Mathematica And Related Systems).

In 1931, neither the content nor the style of argumentation in Gödel's paper was understandable for most mathematicians. It seems therefore an unattainable aspiration to make it accessible for the layman. But it is an intellectual challenge to unravel, even roughly, the ingenuity and the profundity of a reasoning that proves that you cannot prove all true propositions, i.e. that a formal description of a somewhat more complicated field must necessarily remain incomplete. This is a proof for rationality's impotence of getting control of the universal everything, and this knowledge is offered by rationality itself! The post-modernists are probably already sniggering, but nevertheless they will have to work through our explanation in order to be able to justify their pleasure.

We now cast an eye on Gödel's working method, which briefly consists in converting everything into numbers, the so-called Gödel numbers. The concepts, the propositions of the theory, as well as the meta-theoretical propositions about the theory are converted into numbers. A proof, which is an arranged sequence of propositions, is converted into a product of powers of prime numbers of which the exponents are equal to the code numbers of those propositions. In that way, a proof itself corresponds to a number. The most important find is, however, the translation of meta-theoretical propositions into number theoretical relationships, which in their turn are themselves expressed by a number again. By this, they are making part of the theory itself!

For this purpose we introduce a minimum of technical tools, necessary to outline the construction of the argumentation.

First step: arithmetizing the theory

a

Concepts and symbols		→ Numbers
not	$\neg$	1
or	$\vee$	2
if … then	$\rightarrow$	3
there exists a	$\exists$	4
is equal to	$=$	5
zero	0	6
is a successor of	s	7
opening parenthesis	(	8
closing parenthesis	)	9
comma	,	10

For this we use the numbers 1 to 10.

b

Numerical variables	→ Numbers
x	11
y	13
z	17
u	19
…	…

Here we use prime numbers bigger than 10.

c

Propositions in the theory	→ Numbers

Example

'Each natural number y has a successor x'

(	$\exists$	x	)	(	x	=	s	y	)
8	4	11	9	8	11	5	7	13	9

$$2^8 \cdot 3^4 \cdot 5^{11} \cdot 7^9 \cdot 11^8 \cdot 13^{11} \cdot 17^5 \cdot 19^7 \cdot 23^{13} \cdot 29^9 = a$$

Here we use numbers that are the product of powers of successive prime numbers.

The formalized proposition is a product of powers, in which the code numbers of the symbols make the exponents and successive prime numbers make the bases. By factorizing the number a and decoding the sequence of exponents obtained, we regain the original proposition. The number a thus defines the proposition in an unambiguous way.

d	Theorems and proofs (sequence of formulas)	→	Numbers

Example

If 'every natural number y has a successor x'
then 'the number zero has a successor z'

$$\left.\begin{array}{lll}(\exists x)(x = sy) & \rightarrow & \text{code number } a \\ (\exists z)(z = s0) & \rightarrow & \text{code number } b\end{array}\right\} \rightarrow \text{code number } 2^a \,.\, 3^b = c$$

Also here we use numbers that are the product of powers of successive prime numbers.

By factorizing the Gödel number c and next factorizing its exponents, the given theorem can be decoded in an unambiguous way.

Second step: arithmetizing the meta-theory

The ingenuity lies in the finding of translating meta-theoretical propositions in number theoretical relations, by coding them according to the method stated above.

Example

Meta-theoretical proposition:

the sequence of formulas $((\exists x)(x = sy), (\exists z)(z = s0))$ (with code number c)
is a proof for the formula $(\exists z)(z = s0)$ (with code number b).

Because a proof is a sequence of propositions of which the last one is exactly the theorem to be proved, the code number of that theorem will appear as the exponent of the last factor in the product which is the code number of its proof. This last factor is, of course, a divider of the product.

So, the number theoretical relation is:

the number 3^b divides (is a factor of) the number $c = 2^a \cdot 3^b$.
coding: $(\exists u)\ (c = u \cdot 3^b) \rightarrow$ code number d.

Observe that the truth of the meta-theoretical proposition 'is a proof of' now can be checked in the theory itself, just by checking the numerical relationship: 'is a factor of'. And this is expressed by the code number d. To make the last code number d easier to remember, and keeping its meta-theoretical meaning in mind, we will represent it also as proof(c, b).

In this way Gödel succeeded in constructing an arithmetical relationship that interprets the meta-theoretical proposition: 'the sequence of propositions with code number x makes a proof for the proposition with code number z'. In other words 'x is a proof for z'. And next he succeeded in representing this arithmetical relationship by a number, which we call proof(x, z). If we want to express that the arithmetical relationship interpreted by the code number proof(x, z) is not true, we write ¬proof(x, z). On the level of meta-theory this means that the sequence of formulas with code number x is not a proof for the formula with code number z.

The meta-theoretical proposition 'z is unprovable', which can also be formulated as: 'for any sequence of formulas with code number x, it applies that x is not a proof for the formula with code number z', will be represented arithmetically by the formula '(x) ¬proof(x,z)'.

Third step: substitution principle

If we replace a numerical variable in a certain formula with Gödel number a by the number-sign k, then we get a new formula that will generally have a different Gödel number.

Example

$(\exists x)\ (x = sy) \quad \rightarrow \quad$ code number a

$(\exists x)\ (x = sk) \quad \rightarrow \quad$ code number b

Observe that we get the second formula by substituting, in the formula with code number a, the variable with code number 13 (i.e. y) by the number-sign k. To make it easier to remember how the code number b was obtained, we will rather write it as sub(a, 13, k).

So, for example, the following arithmetical formula with code number m:

'(x)¬proof(x, sub(y, 13, y))' $\rightarrow$ code number m (1)

means on the level of meta-theory:

> 'the formula with code number sub(y,13,y) which we got by changing, in the formula with code number y, the variable with code number 13 by the number-sign y, is unprovable'.

Fourth step: construction of the theorem G

Starting from formula (1), with code number m, let us construct the formula with code number sub(m,13,m), so this would be the formula obtained by changing, in the formula with code number m, the variable with code number 13 (i.e. y) into the number sign m.

This leads to:

(G): '(x)¬proof(x, sub(m, 13, m))' $\rightarrow$ code number sub(m, 13, m)

The formula G is thus the arithmetical translation of the meta-theoretical proposition:

'the formula with code number sub(m,13,m) is unprovable'

But this is apparently G itself, regarding its Gödel number!

So, G tells us: 'I am unprovable'.

This resembles to the Cretan who tells us 'I am lying'. It is clear that such a proposition being true as well as being false leads to a contradiction. The paradox of the Cretan can however still be countered by lifting the proposition 'I am lying' onto the meta-theoretical level and by stating that this proposition only applies to a set of earlier stated propositions, to which the new proposition 'I am lying' does not belong.

But the formula G is a different case. G expresses indeed the meta-theoretical proposition 'this proposition is unprovable', but it is in itself an arithmetical formula. This means that either G or the negation of G has to be true in arithmetic. Let us suppose that G would be provable, then depending on its form, G as well as the negation of G would be provable and vice versa, which would mean that arithmetic and its axiom system would not be consistent.

If nevertheless we want to keep on believing in the consistency of arithmetic, then we must admit that neither G nor its negation can be derived from the axioms and that consequently these axioms are incapable of proving all true propositions within the theory. For this reason, this Gödel's theorem G is also called the 'incompleteness theorem'.

In each axiomatic-deductive theory, which is complex enough to include, explicitly or implicitly, the arithmetic of the natural numbers, this incompleteness can always be shown in the same manner. Irrespective of how ever we try to tackle this incapability of the original axioms by completing them with new axioms, such a proposition G can always be constructed. Therefore we know that we will never be able to formalise entirely, in a rational way, certain domains of our experience. This means that there are intrinsic restrictions to the extent of the rational method. Moreover and somewhat ironically, this is shown by this rational method itself. We can prove that we cannot prove everything. This is the deeper philosophical significance of Gödel's theorem

Fifth step: undecidability of the consistency of arithmetic

Starting from the incompleteness theorem and continuing his argumentation, Gödel could also represent the meta-theoretical proposition 'arithmetic is consistent' within arithmetic itself, and show that this formula is also undecidable. The consistency of arithmetic is therefore unprovable starting from the axioms of arithmetic itself. This smashed Hilbert's dream to smithereens. We will also briefly outline this argumentation.

If a theory contains one contradiction, then all its propositions as well as their negations are logically derivable (see section 4.7), i.e. the whole theory is then inconsistent. Consequently (by contraposition): if in a theory at least one proposition is unprovable, then the theory is consistent. Thus the unprovability of propositions also has pleasant sides.

The meta-theoretical proposition 'arithmetic is consistent' is thus equivalent to 'there exists a proposition in arithmetic that is unprovable '. This last sentence we can represent in the usual way in arithmetic itself and thus provide it with a code number:

$$(C): \quad `(\exists y)((x)\neg \text{proof}(x,y))' \quad \rightarrow \quad \text{code number } n$$

It is obvious that this formula C is true in arithmetic. For example, it would be sufficient to substitute y by the code number G of Gödel's theorem. But also the formula C itself cannot have a proof in arithmetic. Because, if 'arithmetic is consistent', then 'there exists a formula that is unprovable' (for example G). This implication can be represented in arithmetic as: $C \rightarrow G$. Now, let C be provable, then from the modus ponens it follows that:

$$C \text{ and } (C \rightarrow G) \Rightarrow G$$

In other words, G would also be provable, which is exactly impossible if arithmetic is to be consistent. So, it is impossible to prove the consistency of arithmetic within arithmetic itself.

In the course of the twentieth century further aspects of incompleteness and undecidability were shown in the field of data processing, of diophantic equations, of the theory of complexity of problems, and several others. We cannot elaborate all these sophisticated areas given the scope of this book. For this we refer the interested reader to the specific professional literature in the bibliography at the end of this book.

Kurt Gödel (1906–1978)

Mathematician and logician, born in Brno, in the present Czech Republic but at that time a part of the twin-monarchy of Austria and Hungary. He undoubtedly provided the most important climax in the study of the fundamentals of mathematics, which had commenced in the second half of the nineteenth century. In 1931, he proved his world famous incompleteness theorem, which implies important philosophical consequences, also outside the field of mathematics. It may seem ironical that he had nevertheless proved just before, in 1929, the completeness of the logic of the first order. From that it is shown that arithmetic, on which the incompleteness theorem applies, can not coincide with this kind of logic. This definitively sunk the project of the *Principia Mathematica* of Russell and Whitehead. Later Gödel became a permanent member of the Institute for Advanced Study in Princeton, with which also Einstein was associated.

Conclusion

In this chapter we have provided a glimpse into certain characteristics of mathematical methodology:

- the origin of the rational way of thinking in the human mind
- the specificity of the mathematical language
- the creation of mathematical concepts
- the construction of the logical-deductive tools
- the starting up of a theory by the choice of an appropriate set of axioms
- the creation of typical structures with their semantic models
- the reflections of the mathematician on his own methodology and the range of philosophical consequences implied by this.

But of course, all this has not yet shown precisely how the 'working mathematician' wrestles with the preceding problems in his daily occupation. Unfortunately, there exists no systematic working method. Not only knowledge and imagination is needed, but also a 'coincidence of circumstances' and some luck. Certain famous conjectures (Goldbach's conjecture, Riemann's hypothesis, the complexity problem of algorithms, the existence of odd perfect numbers, etc.) are still waiting for a proof. Also 'the last theorem of Fermat' was an intriguing unsolved problem for more than three centuries, until in 1995 the English mathematician A. Wiles succeeded in constructing a proof for it, starting from the solution to totally different problems.

At the start of the twentieth century Hilbert still faithfully uttered the hope that every mathematical problem would get a solution sooner or later, because according to him, no 'ignorabimus' can exist in mathematics. In the meantime we know, because of Gödel's incompleteness theorem, that a certain presumption actually can be true without being provable within the context of a certain axiomatic system. Because of this, some mathematicians spend their whole life groping in the dark for their 'philosophers stone', often without ultimate success. But their efforts always generate fruitful mathematics, although not necessary in the field they had in mind. In spite of the perspective offered by Gödel, that mathematical knowledge is a project that can never be completed, mathematical knowledge nevertheless grows steadily and its addicts experience intense pleasure from each newly acquired insight.

"God exists since mathematics is consistent, and the Devil exists since we cannot prove it."

André Weil

Chapter 2

How does mathematics work?

A mathematician is a machine that transforms coffee into theorems.

Paul Erdös

In this chapter we give a number of examples of the way that mathematics develops. In particular, we will highlight the creativity of the mathematical method. These examples can be read independently of each other, so you can dip into them at random.

Starting from simple problems or graphic intuitions, we introduce the basic concepts and elementary tools needed to achieve the intended solutions. Using analogy and generalization, or by slightly modifying the structural elements, we obtain a panorama of new situations that goes far beyond the context of the initial problems. The resulting insights often open the door to unexpected applications.

Let unsuspecting readers be warned that in the ensuing sections they may well be called upon to engage in mathematical exertions, but through such exertions it will then be, with appropriate guidance, just a short stroll to reach a point offering a breathtakingly wide panorama. Those who are not keen on climbing will miss the view

The frontispiece of the first English version of Euclid's Elements, published in 1570 by Sir Henry Billingsley, who was also Lord Mayor of London. The title notice makes erroneous mention of Euclid of Megara instead of Alexandria, something that occasionally happened in the Middle Ages. John Dee assisted Sir Henry with technical advice and contributed by adding annotations and writing the preface.

R. Verhulst, *Under the Spell of Mathematics*, https://doi.org/10.1007/978-3-031-63871-8_2

1 The royal road to multidimensional and non-Euclidean geometry

How can we, starting from simple representations of figures, borrowed from the intuitive two- and three-dimensional Euclidean space, come to an understanding of geometry of a higher and even an infinite dimension?

How can we, starting from our trusted representations of parallelism and of perpendicular lines, penetrate new spaces in which there exist, for example, no parallel line to a given line or in which there exist lines that are perpendicular to themselves?

For this, one doesn't have to do advanced mathematical studies. It is enough to take the little steps of an optimized didactic method.

1.1 The express train of analytic geometry

Step 1 Points, lines and planes

On a cube-shaped block we can recognise the basic figures of planar- and three-dimensional geometry and we can illustrate all the possibilities of their relative positions.

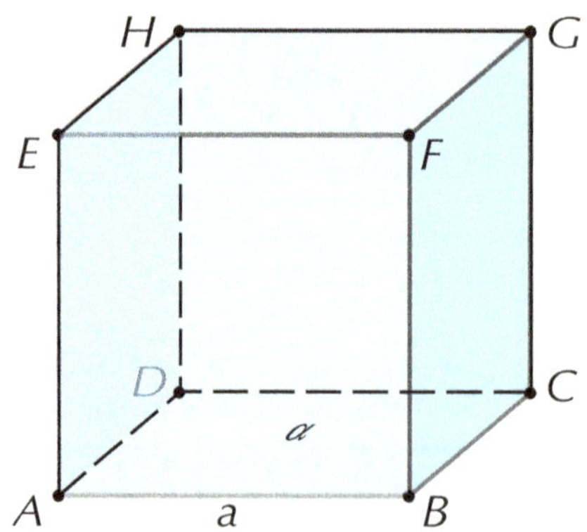

The points A, B, etc. represent the vertices.

The line segments AB, BC, etc. represent the edges. These are parts of lines, for example AB is part of the line a.

The squares $ABCD$, $BCGF$, etc. represent the faces. These are parts of planes, for example $ABCD$ is part of the plane α.

In a plane, for example the one defined by *ABCD*, only two different relative positions for two different lines are possible:

intersecting, like *AB* and *BC*	(one point in common)
parallel, like *AB* and *DC*	(no point in common)

In three-dimensional space, however, three relative positions of two lines are possible:

intersecting, like *AB* and *BC*	(one point in common)
parallel, like *AB* and *DC*	(no point in common and in one plane)
skew, like *AB* and *FG*	(no point in common and impossible to contain in one plane)

In three-dimensional space two different planes can only have two relative positions:

intersecting, like *ABCD* and *ABFE*	(one line in common)
parallel, like *ABCD* and *EFGH*	(no point in common)

Step 2 Coordinates and equations

To make points correspond to coordinates, we have to choose a set of axes and units on them.

Let's do this first in the plane

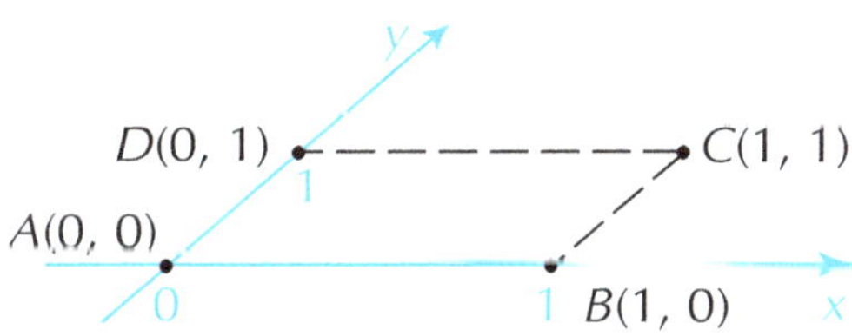

Let's choose the lines *AB* and *AD* of the previous figure respectively as *x*- and *y*- axes.

A is the origin of this set of axes, on which the points *B* and *D* define the units. This gives the following coordinates for the points *A*, *B*, *C* and *D*: *A*(0, 0), *B*(1, 0), *D*(0, 1) and *C*(1, 1).

Calculations with coordinates are done in the same way as calculations with vectors, this means that two operations are defined on them:

an addition: $(a, b) + (c, d) = (a + c, b + d)$

a multiplication by a scalar (real number): $r(a, b) = (ra, rb)$

Geometrically, the addition is no more than the parallelogram rule for the sum of vectors as used in physics, for example

$$\overrightarrow{AB} + \overrightarrow{AD} = \overrightarrow{AC} \text{ corresponds to } (1, 0) + (0, 1) = (1, 1)$$

So, the next figure shows that:

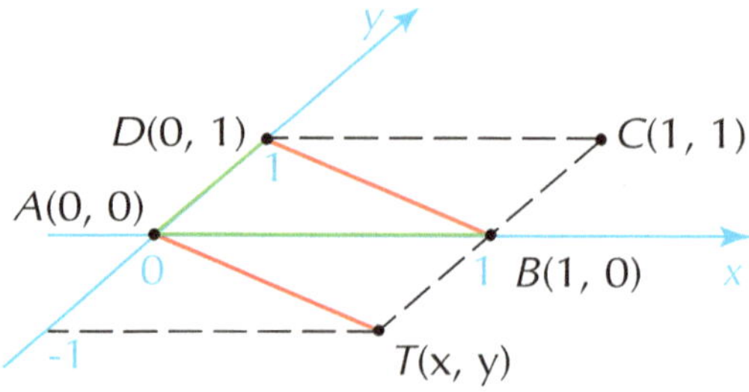

$\overrightarrow{AD} + \overrightarrow{AT} = \overrightarrow{AB}$ corresponds to $(0, 1) + (x, y) = (1, 0)$

And thus $\overrightarrow{AT} = \overrightarrow{AB} - \overrightarrow{AD}$ corresponds to $(x, y) = (1, 0) - (0, 1) = (1, -1)$ (ç).

And since: $\overrightarrow{DB} = \overrightarrow{AT}$ we also have $\overrightarrow{DB} = \overrightarrow{AB} - \overrightarrow{AD}$.

The multiplication of a coordinate with a scalar *r* corresponds geometrically to the (homothetic) magnification of a vector by the same factor *r*, for example:

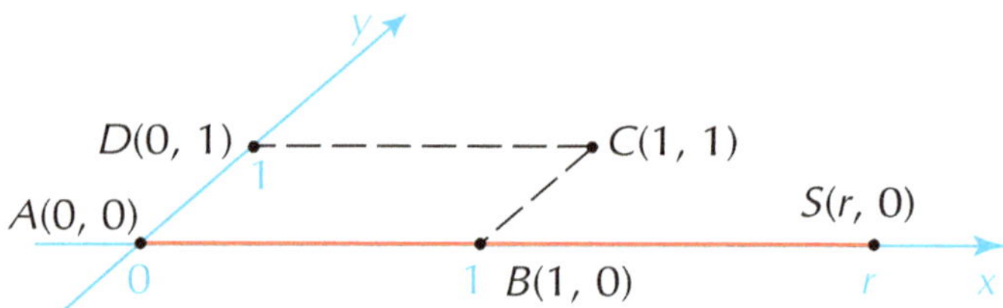

$r\overrightarrow{AB} = \overrightarrow{AS}$ corresponds to $r(1, 0) = (r, 0)$

By means of these operations, we can set up equations of lines respective to this coordinate system.

Example

A point *S* with coordinate (x,y) is on the line *AB* if and only if (x,y) is a real multiple of the coordinate (1,0) of *B*:

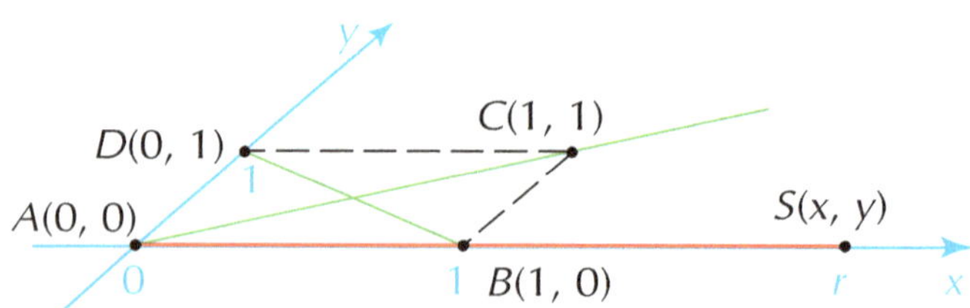

$S(x, y)$ is on $AB \Leftrightarrow$ there exists a real number r such that $(x, y) = r(1, 0)$ (*)

$\Leftrightarrow (x, y) = (r, 0)$

$\Leftrightarrow x = r$ and $y = 0$ (**)

We call (*) a vectorial equation and (**) a set of parametric equations of the line *AB*. In these, *r* is the parameter (variable scalar). By elimination of the parameter from the set of the parametric equations, we get the so-called cartesian equation, which is in this case $y = 0$.

In what follows we will always restrict ourselves to the vectorial equations, because they can be generalized in a trivial way. In science, it is important not to get lost in dead-end streets. Intuition to avoid this is often gained by a historical process of trial and error. But once the safe path has been found, the experienced guide should not without reason mislead the newcomers. Thus, the process of teaching should not explore embryonic historical tracks over and over again.

In an analogous way as for the line *AB*, it can be seen that:

AD has a vectorial equation of the form $(x, y) = r(0, 1)$.

AC has a vectorial equation of the form $(x, y) = r(1, 1)$.

AB, *AD* and *AC* are all lines through the origin *A* and their directions are unambiguously defined respectively by the points *B*, *D* and *C*, alias by the coordinates (1, 0), (0, 1) and (1, 1).

We say that these coordinates define the direction vectors of these lines.

Any multiple of a direction vector of a line is also a direction vector of the same line.

But how can we describe arbitrary lines (not containing the origin) this way?

Example

For the line *DB*:

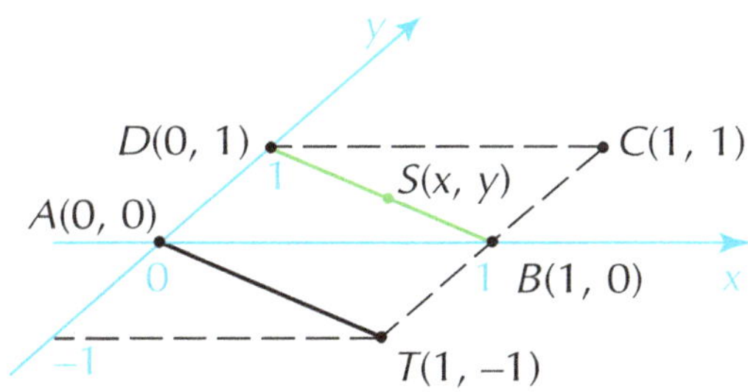

Observe that the line *DB* is parallel to the line *AT*, which contains the origin. Parallel lines have the same direction and hence the same direction vectors. The coordinate (1,– 1) defines a direction vector of the line *AT* and thus also of the line *DB*. Such a direction vector was found in (ç) as the difference of the coordinates of the points *B* and *D*, namely $(1, 0) – (0, 1) = (1, –1)$.

And since (1, 0) determines a point of DB, we can state that:

$$S(x, y) \text{ is on } DB \quad \Leftrightarrow (x, y) = (1, 0) + r(1, -1)$$
$$\Leftrightarrow (x, y) = (1, 0) + r((1, 0) - (0, 1)) \qquad (\S)$$

If we know the coordinates of two different points of a line, then we can write the line's vectorial equation by using them like in (§).

Example

Let $P(a, b)$ and $Q(c, d)$ be two points, then $(x, y) = (a, b) + r((c, d) - (a, b))$ is a vectorial equation of PQ.

This can also be written as $(x, y) = (a, b) + r(c - a, d - b)$.

It takes hardly any more effort to take one step up to three-dimensional space. For this purpose, we just take three axes and units on them. For example AB, AD and AE respectively as x-, y- and z-axis, on which B, D and E determine the units. We notice that the points have now three coordinate numbers each:

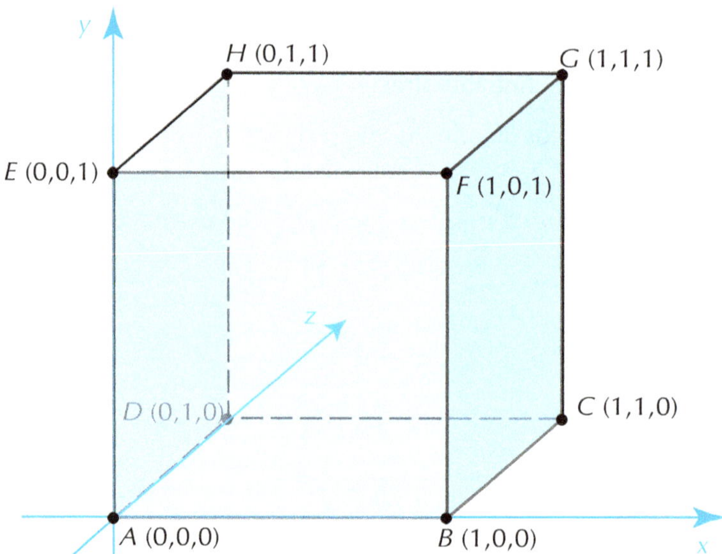

e.g. $A(0, 0, 0)$, $B(1, 0, 0)$, $D(0, 1, 0)$, $E(0, 0, 1)$

$C(1, 1, 0)$, $F(1, 0, 1)$, $H(0, 1, 1)$, $G(1, 1, 1)$

These coordinates can be added up and be multiplied with a real number in an analogous way as we did in the plane:

$$(a, b, c) + (d, e, f) = (a+d, b+e, c+f)$$
$$r(a, b, c) = (ra, rb, rc)$$

The geometrical meaning of these operations is the same in the three-dimensional space as it was in the plane.

In the same way as in the plane, in three-dimensional space we can define the vectorial equation of a line through two points by means of their coordinates.

Example

For $H(0,1,1)$ and $G(1,1,1)$:

$$S(x, y, z) \in HG \Leftrightarrow (x, y, z) = (0, 1, 1) + r((1, 1, 1) - (0, 1, 1))$$
$$\Leftrightarrow (x, y, z) = (0, 1, 1) + r(1, 0, 0)$$

Observe that HG is parallel to AB and that it has a direction vector determined by (1, 0, 0).

Three points, not all three of them on one and the same line, are sufficient to define a plane. So, to define a plane through the origin we only have to choose two more points, not both on one line containing the origin. These last two points determine two different direction vectors of this plane.

Example

The plane ABD has direction vectors $\overrightarrow{AB}$ and $\overrightarrow{AD}$. Each arbitrary point P of this plane determines a vector $\overrightarrow{AP}$ that can be written as the sum of a certain multiple of $\overrightarrow{AB}$ and a certain multiple of $\overrightarrow{AD}$ i.e. $\overrightarrow{AP} = r\,\overrightarrow{AB} + s\,\overrightarrow{AD}$.

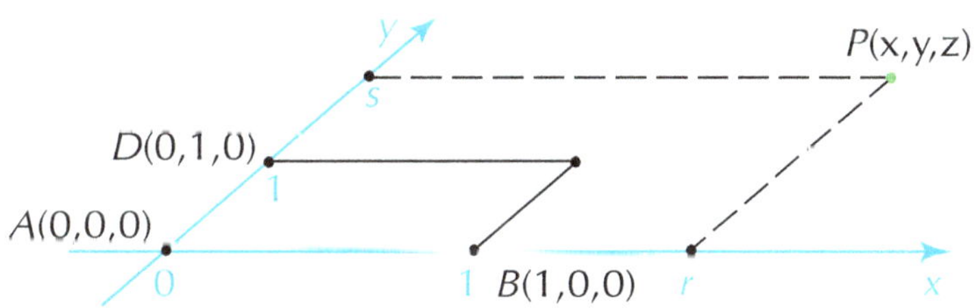

Let $P(x, y, z)$, $B(1, 0, 0)$ and $D(0, 1, 0)$ be three points, then we call $(x, y, z) = r(1, 0, 0) + s(0, 1, 0)$ a vectorial equation of the plane ABD.

For arbitrary planes (planes not containing the origin), we first calculate the direction vectors in an analogous way as for a line not containing the origin (see previous page).

Example

For the plane EFH with $E(0, 0, 1)$, $F(1, 0, 1)$ and $H(0, 1, 1)$, $\overrightarrow{EF}$ and $\overrightarrow{EH}$ are direction vectors.

And since $\overrightarrow{AF} - \overrightarrow{AE} = \overrightarrow{EF}$ or, in coordinates: $(1, 0, 1) - (0, 0, 1) = (1, 0, 0)$

and $\overrightarrow{AH} - \overrightarrow{AE} = \overrightarrow{EH}$ or, in coordinates: $(0, 1, 1) - (0, 0, 1) = (0, 1, 0)$

The plane *EFH* is parallel to the plane *ABD* and indeed, the latter has direction vectors determined by (1, 0, 0) and (0, 1, 0). A vectorial equation of the plane *EFH* is thus:

$$(x, y, z) = (0, 0, 1) + r((1, 0, 1) - (0, 0, 1)) + s((0, 1, 1) - (0, 0, 1))$$
$$(x, y, z) = (0, 0, 1) + r(1, 0, 0) + s(0, 1, 0)$$
$$(x, y, z) = (r, s, 1)$$

This shows that the stringent condition for a point $S(x, y, z)$ to be in the plane *EFH*, is that the third coordinate number z equals 1. After elimination of the parameters r and s, the cartesian equation of this plane in three-dimensional space consequently is $z = 1$.

Step 3 Generalization to a n-dimensional space

What we did in step 2 about interpreting pairs (2-numbers) and triples (3-numbers) as being coordinates of points, we can repeat for n-numbers, in which n is an arbitrary natural number. The addition of such n-numbers and their multiplication with a real number can be defined in the same easy way as for pairs and triples:

$$(a_1, a_2, a_3, a_4, \ldots, a_n) + (b_1, b_2, b_3, b_4, \ldots, b_n) = (a_1+b_1, a_2+b_2, a_3+b_3, a_4+b_4, \ldots, a_n+b_n)$$
$$r(a_1, a_2, a_3, a_4, \ldots, a_n) = (ra_1, ra_2, ra_3, ra_4, \ldots, ra_n)$$

In a space of dimension higher than three, for example four, there are subspaces of every lower dimension, in this case one-, two- and three-dimensional.

A line is determined by two different points and is a one-dimensional subspace depending on one parameter. Its direction is determined by one direction vector.

A plane is determined by three different points, not all of them on one line, and is a two-dimensional subspace depending on two parameters. Its direction is determined by two direction vectors.

A three-dimensional subspace is determined by four points, not all of them in one plane, depending on three parameters. Its 'direction' is determined by three direction vectors.

In an n-dimensional space we can describe subspaces of dimension up to $n-1$.

Because the language is geometrical, we keep on trying to make a visualization also in a dimension higher than three, which apparently seems difficult. Consider, however, that in the case of planar representations of three-dimensional figures, for example of a cube, also some empathizing is asked from our interpreting mind, in order to 'see' certain directions as being independent from each other. So why couldn't some additional effort be asked for higher dimensions so as to interpret certain directions as being independent?

Consider the three-dimensional situation: by shifting (translating) a square along a direction differing from those of the plane of the square, we obtain a three-dimensional cube:

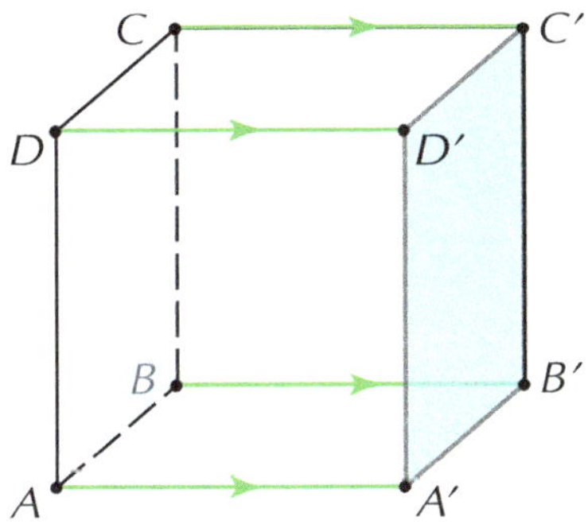

In a four-dimensional space : by translating a cube along a direction differing from those of the three-dimensional space of the cube, we obtain a four-dimensional 'cube'.

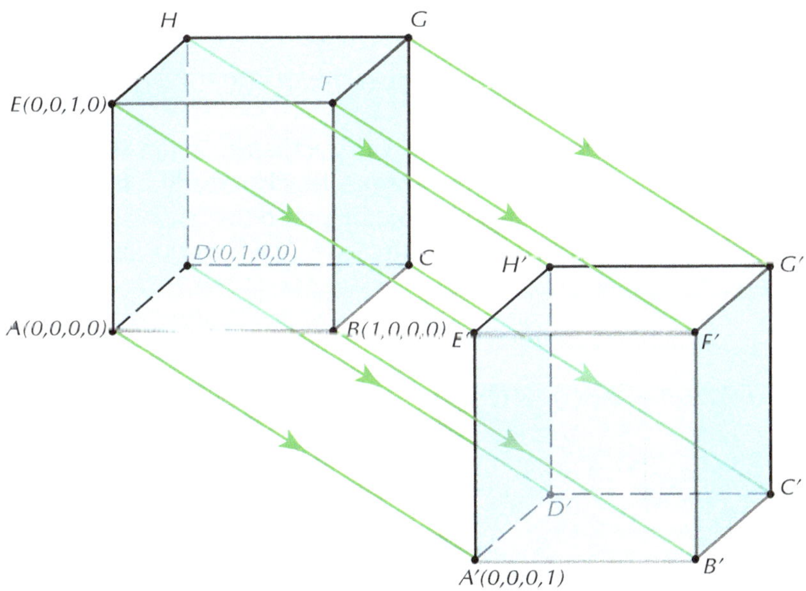

The independence of the different directions is suggested by the coordinates we introduced. In the last figure the directions of *AB*, *AD*, *AE* and *AA′* are independent of each other.

In a four-dimensional space, the vectorial equations of subspaces can be derived in an analogous way as in the three-dimensional space. The coordinate of each point only contains just one more number.

Example

For the line through $A'(0, 0, 0, 1)$ and $E'(0, 0, 1, 1)$:

$S(x, y, z, u) \in A'E'$

$$\Leftrightarrow (x, y, z, u) = (0, 0, 0, 1) + r((0, 0, 1, 1) - (0, 0, 0, 1))$$

$$\Leftrightarrow (x, y, z, u) = (0, 0, 0, 1) + r(0, 0, 1, 0)$$

For the plane through $B(1, 0, 0, 0)$, $F(1, 0, 1, 0)$ and $C(1, 1, 0, 0)$:

$S(x, y, z, u) \in BFE$

$$\Leftrightarrow (x, y, z, u) = (1, 0, 0, 0) + r((1, 0, 1, 0) - (1, 0, 0, 0)) + s((1, 1, 0, 0) - (1, 0, 0, 0))$$

$$\Leftrightarrow (x, y, z, u) = (1, 0, 0, 0) + r(0, 0, 1, 0) + s(0, 1, 0, 0)$$

For the space through $A'(0, 0, 0, 1)$, $B'(1, 0, 0, 1)$, $C'(1, 1, 0, 1)$ and $H'(0, 1, 1, 1)$:

$S(x, y, z, u) \in A'B'C'H'$

$$\Leftrightarrow (x, y, z, u) = (0, 0, 0, 1) + r((1, 0, 0, 1) - (0, 0, 0, 1))$$
$$+ s((1, 1, 0, 1) - (0, 0, 0, 1)) + t((0, 1, 1, 1) - (0, 0, 0, 1))$$

$$\Leftrightarrow (x, y, z, u) = (0, 0, 0, 1) + r(1, 0, 0, 0) + s(1, 1, 0, 0) + t(0, 1, 1, 0)$$

In that way affine geometry, which deals with parallelism in the Euclidean sense, can be practised in every desired dimension. This is already a first step to generalization. In order to involve also the metric properties, such as length, angles and perpendicularity in this generalization, we first have to find an algebraic translation of these concepts.

The OCTACUBE in the Mc Allister building of the mathematics faculty at Pennsylvania State University. This art object, designed by the mathematician Adrian Ocneanu, shows three-dimensional projections (octahedra) of a four-dimensional solid. For its construction a computer programme had to be written that designed the 96 shapes, which had to fit at 23 points of intersection of the axes. A 24th point of intersection is situated in the infinite. Ocneanu studies multidimensional models for the benefit of the theory of quantum-fields.

Step 4 The scalar product of vectors in an analytical form

We first analyse the concepts 'length' and 'perpendicularity' in the plane provided with a cartesian reference system.

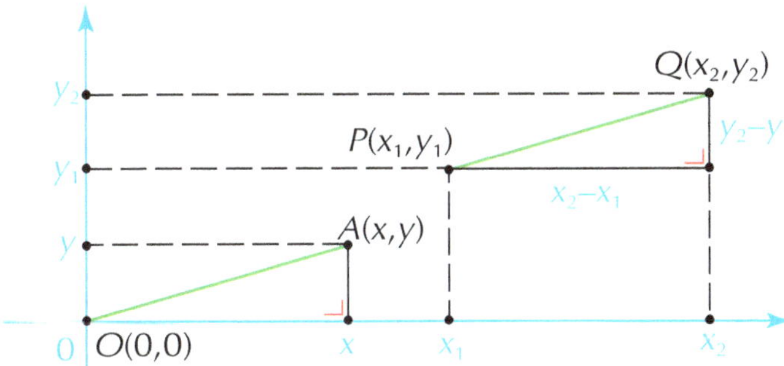

The distance between the origin $O(0, 0)$ and the point $A(x, y)$ is, according to the Pythagorean theorem, given by: $\sqrt{x^2+y^2}$

In an analogous way we find the distance between the point $P(x_1, y_1)$ and the point $Q(x_2, y_2)$ by the formula: $\sqrt{(x_2-x_1)^2+(y_2-y_1)^2}$

The lines OP and OQ in the next figure intersect perpendicularly in the origin if and only if in the triangle OPQ, having PQ as hypotenuse, the Pythagorean theorem applies:

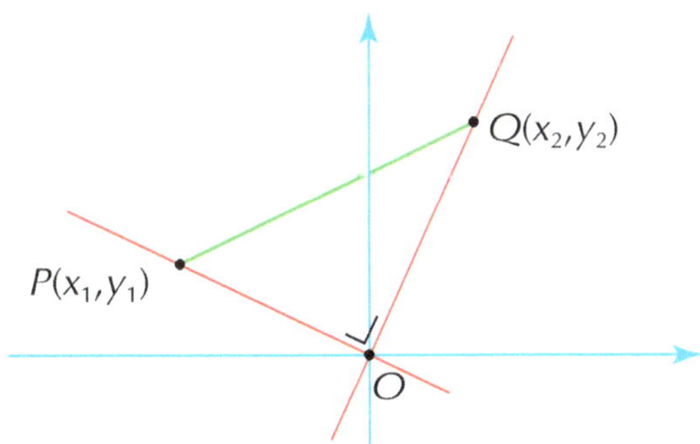

$$
\begin{aligned}
OP \perp OQ &\Leftrightarrow \overline{OP}^2 + \overline{OQ}^2 = \overline{PQ}^2 \\
&\Leftrightarrow \left(\sqrt{x_1^2+y_1^2}\right)^2 + \left(\sqrt{x_2^2+y_2^2}\right)^2 = \left(\sqrt{(x_2-x_1)^2+(y_2-y_1)^2}\right)^2 \\
&\Leftrightarrow x_1^2+y_1^2+x_2^2+y_2^2 = x_2^2-2x_2x_1+x_1^2+y_2^2-2y_2y_1+y_1^2 \\
&\Leftrightarrow 2(x_2x_1+y_2y_1)=0 \\
&\Leftrightarrow x_1x_2+y_1y_2=0 \qquad (*)
\end{aligned}
$$

The expression left of the equal-sign '=' in (*) is called the scalar product of (x_1, y_1) and (x_2, y_2) and we write this as: $(x_1, y_1) \bullet (x_2, y_2)$.

So: $(x_1, y_1) \bullet (x_2, y_2) = x_1x_2 + y_1y_2$.

This implies that the product of two pairs of real numbers (considered as being vectors) is not a vector but a real number (scalar).

From (*) it can be seen that: two lines containing the origin are perpendicular if and only if the scalar product of their direction vectors equals zero. Since parallel lines do have the same direction vectors, this condition also applies to arbitrary lines, not containing the origin.

In particular, we can calculate the following scalar products:

$(1, 0) \bullet (1, 0) = 1 \cdot 1 + 0 \cdot 0 = 1$ (square of the first basic vector)

$(1, 0) \bullet (0, 1) = 1 \cdot 0 + 0 \cdot 1 = 0$ (product of first and second basic vector)

$(0, 1) \bullet (1, 0) = 0 \cdot 1 + 1 \cdot 0 = 0$ (product of second and first basic vector)

$(0, 1) \bullet (0, 1) = 0 \cdot 0 + 1 \cdot 1 = 1$ (square of the second basic vector)

By means of these four numbers the scalar product of (x_1, y_1) and (x_2, y_2) can be expressed using matrices as follows:

$$\begin{bmatrix} x_1 & y_1 \end{bmatrix} \begin{bmatrix} 1 & 0 \\ 0 & 1 \end{bmatrix} \begin{bmatrix} x_2 \\ y_2 \end{bmatrix} = \begin{bmatrix} x_1x_2 + y_1y_2 \end{bmatrix}$$

By means of this scalar product, all metric properties concerning length, angle and perpendicularity that appear in planar analytic geometry can be transferred into an algebraic context. The four numbers 1, 0, 0 and 1, in the 2x2-matrix of the previous formula hence fully characterise this kind of geometry.

This approach can be generalized to the three- as well as to the *n*-dimensional space.

In the three-dimensional space the following formulas apply:

the length of a vector with coordinate (x, y, z) is: $\sqrt{x^2 + y^2 + z^2}$

the distance between two points with coordinates (x_1, y_1, z_1) and (x_2, y_2, z_2) is:

$$\sqrt{(x_2 - x_1)^2 + (y_2 - y_1)^2 + (z_2 - z_1)^2}$$

the scalar product of two vectors with coordinates (x_1, y_1, z_1) and (x_2, y_2, z_2) is:

$$x_1x_2 + y_1y_2 + z_1z_2$$

or in matrix form: $\begin{bmatrix} x_1 & y_1 & z_1 \end{bmatrix} \begin{bmatrix} 1 & 0 & 0 \\ 0 & 1 & 0 \\ 0 & 0 & 1 \end{bmatrix} \begin{bmatrix} x_2 \\ y_2 \\ z_2 \end{bmatrix} = \begin{bmatrix} x_1x_2 + y_1y_2 + z_1z_2 \end{bmatrix}$

The nine numbers 1, 0, 0, 0, 1, 0, 0, 0 and 1, in the 3 x 3 matrix of the previous formula fully characterise the three-dimensional Euclidean space.

In analogy to what was stated in the plane, we say that in the three-dimensional space two lines (possibly also skew lines) are perpendicular if and only if the scalar product of their direction vectors equals zero.

Example

In the following figure the skew lines *BF* and *HG* are perpendicular:

since: a direction vector of *BF* can be found as (1, 0, 1) – (1, 0, 0) = (0, 0, 1)

a direction vector of *HG* can be found as (1, 1, 1) – (0, 1, 1) = (1, 0, 0)

and: $(0, 0, 1) \cdot (1, 0, 0) = 0 \cdot 1 + 0 \cdot 0 + 1 \cdot 0 = 0$

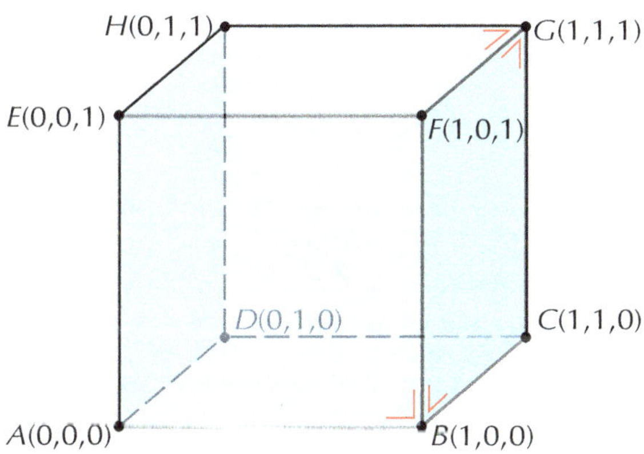

By means of perpendicularity of lines we can also characterise perpendicularity of a line and a plane, and of two planes. But it is not our intention to develop the whole three-dimensional geometry here, but to show the way to generalization. For analogous formulas in an *n*-dimensional space we only have to write *n* coordinate numbers. This way the breakthrough to any dimension is made possible. It doesn't require any other effort than to write longer coordinates. So we are truly on the right track on our highway!

Now the question may arise whether such higher dimension geometries have any sense or use at all. The answer is positive: they are used in applications in which the number of variables (degrees of freedom) is more than three. For example, the time space in relativity theory is four-dimensional (three spatial and one time dimension). The string theory (the theory about the most elementary components of the universe) takes, according to the mathematician and physicist Schwarz, 'only' ten dimensions and not twenty-six, as was initially suggested. Thus applied sciences have no longer the same qualms, as Greek geometry had, to exceed the barrier of three dimensions if there is a need to.

Step 5 Geometry related to a symmetrical bilinear form

Now we have arrived at a tempting situation, like it always happens in mathematics at a moment of extreme abstraction, the tickling urge to mess about a bit with the original model and to see what kind of innovations it generates. We will restrict ourselves to a simple example in the plane.

The metric characteristics of the Euclidean plane can be fully expressed by means of the scalar product linked to the matrix:

$$\begin{bmatrix} 1 & 0 \\ 0 & 1 \end{bmatrix}$$

Now, let us change this product by taking $\begin{bmatrix} 1 & 0 \\ 0 & -1 \end{bmatrix}$ as the defining matrix.

This leads to::

$$(x_1, y_1) \blacklozenge (x_2, y_2) = \begin{bmatrix} x_1 & y_1 \end{bmatrix} \begin{bmatrix} 1 & 0 \\ 0 & -1 \end{bmatrix} \begin{bmatrix} x_2 \\ y_2 \end{bmatrix} = \begin{bmatrix} x_1x_2 - y_1y_2 \end{bmatrix}$$

This new product keeps the essential properties of the previous one, namely it is still 'symmetrical and bilinear', in other words: it is a scalar product as well. Let's take a look at the consequences for the orthogonality (i.e. perpendicularity relative to the scalar product) of lines in particular.

In the usual Euclidean plane the coordinate axes were perpendicular, as shown from their direction vectors (1, 0) and (0, 1):

$$(1, 0) \bullet (0, 1) = 1 \cdot 0 + 0 \cdot 1 = 0$$

So were the bisectors of the two axes, as shown from their direction vectors (1, 1) and (–1, 1):

$$(1, 1) \bullet (-1, 1) = 1 \cdot (-1) + 1 \cdot 1 = 0$$

In the plane provided with the new scalar product, the coordinate axes are still orthogonal:

$$(1, 0) \blacklozenge (0, 1) = 1 \cdot 0 - 0 \cdot 1 = 0$$

but the bisectors are not orthogonal any more, since:

$$(1, 1) \blacklozenge (-1, 1) = 1 \cdot (-1) - 1 \cdot 1 = -2 \neq 0$$

Moreover, we notice some strange particularities:

- each bisector is orthogonal respective to itself, since:

$$(1, 1) \blacklozenge (1, 1) = 1 \cdot 1 - 1 \cdot 1 = 0 \quad \text{and}$$

$$(-1, 1) \blacklozenge (-1, 1) = (-1) . (-1) - 1 \cdot 1 = 1 - 1 = 0.$$

– all pairs of lines that are symmetrical respective to one of the bisectors, like those with direction vectors (3, 2) and (2, 3), are orthogonal:

$$(3, 2) \blacklozenge (2, 3) = 3 \cdot 2 - 2 \cdot 3 = 0$$

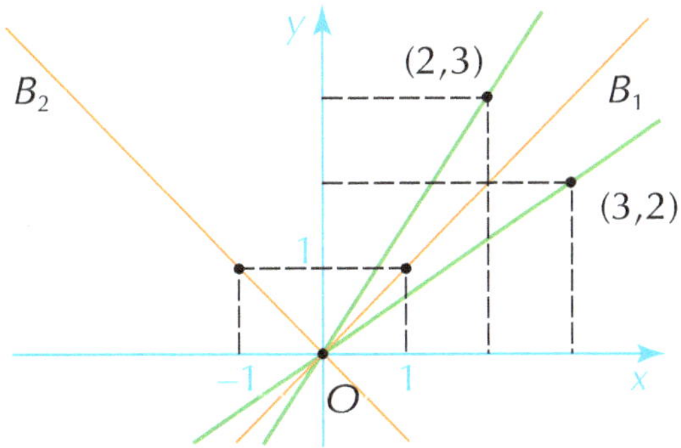

Anyone who is pondering that all these extravagances are no more than just a chaste pastime, will be surprised to hear that exactly such spaces (of Minkowski) appeared to be suitable for providing a theoretical basis for certain aspects of relativity theory. The shrinking effect of the transition from the original rectangular coordinate system on a pair of 'perpendicular' lines that are 'closer to' the bisector, can be understood as an adaptation of the coordinate system that must be taken into account when the speed of the observer relative to the observed object approaches the speed of light (the bisector). You may be ready to find some solace in the witticism of Einstein who seems to have said on this: "since mathematicians threw themselves into my theory, I for myself don't understand anything of it any more".

This is the model of the time space of Einstein and Minkowski represented by the light cone. The upper half of the cone represents the points from which the rays of light are emitted (the future), the bottom half represents the points in which the rays of light are received (the past). The figure on the right shows how different the set of simultaneously experienced events is for two different observers, depending on their speed relative to the speed of light (the position of the arrows with regard to the light cone). In this time space both the pink and the blue arrow are respectively 'perpendicular' to the equal coloured planes of the events that are simultaneous to them.

But Einstein's witticism cannot temper the creativity of the mathematician. One can, for example, fill in four random numbers a, b, c and d into the defining matrix of a scalar product of the plane. If we want the associated relationship of lines being orthogonal to be symmetrical, i.e. that the fact that AB is orthogonal to CD implies that also CD is orthogonal to AB, it is necessary for the numbers b and c to be equal. The defining matrix thus has the form $\begin{bmatrix} a & b \\ b & d \end{bmatrix}$

Three numbers thus define a geometry of the plane, linked to a symmetrical bilinear form (scalar product). A lot of these geometries resemble each other and can thus be categorized in classes, by identification based on, among other things, the characteristics of the defining matrix.

Of course this can be reproduced in higher dimensions, in the three-dimensional space by a 3 x 3-matrix and in an n-dimensional space by an n x n-matrix. But already another generalization shows up; let's call it a slip road to an even broader highway.

M.C. Escher (1898–1972)

In the 'Waterfall', the graphic artist M. C. Escher plays a trick on our illusion of three-dimensional interpretations of planar drawings. After falling, the water runs back to the apparent highest level. In a lot of his works, Escher is flirting with profound mathematical concepts.

Step 6 Geometry of a vector space provided with a scalar product.

In the previous steps, the concepts point, line, plane and space have still preserved some intuitive geometrical meaning, in spite of the algebraic transcriptions. We now take a step up to an even higher level of abstraction by passing on from geometrical concepts to random objects for which an analogous structure can be found, i.e. objects that we can add up and multiply with a real number. Those objects we will then also call vectors. The vector space structure is already sufficient to repeat steps 1,2 and 3 of affine geometry. If, moreover, we can define a scalar product on those vectors (in which the product of two such vectors is a real number) then we can also repeat step 4 for the metric properties.

As an illustration, we give an important example from function analysis, in particular the class of continuous functions on a closed interval, e.g. [0, 2]. The elements of that class are presented by (uninterrupted) graphs, as in the next figure.

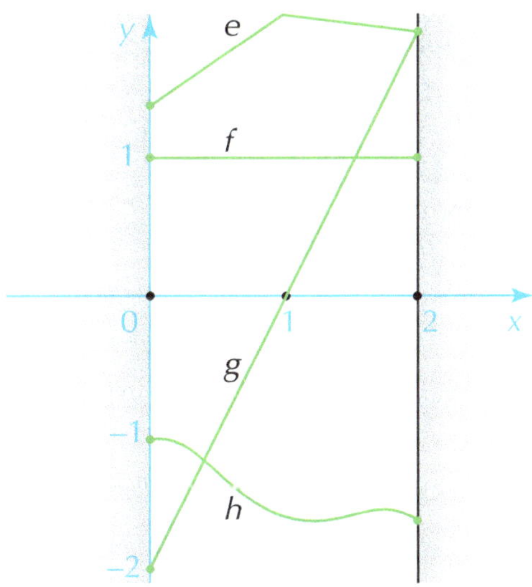

Because the sum of two continuous functions is a continuous function and the product of a continuous function with a real number also a continuous function, the continuous functions can be taken as vectors. The 'line' through two different elements f and g of this space is then written in the same analytical way as we did for the coordinate space:

the function x is on the 'line' through the functions f and g

$$\Leftrightarrow x = f + r(g - f)$$

In this, $g - f$ is a direction vector of that 'line'.

Two of those 'lines' are 'parallel' if their direction vectors (and these are functions!) are real multiples of each other.

Since the vector space of the continuous functions has infinite dimension, we can set up subspaces of whatever dimension by means of vectorial equations.

Since the definite integral of a continuous function on a closed interval is a real number, we can use the definite integral as a means to define a scalar product on continuous functions:

$$f \blacklozenge g = \int_0^2 f \cdot g\,dx$$

By means of this we can now also apply the metric concepts like 'length', 'perpendicularity, etc. to these functions.

Examples

Let $f(x) = 1$ and $g(x) = 2x - 2$ (see previous figure).

The length of f and g (in relation to the scalar product $\blacklozenge$) is calculated as follows:

the length of $f = \sqrt{f \blacklozenge f} = \sqrt{\int_0^2 f(x) \cdot f(x)\ dx} = \sqrt{\int_0^2 1 \cdot 1\ dx} = \sqrt{\int_0^2 1\ dx}$

$$\overset{(1)}{=} \sqrt{x\Big|_0^2} = \sqrt{2-0} = \sqrt{2}$$

the length of $g = \sqrt{g \blacklozenge g} = \sqrt{\int_0^2 g(x) \cdot g(x)\ dx} = \sqrt{\int_0^2 (2x-2)^2\ dx}$

$$= \sqrt{\int_0^2 \left(4x^2 - 8x + 4\right)\ dx}$$

$$\overset{(2)}{=} \sqrt{\frac{4}{3}x^3 - 4x^2 + 4x\Big|_0^2} = \sqrt{\frac{32}{3} - 16 + 8} = \sqrt{\frac{8}{3}}$$

For those who might have 'forgotten' how a definite integral can be calculated, we give some hints so as to make the results in (1) and (2) comprehensible.

The definite integral of a function h over the interval [a, b] i.e. $\int_a^b h\ dx$ can be calculated by the formula $H(b) - H(a)$ (shortly written as $H\Big|_a^b$)

in which H is a function of which the derivative equals h. We say that H is a primitive function of h.

Since the derivative of x equals 1, x is a primitive function of 1. This explains formula (1).

Because the derivative of $\frac{4}{3}x^3 - 4x^2 + 4x$ equals $4x^2 - 8x + 4$, $\frac{4}{3}x^3 - 4x^2 + 4x$ is a primitive function of $4x^2 - 8x + 4$, which leads to formula (2).

But naturally, simple faith would be more than adequate for this.

Observe that the 'length' calculated this way has no connection to the usual Euclidean length of the graphs of f and g, because these are respectively 2 and $2\sqrt{5}$. The 'geometry' that we practise here is a sophisticated geometry in which the vector points are functions.

Also the fact that two functions are orthogonal depends on the stated scalar product $\blacklozenge$.

Example

The functions f and g where $f(x) = 1$ and $g(x) = 2x - 2$ are orthogonal, since their scalar product equals zero:

$$f \blacklozenge g = \int_0^2 f(x) \cdot g(x)\, dx = \int_0^2 1 \cdot (2x-2)\, dx = \int_0^2 (2x-2)\, dx$$

$$= x^2 - 2x \Big|_0^2 = (4-4) - (0-0) - 0$$

It is obvious that starting from the Insight reached in step 6 such a kind of algebraic geometry can be built from primitive elements that can be something quite different than points intuitively seen as 'dots' at the start of step 1. For example, cone sections, curves of higher degree, spheres, etc. can be considered as 'points' of a particular kind of geometry. With this fertile idea in mind we can leave algebra again and go on a purely synthetic geometrical tour.

1.2 The world of incidence-geometry

Let's return for a while to the dear little 'plane' of our youth with its familiar points and lines, which offered us such a lot of the 'pleasure' that we shared with the ancient Greeks. In that, lines are considered to be sets of points and points are primitive elements. A point may belong to a certain line and a line may contain a certain point: this relationship between points and lines is called an incidence relation. For the purpose of generalization we now make abstraction of the specific meaning of points and lines and, to simplify the matter for the less experienced abstracter, we will speak of 'primitive elements' instead of 'points' and of 'subspaces' instead of 'lines'.

We call an incidence space each non-empty set of primitive elements in which we select a particular set of subsets as subspaces, and in which the incidence relation means that the primitive elements can appear as elements of those subspaces.

Example

We take as a set of primitive elements $\{1, 2, 3, 4\}$ and as a set of subspaces the set of pairs that can be made with these, i.e.

$$\{\{1, 2\}, \{1, 3\}, \{1, 4\}, \{2, 3\}, \{2, 4\}, \{3, 4\}\}.$$

The following incidence relations apply:

$$1 \in \{1, 2\} \text{ and } 3 \notin \{1, 2\}.$$

Nikolai Ivanovitsj Lobachevsky (1793–1856)

Russian mathematician, professor at the university of Kazan. He developed a geometry in which the parallel postulate of Euclid does not apply, such as J. Bolyai and K.F. Gauss had also done, independently from each other. In 1826, being the first to publish his insights about a hyperbolic geometry, his ideas fell on deaf ears. Only many years after his death, was non-Euclidean geometry commonly accepted as an alternative to the Euclidean.

Nevertheless, in order to create some geometrical context in an incidence space, we have to enrich the structure with axioms that suggest geometry. This means that they are inspired by what we think we know about points and lines. We must deal with axioms carefully. An absolute requirement is that they are consistent, i.e. that already from the start they do not mutually contradict and that sooner or later we cannot simultaneously derive from them a proposition as well as its negation. This last condition is not always obvious. For this reason mathematicians have introduced a technique of implementing the theory in a concrete model, in which the axioms can be verified. Furthermore, it's also elegant, although not necessary, that the axioms are mutually independent. This means that any axiom is not derivable from the other axioms.

The incidence relation in an incidence space is set up by characteristic axioms, of which the most essential is the following:

Axiom 1 Two different primitive elements belong to exactly one subspace.

In order to experience at least a bit of excitement in such a space and not to end up in trivial situations, we add the following, rather banal axiom:

Axiom 2 There exist at least four different primitive elements, of which no three belong to the same subspace.

Observe that the example given earlier (model) satisfies these axioms. This proves the consistency of the two axioms.

Because the number of concepts is not very large we can supplement or refine them slightly, e.g. with the definition of parallel subspaces.

Definition: 'Two subspaces are parallel if and only if they are equal or have no primitive elements in common'

In the given example $\{1, 2\}$ and $\{3, 4\}$ are thus parallel, but not $\{1, 2\}$ and $\{1, 3\}$.

From what precedes it is easy to prove that the relation 'parallel' has the following properties: it is

- reflective, i.e. each subspace is parallel to itself
- symmetrical, i.e. if a subspace A is parallel to a subspace B, then B is also parallel to A.

These axioms are, however, not sufficient to indicate whether the parallel-relation is also transitive, i.e. from A parallel to B and B parallel to C it follows that A is also parallel to C. Neither is it possible to infer how many subspaces exist, parallel to a given subspace and containing a primitive element not situated in that given subspace.

First we give some examples to show that there are several situations possible. Then we will formulate additional axioms characterizing some of these situations and thus lead to a classification of incidence spaces.

Example 1

primitive elements: {1, 2, 3, 4}

subspaces : {{1, 2}, {1, 3}, {1, 4}, {2, 3}, {2, 4}, {3, 4}}.

We take a subspace, e.g. {1, 2}, and a primitive element not included in it, e.g. 3, it can then be seen that there is precisely one subspace that contains 3 and that is parallel to {1, 2}, i.e. {3, 4}. You can check this for any choice of a subspace and of a primitive element not included in it. This model shows that we can confidently add a third axiom that states:

Axiom 3 For every primitive element not included in a given subspace, there exists precisely one subspace containing that primitive element and parallel to the given subspace.

An incidence space that satisfies axioms 1, 2 and 3 is called an affine space.

The given example is thus an affine space, just as the Euclidean plane with its usual points and lines is.

Example 2

primitive elements: the internal points of a disc (without the points on the edge)

subspaces: the chords in this disc (without their extremities)

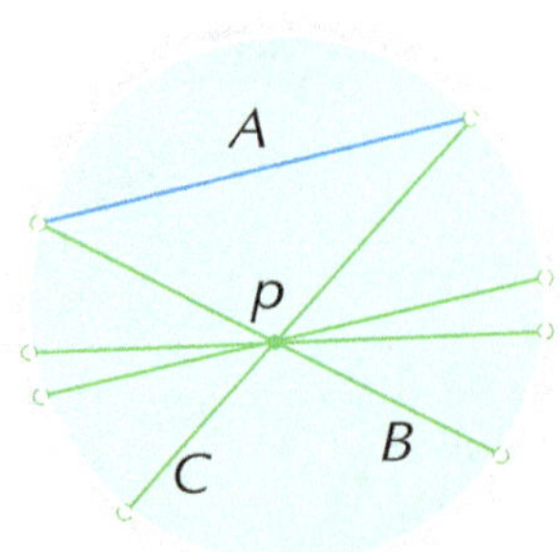

This example satisfies axioms 1 and 2, but not axiom 3: if we take a subspace, e.g. the chord *A*, and a primitive element not on it, e.g. *p*, then apparently there is an infinite number of subspaces that contain *p* and that are parallel to *A*, i. e. the subspaces *B* and *C* and all the subspaces 'in between'. This applies to any situation chosen for *A* and *p*.

This model shows that axiom 3 is indeed independent of axioms 1 and 2, and therefore can be freely replaced by axiom 3bis below. This new axiom will then also be consistent with axioms 1 and 2.

Axiom 3bis For every primitive element not included in a given subspace, there exist at least two different subspaces containing that primitive element and parallel to the given subspace.

An incidence space that satisfies axioms 1, 2 and 3bis is called a hyperbolic space.

The given example, the Cayley-Klein model, is thus a hyperbolic space and not an affine space. This model has played a historical role in the acceptance of the consistency of non-Euclidean geometries. The ingenious idea of linking the consistency of hyperbolic geometry to a model that was borrowed from Euclidean geometry, i.e. an open disc and its chords, has faced the sceptics with a choice: either to accept that the new geometry is indeed consistent, or to also doubt the consistency of Euclidean geometry, which would be akin to throwing out the baby with the bath water.

The hyperbolic spaces were the first non-Euclidean spaces that were invented, independently from each other, by the Hungarian Janos Bolyai (1802–1860), the Russian Nicolai Ivanowitsj Lobatchevsky (1793–1856) and the German Karl Friedrich Gauss (1777–1855).

Example 3

primitive elements: pairs of (opposite) points of intersection of a sphere and one of its diameters (one pair is only one primitive element!).

subspaces: the great circles on the sphere (circles of which the centre coincides with the centre of the sphere)

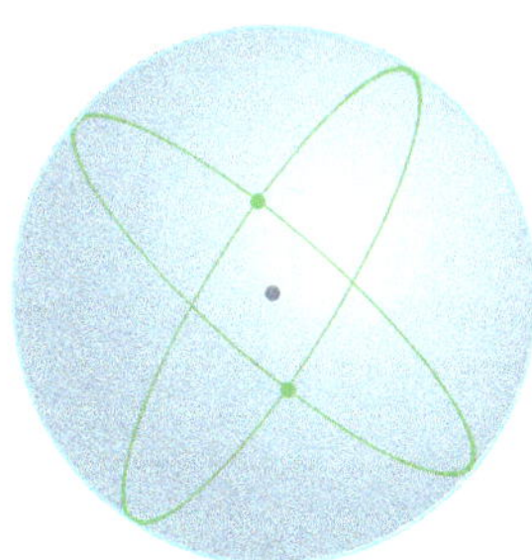

This model satisfies axiom 1 and 2, but not axiom 3, neither axiom 3bis, since two different subspaces (great circles) always intersect in precisely one primitive element (one pair of opposite points). So we can as well keep the consistency if we add the following axiom 3tris to the axioms 1 and 2:

Axiom 3tris For every primitive element not included in a given subspace, there exists no subspace containing that primitive element and parallel to the given subspace.

An incidence space that satisfies axioms 1, 2 and 3tris is called an elliptic space.

The given example is thus an elliptic space and not an affine or hyperbolic space.

The first example of an elliptic space was created by the German mathematician Bernhard Riemann (1826–1866), a pupil of Gauss.

Example 4

primitive elements: the internal points of the first three quadrants of the plane, provided with a cartesian reference system

subspaces: the traces (remainders) of lines onto these quadrants

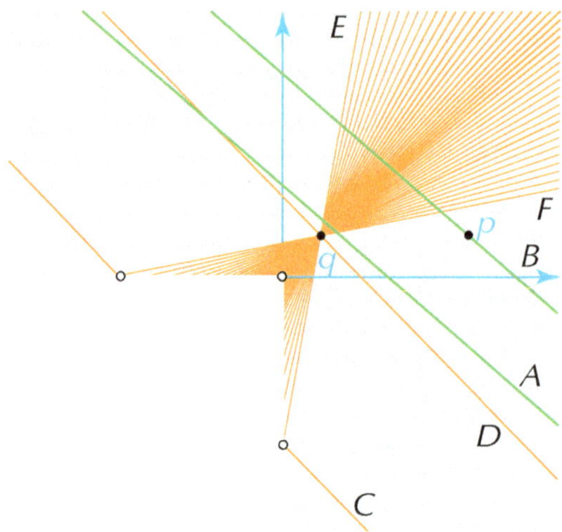

This model satisfies axioms 1 and 2, but not axiom 3, neither axiom 3bis nor axiom 3tris, since for the subspace A and the primitive element p there is only one subspace parallel to A, namely B, but for subspace C and the primitive element q there is an infinite number of subspaces parallel to C, namely D, E and F and all subspaces 'in between'.

The given example is thus an incidence space that is neither affine, nor hyperbolic, nor elliptic.

In this item we have studied, by the criterion of parallelism, the alternatives for incidence spaces in a synthetic way. In the next item we will deal with spaces by the criteriòn of metric properties, again in a synthetic way.

1.3 Geodesics on a curved surface

In the usual Euclidean plane, the shortest distance between two points is the one measured along a straight line connecting those two points, and the sum of the internal angles of a triangle equals exactly 180°. The proof of that last property is based on the axiom of the unique parallel to a given line containing a point not on that line.

In another type of space those properties are not necessarily still valid. They may well then be replaced by adapted versions, appropriate to that space. Let's have a look at those aspects, for example for the model of the sphere in example 3.

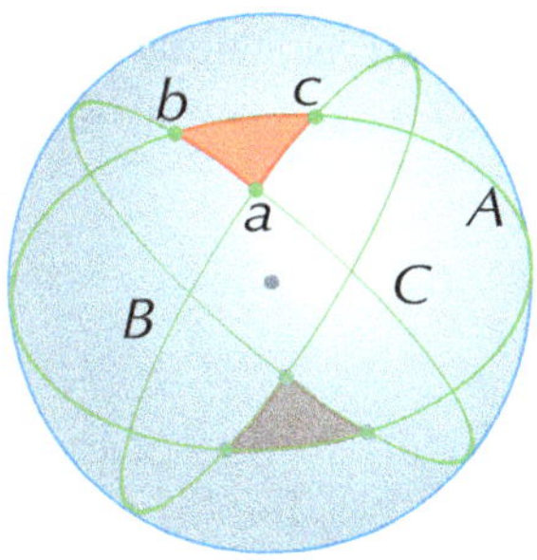

A, *B* and *C* are geodesics

abc is a spherical triangle

The shortest distance between two points on the sphere is measured along a great circle containing those two points, and is known as the great-circle distance. That is why, in this context, we call the great circles on the sphere geodesics. If we connect three points on the sphere, not all three on the same great circle, by arcs of the great circles containing those points then we call the result a spherical triangle. By drawing, in a vertex of this spherical triangle, tangent lines to the arcs in this vertex, we obtain an angle that we call an angle of the spherical triangle. The sum of the three angles of such a spherical triangle in this model exceeds 180°. A sphere is a curved surface with a constant positive curvature.

It was one of Riemann's ideas to study the geometry of curved surfaces with constant positive or negative curvature. A famous example of a curved surface with constant negative curvature is a pseudo-sphere. It can be generated as the surface of revolution, i.e. by rotating a curve K having the shape of a bird around its asymptote A. This surface looks like two trumpets glued together.

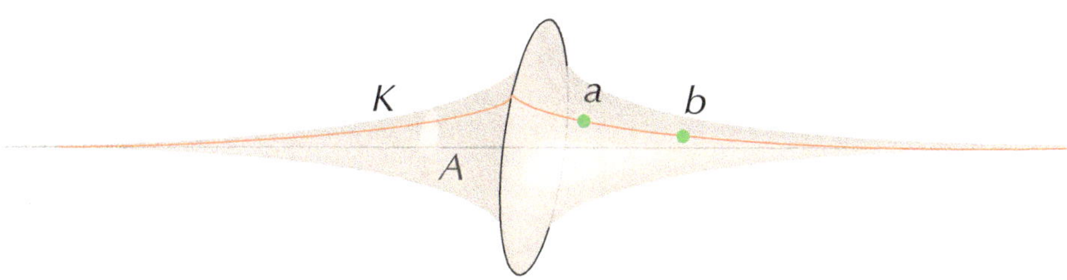

The shortest distance between two points on this pseudo-sphere is measured along a specific curve, which is a geodesic of that space. If we connect three points of this space, not all three on the same geodesic, with the arcs of the geodesics containing two of these points, then we obtain a pseudo-spherical triangle. The angles are also defined by means of tangent lines to the geodesics, just like on a sphere. In this space with negative curvature, the sum of the angles of such a triangle is less than 180°.

To calculate distances and areas in such curved spaces, it takes 'big theoretical canons' that belong to the domain of differential geometry. The problems of shortest distance, area, volume, etc. in higher dimensional spaces are the subject of the so-called calculus of variations. Since the formulation of the general relativity theory, in which gravitation is considered as a distortion of space, the theory of curved spaces is a suitable tool to deal with those aspects of relativity.

Summary

This section showed how, by hanging on to a certain line of thought, creative ways opened up that initially seemed eccentric but finally also had practical applications in their scope. Due to this line of thought, it is not astonishing that at the end of the nineteenth century mathematicians started to wonder what could still be the meaning of a 'geometry'. It was the German mathematician Felix Klein, professor at Erlangen, who in an article in 1872 highlighted the importance of group theory for the classification of geometrical spaces. This article is since then known under the name 'Erlanger program'. He suggested seeing a geometry as a set of primitive elements on which operates a group of transformations and wherein the attention goes to those properties that remain unchanged (invariant) under that group of transformations. Thus the affine geometry can be considered as that type of geometry in which parallelism remains under specific transformations like translations (shifts) and homotheties (scale changes). And metric geometry can be considered as that geometry in which length and angle are invariant under transformations, such as reflections and rotations.

The tale of geometry is of course still not fully unravelled by this. Nothing has been said about projective geometry, the geometry that studies what is invariant from a perspective point of view, or about topology, the geometry that studies what is invariant under smooth deformations such as 'stretching', 'folding', or 'denting' (without breaking). But it is not our intention to be exhaustive in this book.

2 From the empty nothingness to weird infinities

Both the absolute nothing as well as the infinite have fascinated a lot of philosophers and scientists throughout the centuries. The rejection of the actual infinite by Aristotle and later by the intuitionists turned out to be an enormous impediment to the creation of many subtle mathematical concepts. In number theory and in geometry the concept of infinity has adopted many aspects. On the other hand, in set theory there exists only one empty set. We show in this section how, starting from this empty set, one can rise to an inconceivable variety of infinite sets with different numerousness.

2.1 The infinite sequence of the natural numbers

In the beginning there was nothing. Mathematicians called this the emptiness, the empty set. As soon as nothing gets its name, at once there is something, i.e. the name of that nothing. Let's call it null and represent it by 0, this makes: $\{\} = 0$.

If we put this name in a set, then we get a set with something in it, i.e. $\{0\}$. This new set differs from the empty set $\{\}$, so it has to get a different name, i.e. one. This makes: $\{0\} = 1$. We can continue this forever: $\{0, 1\} = 2$, $\{0, 1, 2\} = 3, \ldots$, $\{0, 1, 2, 3, \ldots, n-1\} = n$, etc. This way we build the infinite set of the natural numbers $\{0, 1, 2, 3, \ldots, n, \ldots\}$. This set we represent by $\mathbb{N}$. Because of its construction itself, the set $\mathbb{N}$ can be arranged as an infinite sequence.

Georg Cantor (1845–1918)

German mathematician, founder of modern set theory. The surprising results in the field of the transfinite cardinal and ordinal numbers and the paradoxes that arose from his way of argumentation, split the mathematical world into protagonists and antagonists. Among the sceptics were eminences like Herman Weyl, Luitzen E.J. Brouwer, Henri Poincaré and his own teacher Leopold Kronecker. Nevertheless, David Hilbert strongly defended Cantor and went so far in this as to state: "Nobody will ever chase us from the paradise that Cantor has created for us.".

Observe that according to this construction, each natural number is equal to the set of its predecessors, e.g. 6 = { 0, 1, 2, 3, 4, 5}. Using the natural numbers as model sets and model sequences, we can count and arrange arbitrary sets. For this purpose we have to define relations between sets, namely the relation of having the 'same number' of elements and the relation of having 'fewer' elements than the other set.

2.2 An infinite procession of cardinals

2.2.1 The natural numbers as cardinal numbers of finite sets

We say that two sets *A* and *B* have the same number of elements (are equinumerous) if and only if there exists a one-to-one correspondence between their elements; in other words: if every element of *A* corresponds to exactly one element of *B* and vice versa. We then say that there exists a bijection between *A* and *B*.

Example

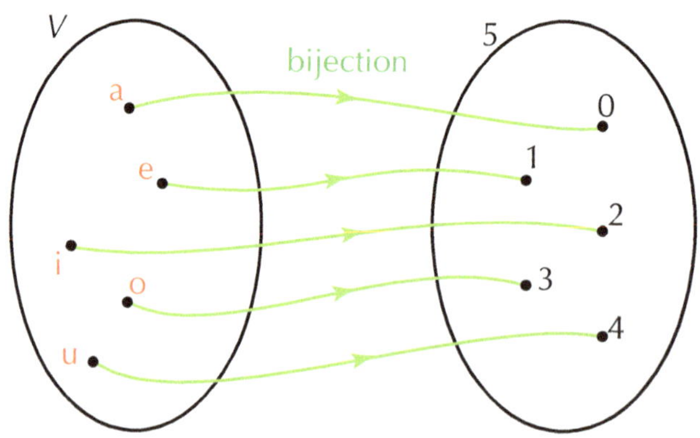

Since the set of vowels *V* = {a, e, i, o, u} contains the same number of elements as the natural number 5 = {0, 1, 2, 3, 4} we state that the cardinal number of the set *V* equals 5. There exists a bijection from *V* onto 5, namely:

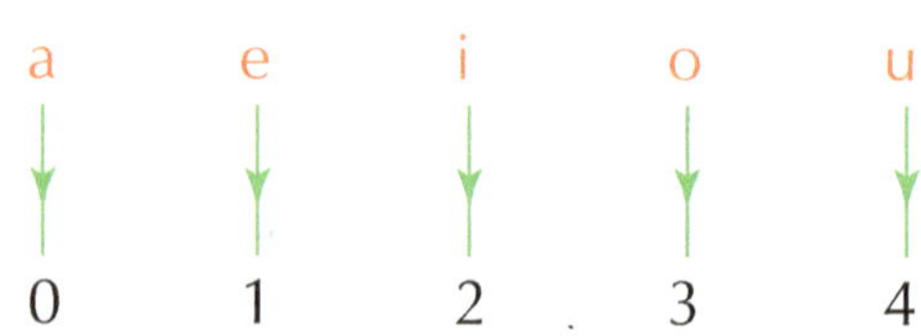

This way, the natural numbers serve as cardinals of finite sets.

The cardinal of a set *S* is symbolized as #*S*.

So, for our set *V*, we note: #*V* = 5.

2.2.2 *The cardinal of* $\mathbb{N}$*, representing the countably infinite*

We already have an infinite set available, namely:

$$\mathbb{N} = \{0, 1, 2, 3, \ldots, n, \ldots\}.$$

We call the cardinal of $\mathbb{N}$ 'aleph-null' and represent it by $\aleph_0$. Every infinite set that is equinumerous to $\mathbb{N}$ thus has the same cardinal number $\aleph_0$. Such sets we call 'countably infinite sets', because their elements can be put in correspondence with the elements of the set of the natural numbers. They make the first kind of infinite sets.

Examples

The set of the even natural numbers $E = \{0, 2, 4, 6, 8, \ldots, 2n, \ldots\}$ contains as many elements as the set of the natural numbers $\mathbb{N} = \{0, 1, 2, 3, \ldots, n, \ldots\}$, because there exists a bijection between $\mathbb{N}$ and E, namely:

0	1	2	3	4	...	n	...
↓	↓	↓	↓	↓		↓	
0	2	4	6	8	...	$2n$	...

This bijection maps every natural number onto its double. E thus has cardinal number $\aleph_0$.

In the same way, the set $S = \{0, 1, 4, 9, 16, \ldots, n^2, \ldots\}$ of the second powers of the natural numbers also has cardinal number $\aleph_0$, since there exists a bijection between $\mathbb{N}$ and S, namely:

0	1	2	3	4	...	n	...
↓	↓	↓	↓	↓		↓	
0	1	4	9	16	...	n^2	...

This bijection maps every natural number onto its square.

We can share Galileï 's surprise about the fact that a proper subset of $\mathbb{N}$, like E or S, in which even an infinite number of elements of $\mathbb{N}$ has been left out, still keeps on having the same number of elements as $\mathbb{N}$ itself. This, however, is exactly the characterizing property of infinite sets. For this reason Richard Dedekind (1831–1916) used the characteristic of having an equinumerous proper subset for the definition of an infinite set.

2.2.3 Transfinite cardinals. Cantor's theorem.

Obviously, not all sets are equinumerous. A set *A* contains less elements than a set *B* if every element of *A* corresponds to an element of *B*, but still some elements of *B* are left over. Then we say that the cardinal of *A* is (strictly) less than the one of *B* and that the one of *B* is (strictly) more than the one of *A*. We write this as $\#A < \#B$.

Example

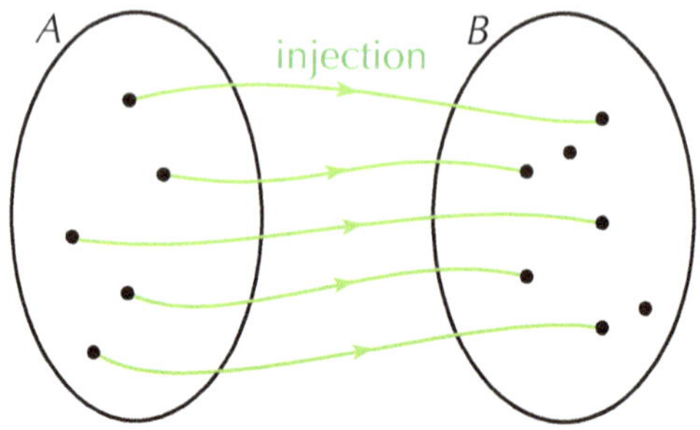

A mapping from the set *A* into the set *B*, which maps different elements of *A* onto different elements of *B*, and by which not necessarly but possibly all the elements of *B* are reached, is called an injection of *A* into *B* (as seen from the example above). Because in particular also bijections are injections, we say that the cardinal of *A* is (strictly) less than the one of *B* if and only if no injection of *A* into *B* is also a bijection.

Now we come to a fact in the history of mathematics that really caused a sensation and did not pass by without controversies, namely the amazing Cantor theorem and its proof.

For any set *S*, finite or infinite, we can construct a new set, namely the set of its subsets ('parts') $\mathcal{P}S$.

Example

Let $S = \{a,b,c\}$, then: $\mathcal{P}S = \{\{\ \}, \{a\}, \{b\}, \{c\}, \{a, b\}, \{a, c\}, \{b, c\}, \{a, b, c\}\}$.

Observe that *S* contains three elements and $\mathcal{P}S$ eight: we can consider a subset of *S* being made by deciding for each of the three elements of *S* between the two alternatives: including it in the subset or not. This explains that we can choose $2 \cdot 2 \cdot 2 = 2^3 = 8$ subsets from a set of 3 elements. Likewise, a set with 4 elements has $2^4 = 16$ subsets, and by generalizing we can conclude that a set with *n* elements has 2^n subsets.

The given example shows that the cardinal number of *S* is (strictly) less than that of $\mathcal{P}S$.

Cantor succeeded in proving that this is true for any set *S* and, very surprisingly, also for infinite sets. This theorem of Cantor's, namely $\#S < \#\mathcal{P}S$, implies that one can construct sequences of infinite sets of which the next one has higher cardinality than the previous one and that there cannot be a highest cardinality for sets. According to this theorem, cardinals in the following sequence only get bigger and bigger:

$$\#\mathbb{N} < \#\mathcal{P}\mathbb{N} < \#\mathcal{P}\mathcal{P}\mathbb{N} < \#\mathcal{P}\mathcal{P}\mathcal{P}\mathbb{N} < \ldots$$

We define: $\#\mathcal{P}\mathbb{N} = \aleph_1$, $\#\mathcal{P}\mathcal{P}\mathbb{N} = \aleph_2$, $\#\mathcal{P}\mathcal{P}\mathcal{P}\mathbb{N} = \aleph_3$, etc. We call them transfinite cardinals, and they characterise infinite sets of higher cardinality than the set of natural numbers. One can prove that the set of real numbers $\mathbb{R}$, in other words the set of points on a line, is equinumerous to $\mathcal{P}\,\mathbb{N}$. So $\mathbb{R}$ has cardinal $\aleph_1$. That is why $\aleph_1$ is also called the continuum.

Because Cantor's theorem and its consequences caused such a lot of consternation and disbelief, we give a sketch of the proof of this theorem. Our argumentation is general, but we illustrate each step with the example $V = \{a,b,c\}$.

We have to prove that no injection of *V* into $\mathcal{P}V$ can be a bijection, in other words, that no injection *i* of *V* into $\mathcal{P}V$ reaches all the elements of $\mathcal{P}V$. So, we have to show that for any injection *i*, there will always be at least one subset in $\mathcal{P}V$ that is not reached by *i*.

Let's take an arbitrary injection *i* of *V* into $\mathcal{P}V$. In our example this could be:

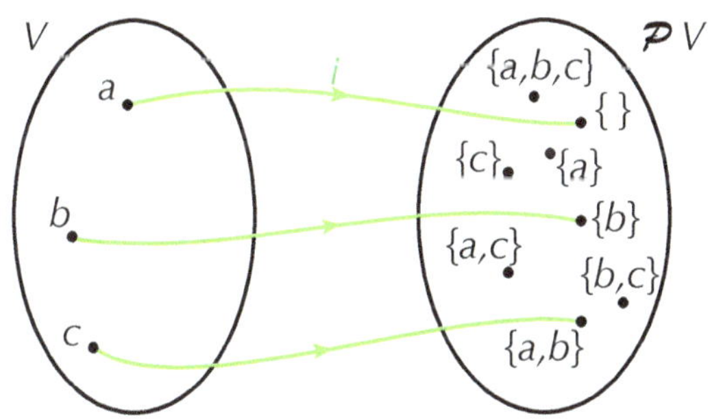

i maps *a* onto { }, *a* is not an element of its image set, we call *a* a 'traitor'

i maps *b* onto {*b*}, *b* belongs to its image set, we call *b* an 'accomplice'

i maps *c* onto {*a*, *b*}, *c* does not belong to its image set, thus *c* is also a 'traitor'.

In this way each element of *V* is either an 'accomplice' or a 'traitor' for the given injection *i*.

Let's have a look at the set of the 'traitors'. This is again a subset of *V* and hence an element of $\mathcal{P}V$. In our example it is {*a*,*c*}.

We now show that it is impossible that *i* reaches the subset of 'traitors'.

- either an 'accomplice' is mapped onto this subset, but then this one should belong to its image set, which is impossible since the image set only contains 'traitors';
- either a 'traitor' is mapped onto this subset, but then this one would belong to its image set that contains all 'traitors', which implies that it would be an 'accomplice'.

Since *V* only contains 'accomplices' and 'traitors', no element of *V* is mapped onto the subset of its traitors. Thus the injection *i* is not a bijection.

Indeed, in our example, *i* doesn't map any element of *V* onto {a,c}.

This argumentation can be carried out for all the injections of *V* into $\mathcal{P}V$.

This proves the theorem: the cardinal of a set *V* is always (strictly) less than the cardinal of the set of its subsets $\mathcal{P}V$!!!

The argumentation on which this proof is based bewildered mathematicians at the end of the nineteenth century. Many of them even rejected it as being an aberration of the set theory, which moreover had to cope with even more dramatic paradoxes. Because of these strange results, subsequent research into the fundamentals led to different philosophical schools within mathematics. How this happened will be explained briefly in chapter 3.

2.2.4 Calculating with cardinals

With cardinals we can calculate by means of sets that represent these cardinals, similarly to how it is done, for instance, on an abacus. To add 3 to 2, one first moves three beads (a set with three elements), next one moves two other beads (a set with two elements, separate from the first set), then one puts both sets of beads together to make one set and one counts the number of them. This makes the number 5. The cardinal 5 is thus the sum of the given cardinals 3 and 2.

In this way, the sum of two cardinals is made by means of the union of two separate sets that are representations of those cardinals. Since in a set the elements are not arranged in a certain order, the union of sets is commutative (independent of the order of the sets) and hence the addition of cardinals is. The union of two sets *A* and *B* is written as $A \cup B$.

So the definition of the addition of cardinals can be expressed symbolically as:

$$\#A + \#B = \#(A \cup B) \qquad \text{(where } A \text{ and } B \text{ are disjunct sets)}$$

Example

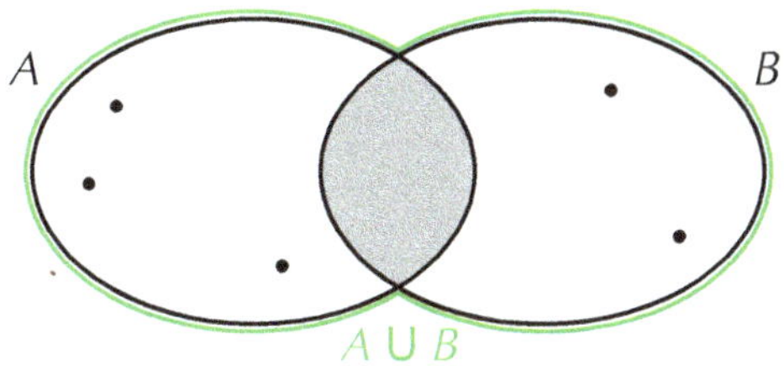

Also: $3 + 2 = 5 = 2 + 3$

We have constructed the natural numbers by considering them as being the set of their predecessors, i.e. $\{0, 1, 2, 3, \ldots, n-1\} = n$. By adding n to the set that defines it, we get its successor, i.e. $n \cup \{n\} = n+1$. This formula expresses a special case of the addition of cardinals.

By means of this successor technique we can construct the infinite set of the natural numbers. But can this technique get us even beyond $\mathbb{N}$, in other words, does $\mathbb{N} \cup \{\mathbb{N}\}$ define a set of higher cardinality than $\mathbb{N}$? This question can also be put as: is $\aleph_0 + 1$ more than $\aleph_0$? We will show that this is not the case and that: $\aleph_0 + 1 = \aleph_0 = 1 + \aleph_0$. For this purpose, it is sufficient to construct a bijection from $\{\mathbb{N}\} \cup \mathbb{N}$ onto $\mathbb{N}$.

Well,

$$\begin{array}{rcccccccccc} \{\mathbb{N}\} \cup \mathbb{N} = & \{\mathbb{N}, & 0, & 1, & 2, & 3, & 4, & \ldots, & n, & \ldots\} \\ & \downarrow & \downarrow & \downarrow & \downarrow & \downarrow & \downarrow & & \downarrow & \\ \mathbb{N} = & \{0, & 1, & 2, & 3, & 4, & 5, & \ldots, & n+1, & \ldots\} \end{array}$$

is a bijection.

We can even prove that $\aleph_0 + \aleph_0 = \aleph_0$.

Let's take, for instance, $A = \{a_0, a_1, a_2, \ldots, a_n, \ldots\}$ and $B = \{b_0, b_1, b_2, \ldots, b_n, \ldots\}$,

So both A and B have cardinal $\aleph_0$.

Since A and B are separate, it applies that: $\#(A \cup B) = \#A + \#B = \aleph_0 + \aleph_0$.

But

$$\begin{array}{rcccccccccc} A \cup B = & \{a_0, & b_0, & a_1, & b_1, & a_2, & b_2, & \ldots, & a_n, & b_n, & \ldots\} \\ & \downarrow & \downarrow & \downarrow & \downarrow & \downarrow & \downarrow & & \downarrow & \downarrow & \\ \mathbb{N} = & \{0, & 1, & 2, & 3, & 4, & 5, & \ldots, & 2n, & 2n+1, & \ldots\} \end{array}$$

is a bijection.

Which shows that: $\aleph_0 + \aleph_0 = \aleph_0$.

For making bigger cardinals starting from the cardinal $\aleph_0$ of $\mathbb{N}$, we apparently need the boosts offered by Cantor's theorem, i.e.

$$\#\mathbb{N} < \#\mathcal{P}\mathbb{N} < \#\mathcal{P}\mathcal{P}\mathbb{N} < \ldots$$

2.2.5 *Hotel Hilbert*

A pleasant story to illustrate the properties of the countably infinite, aleph null, is the story of Hotel Hilbert, which unlike a normal hotel, has an infinite number of rooms, namely the countably infinite number $\aleph_0$.

A minibus arrives, loaded with ten VIP's asking for rooms. Unfortunately all rooms are occupied. The receptionist reports the problem to his boss, actually the famous mathematician Hilbert. He solves the problem in a wink: let all hotel guests move to a room with a number 10 more than that of their present room, this means 1 goes to 11, 2 to 12 etc... ; and then offer the rooms with numbers 1 to 10, which are now free, to the newly arrived guests.

The next day a group of a countably infinite number of participants in a mathematics symposium reports at the hotel desk. In despair, the receptionist calls in Hilbert again, who once more has a solution: let all hotel guests move to a room the number of which is exactly twice their present room's number, so 1 goes to 2, 2 to 4 etc.; in this way only the rooms with an even number remain occupied but a countably infinite number of rooms with odd numbers are freed for the mathematicians.

Of course, the story doesn't end here. Next, a countably infinite number of groups with each a countably infinite number of members report at the hotel desk. Unfortunately Hilbert himself went on his holidays, so it is your turn to assist the receptionist!

An ode to aleph-null by the club of inveterate beer drinkers. If on an evening in this club one of the aleph-null bottles fell from the shelf and broke, nobody would have to do with one pint less. Even if the Hotel Hilbert, with its aleph-null rooms, was fully booked.

2.2.6 *The continuum hypothesis*

If we use Cantor's theorem to create, starting from the empty set, a sequence of sets of step by step higher cardinality, we progress along the following types of finite sets:

$$\{\} = 0,$$
$$\mathcal{P}\{\} = \{\{\}\} = \{0\} = 1,$$
$$\mathcal{PP}\{\} = \mathcal{P}\{0\} = \{\{\}, \{0\}\} = \{0, 1\} = 2,$$
$$\mathcal{PPP}\{\} = \mathcal{P}\{0,1\} = \{\{\}, \{0\}, \{1\}, \{0,1\}\}.$$

Observe that the set $\mathcal{PPP}\{\}$ already contains four elements, so by jumping in this way we missed the type with three elements. This method makes us skip the cardinal type 3 and many more. In fact, doing so, we only reach the sets successively with the following cardinal numbers:

$$0,\ 2^0 = 1,\ 2^1 = 2,\ 2^2 = 4,\ 2^4 = 16, \text{ etc.}$$

This sequence of sets would make us skip quite a lot of natural numbers.

So the question arises whether we suffered the same shortcoming with the sequence of the transfinite cardinal numbers. Wouldn't there exist types of infinite sets with a cardinal number between, for instance, $\aleph_0$ and $\aleph_1$?

In other words, do there exist sets of higher cardinality than the set of the natural numbers, having cardinal $\aleph_0$, but of lower cardinality than the set of real numbers, having cardinal $\aleph_1$ (the continuum)? The conjecture that such an intermediate set does not exist therefore carries the name: continuum hypothesis. It was the first of the 23 famous problems that David Hilbert proposed to the mathematical world, in his speech during the international congress of 1900 In Paris. The problem was solved in different ways by Kurt Gödel and Paul Cohen. In 1939, Gödel proved that the axiom system of the 'usual' set theory remains consistent if the continuum hypothesis is added. In 1963 P. J. Cohen managed to prove that the continuum hypothesis is independent of the other axioms of set theory and hence that the negation of the continuum hypothesis could be added as well, keeping the axiom system consistent. The continuum hypothesis appears to have a status in set theory that is analogous to the status of the parallel postulate in geometry.

2.3 Transfinite ordinals

2.3.1 The natural numbers as prototypes of well-ordered sets

The sequence of natural numbers is ordered due to its construction. Moreover, this order is special because of the property that all non-empty subsets of $\mathbb{N}$, finite or infinite, have a least element. This property characterizes the so-called 'well-ordered sets'. That is why the natural numbers can also serve as prototypes for the finite well-ordered sets. In this context we call them ordinal numbers. We represent the natural numbers, considered as being ordinals, by sequences and indicate the order by arrows.

Example

The ordinal 4 is represented as:

As an ordinal, every natural number also equals the set of its predecessors.

In the empty set there is nothing to arrange, for indeed not much is to be seen there. Also in a set with only one element ordering is not needed, a mere dot suffices to note its presence.

The set of the natural numbers $\mathbb{N}$ is a well-ordered set itself, though infinite. We call the ordinal of $\mathbb{N}$ 'omega' and write it as ω. Since there is no greatest natural number we represent ω by the following infinitely progressing series:

This implies also that ω is equal to the set of all its predecessors, i.e. all natural numbers. So as to be able to represent ω in the sequence of ordinals we introduce a special symbol, i.e. an 'arrowhead' ≻, to indicate that this element has no direct predecessor but is preceded by all elements to the left. So we get the following sequence of ordinals with their order type shown under each of them:

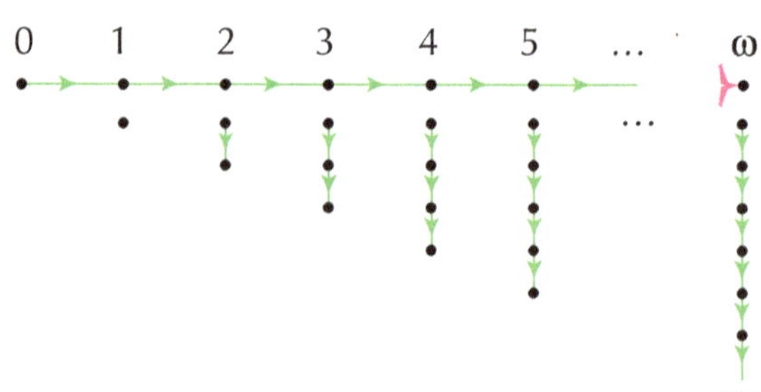

2.3.2 Calculating with ordinals

We add ordinals by writing the order type of the second next to the order type of the first (put them in juxtaposition) and next determining the order type of this joined sequence.

Example

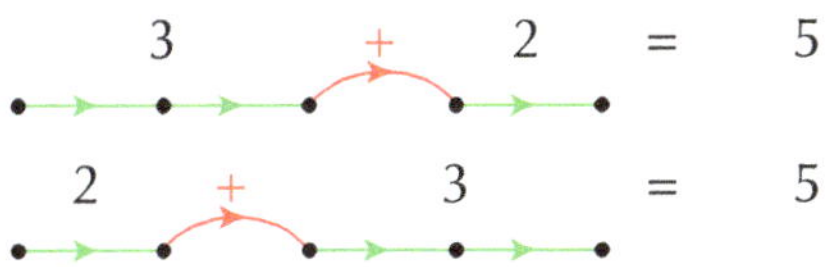

For finite ordinals we thus find the same results and properties as for the cardinals. But if we bring in the ordinal ω as a term, this doesn't hold true anymore.

Example

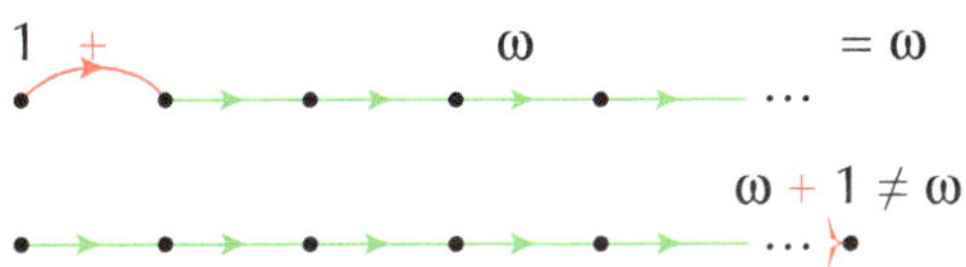

An infinite sequence with one in front of it is still nothing more than an infinite sequence, but an infinite sequence with one behind it is suddenly a different order type because it has a last element! We notice that $\omega + 1$ is another order type than ω and that $\omega + 1$ directly follows ω in the sequence of the ordinals. Moreover: $1 + \omega \neq \omega + 1$.

In an analogous way we notice that $2 + \omega = \omega$ but $\omega + 2 \neq \omega$ and $\omega + 2 \neq \omega + 1$ since:

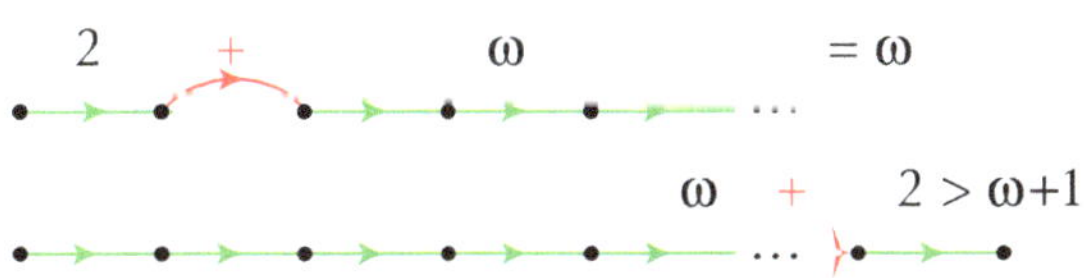

In this way we can construct bigger and bigger ordinals. The conclusion is that there exists no greatest ordinal, just as there exists no greatest cardinal either.

Summary

It is remarkable that cardinals and ordinals went different ways as soon as we crossed the border of the transfinite territory. Also the properties of the operations are no longer the same from that border onwards. Using well-ordered infinite abacuses gives different results than using un-ordered ones. This is a nice illustration of the fact that mathematical concepts and results are not just intuitively evident but always depend on the method by which they are generated.

3 Amazing birthdays and cloakroom situations

A lot of events in life are not just calculable or predictable: the day of our birth, finding a partner, the sex of our children, how long we will live, etc. That is why we say that they are determined by 'chance'. Already from the invention of the so-called games of chance onwards, people have been fascinated by the thrilling fancies of the wheel of fortune. Although the probability of profit may appear to be small, many people cannot resist testing their luck on the lottery, the jackpot or the roulette. The winning outcome is always uncertain, but the addictive gambler believes that chance does somehow not entirely work at random and that in an honest game 'in the long-term' a just distribution between all possible outcomes will show up, which he hopes to be advantageous for him. When tossing a coin once, you cannot know whether the result will be a head or a tail, but when tossing it 10000 times the number of times that head or tail appears will indeed come close to 5000. Still we know that, even if at a given moment the numbers would be exactly equal, the following toss will break that equilibrium again and there is definitely not a kind of limit tendency. In principle, also a sequence of 10000 times heads is possible, but the 'law of large numbers' states that this is very 'improbable'. As long as it concerns a game or a lottery, there would be limited damage. But when dealing with life insurance, surgery or a parachute jump, we would nevertheless like to have some more reassuring indications, although they obviously cannot be guaranteed.

Roulette
is one of the most popular games of chance that has its special place in all casinos of the world. It contains 38 numbers, namely 0, 00, 1, 2, ... , 36. You can only bet on the numbers 1 up to 36. Once the wheel is put in motion the dealer announces 'rien ne va plus' (no more bets) and the tension increases as the speed of the ball decreases.

To what extent does chance show its 'cards', and what can statistical research reveal? The start of probability theory is generally situated in the seventeenth century, when Blaise Pascal and Pierre de Fermat were corresponding on problems that Chevalier De Méré presented to them, concerning a game of dice. In the following years and centuries, probability calculus was further developed by the mathematicians Christiaan Huygens, Jakob Bernoulli and Laplace among others. But since the Russian mathematician Andrei Kolmogorov gave probability theory a rigorous axiomatic basis in 1933, several domains of nature and human sciences cannot be imagined without the methods of mathematical statistics, which are based on this probability theory.

Within the scope of this book, it is impossible to give even a brief outline of the vast spectrum covered by the current probability calculus. We restrict ourselves to some striking examples of chance phenomena, which seemingly go against intuition, but which are always confirmed by statistical samples.

3.1 Happy birthday to you and to you

If, as a simplification, we leave out February 29th (of leap years), then we can state that any person can have 365 different days as their birthday. We number them 1 up to 365 instead of January 1st up to December 31st. In a company of several persons it is possible that some persons do have the same birthday, leaving the year of birth out of consideration. We examine the next problem:

> *'What is the probability that in a company of n persons, at least two of them have the same birthday?'*

According to probability theory an answer to a question like this is only significant within the framework of a sample space. That is the set of all possible outcomes, in which to each outcome a number between 0 and 1 is assigned as being the probability that it would occur, and to which some elementary properties apply.

The first property is that the sum of the probabilities of all possible outcomes must equal 1.

So, in the case that an equal probability is assigned to all outcomes, the probability of any single outcome equals 1 divided by the number of all possible outcomes. In this case we say that the sample space is a uniform sample space.

In the posed problem the possible outcomes are the various possible assignments of the birthdays to the n persons. If we symbolise the n persons by p_1, p_2, p_3, ... ,p_n, then we can consider such an assignment as a mapping of the set of persons $P = \{p_1, p_2, p_3, \dots ,p_n\}$ into the set of numbers $B = \{1, 2, \dots , 365\}$, mapping every person onto the number of their birthday. It is possible that two persons will be mapped onto the same number, i.e. they have the same birthday.

Example:

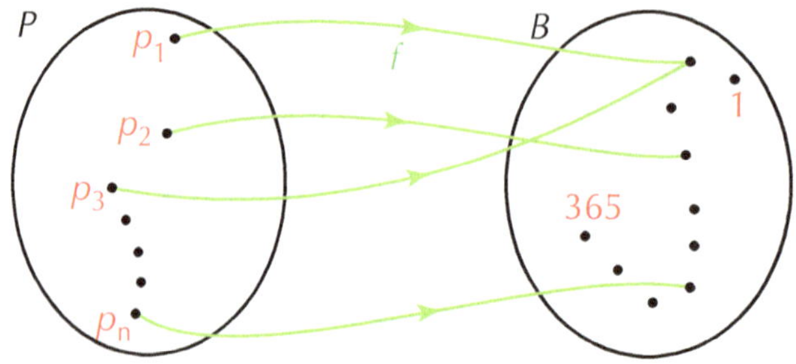

Since for each person we have the choice from 365 days, the number of mappings possible is 365 · 365 · 365 · ... · 365 (n factors), which makes 365^n. In this birthday problem the possible outcomes (outcome set) are the possible mappings of P into B. We can assume this to be a uniform sample space. The probability of each mapping is thus $\frac{1}{365^n}$.

A second property states that the probability of a subset of a (finite) outcome set equals the sum of the probabilities of its individual elements. In this context we call a subset of the outcome set an 'event' in the sample space. In the birthday problem we want to find the probability of the 'event' of all mappings f of P into B that do not map all persons onto different birthdays, i.e. mappings like in the example above. Opposite to this, a mapping of P into B that maps all persons onto different birthdays would be an injection of P into B. The set of the injections is also an event in the sample set of our problem, namely the event complementary to the one we are looking for. In making an injection of P into B for the first person we have a choice of 365 days, but for the second only a choice of the 364 days left, for the third 363 days,... and for the n-th person a choice of $365-n+1$ days left. So the number of injections equals $365 \cdot 364 \cdot 363 \cdot ... \cdot (365-n+1)$.

In the picture on the left, a sample of a given number of persons (indicated by the white frame) is taken at random from a mass. In the graph on the next page the probability of at least two persons in that sample having the same birthday can be read, depending on the number of persons in the sample..

An arbitrary mapping of *P* into *B* is either an injection or not an injection. If we call the set of all injections *I* and the set of all non-injections *N*, then according to the first property:

probability of *N* + probability of *I* = 1

So probability of *N* = 1 – probability of *I*

$$= 1 - \frac{365 \cdot 364 \cdot 363 \cdot \ldots \cdot (365 - n + 1)}{365^n}$$

In this formula *n* can vary, so for any number *n* of persons we can calculate the probability that at least two of them have the same birthday. This leads to surprising results, as can be seen in the following table and graph:

n	10	20	22	23	30	50	60
probability of *N*	0.12	0.41	0.48	0.51	0.71	0.97	0.99

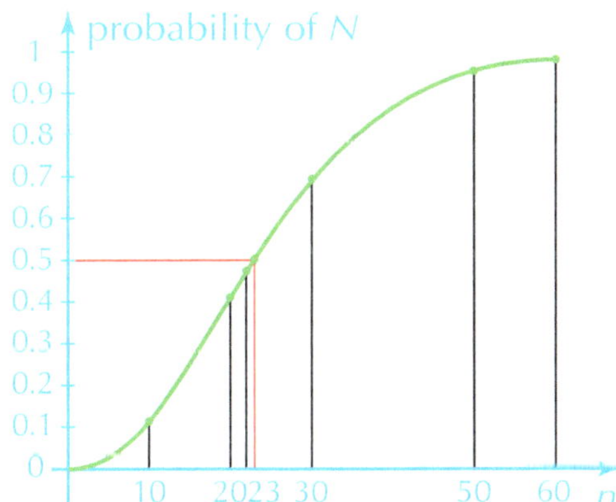

This shows that in a group of only 23 persons the probability of two of them having the same birthday already exceeds 50%, and that in a group of sixty this probability is about 99%, which means that it is almost certain. However, it is possible to select a group of 365 persons, all with different birthdays. The example above, however, shows that this will not be the case if the persons are gathered at random. It is an amusing statistical experiment to check this at parties or meetings of fifty people and more. It will make a sensational experience for the doubting Thomas's. Not a single value in the table corresponds to intuition. Only the formula shows us that we can expect surprising results.

3.2 Is this my hat or is it Euler's?

Another classic of probability theory is the famous exchange problem. We'll formulate it more precisely in the following hat problem:

> *'Four persons hand in their hats at the cloakroom desk. If later the hats are handed back at random to those four persons, what is the probability that at least one of them gets his own hat back?'*

Let us number the persons from 1 to 4, and let us give the same numbers respectively to their hats, then the activity of handing back the hats at random can be considered as a one-to-one mapping of the set {1, 2, 3, 4} in itself. This we call a permutation of the set.

Example:

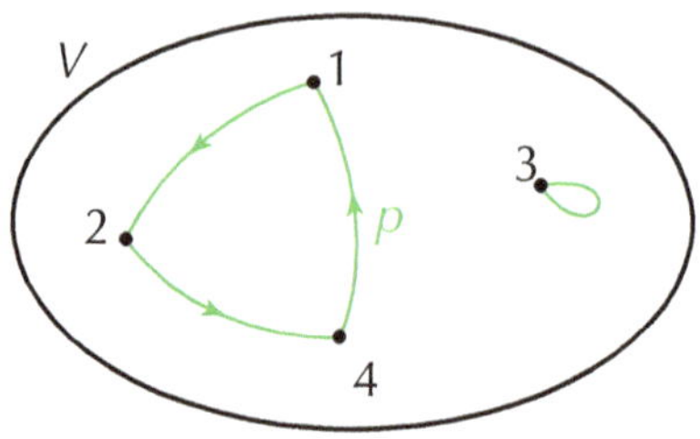

The arrow from 4 to 1 means that number 4's hat has been handed to number 1.

A loop-shaped arrow, like at element 3 in the given example, means that the person with that number got back his own hat, in this case only person 3.

The outcome set of the sample space in this example is the set of all possible permutations of {1, 2, 3, 4}. They number $4 \cdot 3 \cdot 2 \cdot 1$, since the first hat can be returned to four persons, the second only to three, the third only to two, while the last can only be returned to the last person remaining. We symbolise $4 \cdot 3 \cdot 2 \cdot 1$ as 4!, thus

$$4! = 4 \cdot 3 \cdot 2 \cdot 1 = 24.$$

We assume that, when giving back at random, every permutation has the same probability. Hence, the probability of one permutation is:

$$\frac{1}{4!} = \frac{1}{24}$$

In this problem, the event we want to find the probability of is the set *A* of all permutations of {1, 2, 3, 4} with at least one loop. Because it is easier to count the permutations without any loop at all, we will first try to count the elements of the set *B* containing all permutations of that kind.

If we want to construct a permutation that contains no loop at all, then we have to take care that every hat goes to someone else than its owner. If the hats of two persons are exchanged, then also the hats of the other two persons are exchanged. In our example there are three permutations of that type, since person 1 can have his hat exchanged with each one of the other three persons 2, 3 or 4. Like this:

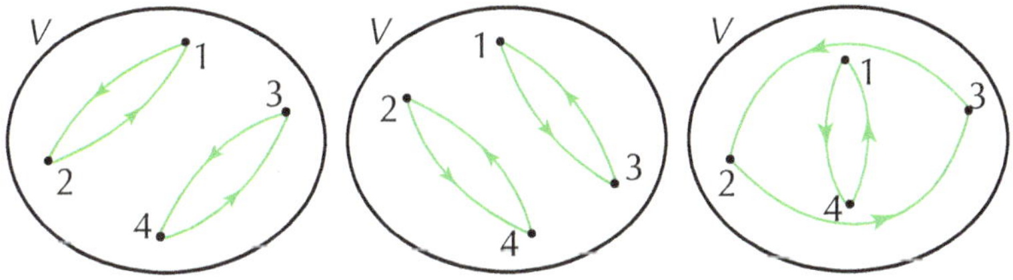

If a permutation of {1, 2, 3, 4} contains no loop and no exchanging pairs like the pairs in the examples above, then it is a circuit (cyclic permutation). There exist exactly $3 \cdot 2 = 6$ of this type, because for returning the first hat we have a choice of three persons, but for the person getting that hat we can only choose between the two remaining persons, and that determines the circuit. Like this:

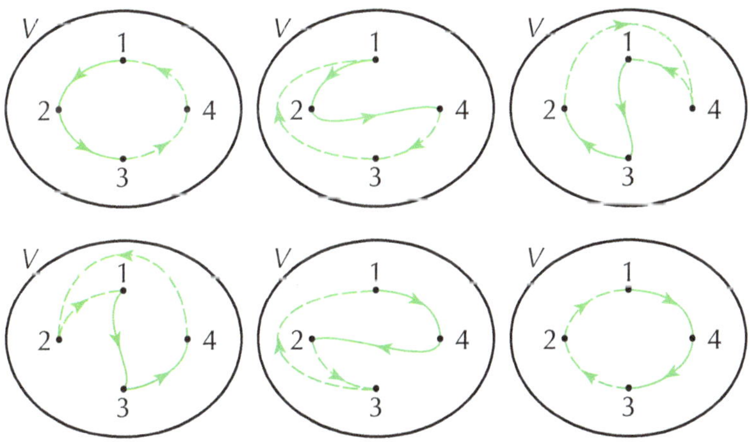

The set *B* of all permutations of {1, 2, 3, 4} without any loop thus contains nine elements. The probability of *B* is thus $\frac{9}{24}$. By means of this result we can now calculate the probability of the set *A* of all permutations with at least one loop. Since from the first property of a sample space it follows that:

probability of *A* + probability of *B* = 1

probability of *A* = 1 − probability of *B* $= 1-\frac{9}{24}=\frac{15}{24}=\frac{5}{8}=0.625$

A person getting back his own hat is almost twice as probable as no one getting back his own hat.

Can we also find a general formula for this exchange problem for n persons? How will the probability change? Will it increase with the number of persons? What would the probability become if the number of persons tends to the infinite? Because the answer to this last question is so surprising we will continue our exploration.

We can calculate that the set B_n of all permutations without any loop of a set with n elements, where n is respectively 1, 2, 3 and 4, contains the following number of elements:

B_1 contains zero elements, because the only permutation of the set

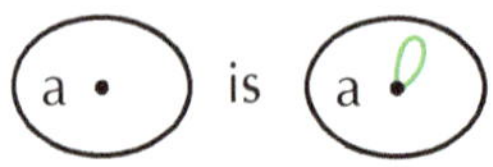

B_2 contains one element, because the only permutation without any loop of

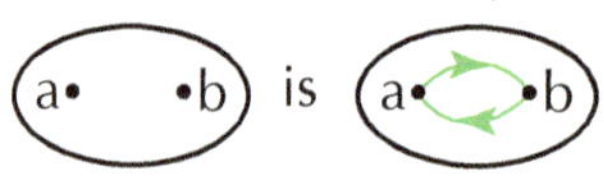

B_3 contains two elements because the permutations without any loop of

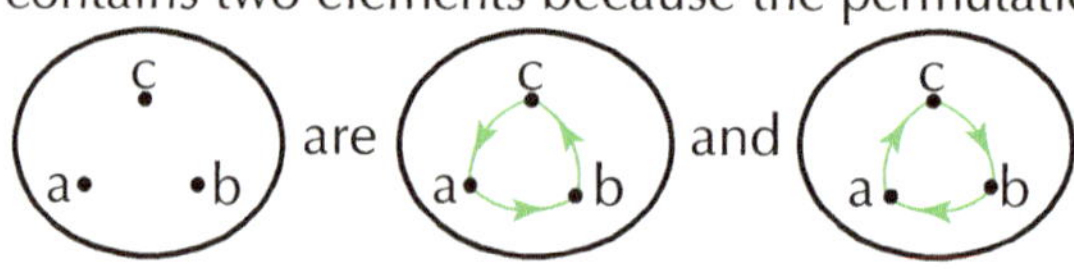

B_4 contains nine elements (see the problem of exchanging four persons' hats). The number of elements of B_n can be calculated recursively, this means: there is a relationship between the number of elements of B_n and that of B_{n-1}, i. e.

$$\#B_n = n \cdot \# B_{n-1} + (-1)^n.$$

This formula indeed applies for the already mentioned B_1 till B_4, because:

$$\#B_2 = 2 \cdot \#B_1 + (-1)^2 = 2 \cdot 0 + 1 = 1$$
$$\#B_3 = 3 \cdot \#B_2 + (-1)^3 = 3 \cdot 1 - 1 = 2$$
$$\#B_4 = 4 \cdot \#B_3 + (-1)^4 = 4 \cdot 2 + 1 = 9$$

One can prove that this formula applies for any n. This results in:

$$\#B_5 = 5 \cdot \#B_4 + (-1)^5 = 5 \cdot 9 - 1 = 44 \quad \text{and}$$
$$\#B_6 = 6 \cdot \#B_5 + (-1)^6 = 6 \cdot 44 + 1 = 265$$

By means of these numbers we can calculate all probabilities for higher n. Since B_n is the set of all permutations without any loop of a set with n elements, and A_n the set of all permutations with at least one loop of a set with n elements, this results in:

prob. of $B_1 = \frac{0}{1!} = \frac{0}{1} = 0$ and prob. of A_1 = 1 – prob. of B_1 = 1 – 0 = 1

prob. of $B_2 = \frac{1}{2!} = \frac{1}{2}$ and prob. of A_2 = 1 – prob. of $B_2 = 1 - \frac{1}{2} = \frac{1}{2}$

prob. of $B_3 = \frac{2}{3!} = \frac{2}{6}$ and prob. of A_3 = 1 – prob. of $B_3 = 1 - \frac{2}{6} = \frac{4}{6}$

prob. of $B_4 = \frac{9}{4!} = \frac{9}{24}$ and prob. of A_4 = 1 – prob. of $B_4 = 1 - \frac{9}{24} = \frac{15}{24}$

prob. of $B_5 = \frac{44}{5!} = \frac{44}{120}$ and prob. of A_5 = 1 – prob. of $B_5 = 1 - \frac{44}{120} = \frac{76}{120}$

prob. of $B_6 = \frac{265}{6!} - \frac{265}{720}$ and prob. of A_6 = 1 – prob. of $B_6 = 1 - \frac{265}{720} = \frac{455}{720}$

One can prove that in general the probability of B_n is given by the following formula:

$$\text{probability of } B_n = 1 - \frac{1}{2!} + \frac{1}{3!} - \frac{1}{4!} + \frac{1}{5!} - \frac{1}{6!} + \ldots + \frac{(-1)^{n-1}}{n!} \qquad (*)$$

Example: $$\text{probability of } B_6 = 1 - \frac{1}{2!} + \frac{1}{3!} - \frac{1}{4!} + \frac{1}{5!} - \frac{1}{6!}$$
$$= 1 - \frac{1}{2} + \frac{1}{6} - \frac{1}{24} + \frac{1}{120} - \frac{1}{720}$$
$$= \frac{265}{720}$$

The limit of formula (*) for n tending to the infinite exists and equals the reciprocal of Euler's constant e. The number e is the base of the natural logarithms.

e = 2.7182818... and $\frac{1}{e} = 0.3678794...$

Observe that for the probability of B_6 we got $B_6 = \frac{265}{720} = 0.3680555...$

Rounded to the third decimal, already for $n = 6$ the probability B_6 almost equals $\frac{1}{e}$

This means that the exchanging problem in a set of six persons hardly has a different probability than the same problem in an 'infinite set of persons'. It certainly remains beyond the reach of intuition that even in an infinite set the probability nobody gets back his own hat is more than one third. The probability that in a large population at least one person gets his own hat back will always be about

$$1 - \frac{1}{e} = 0.6321205...$$

The formula (*) suggests the possibility of finding a good approximation of the constant e by statistical experiments, in which the probability of B_n is measured by repeating the experiment many times (Monte Carlo method). That this transcendent number also appears in the exchange problem is already bewildering in itself. But it is not an isolated case. The experiment of dropping a needle many times on a parquet floor and counting the proportion of times that it crosses the seam of two parallel boards (Buffon-experiment), can be linked to the transcendent number $\pi = 3.14159...$, which determines the ratio of the perimeter of a circle to its diameter.

In all this we must not forget that the probability theory is a deductive theory, based on the axioms of a sample space. All deductions and calculations are related to this structure. Before applying them to a concrete problem we will have to test the extent to which a certain probability model is suitable for describing the problem successfully. This combination of probability theory and statistics is essential in the domain of mathematical statistics.

The Monte Carlo Method is often used for setting up emergency scenarios and for being able to anticipate calamities. In the case of avalanches, for example, it is not an option to wait for the actual experience itself before taking action. On the basis of a computer model of the mountain landscape, people can simulate the possible starting points of avalanches and follow their course. This way, for example, they can decide on locations where it is not such a good idea to build houses.

3.3 Paradoxes in probability calculus

Is it possible that, by using different models, one can get different results for the probability of a certain event? Yes, with the next example we illustrate how.

> *If we draw at random a chord in a circle, what is the probability that the length of that chord is more than the length of an inscribed equilateral triangle?*

3.3.1 First solution

A chord through a fixed point *A* of a circle is uniquely determined by the choice of another point *P* on the circle. In this, all points of the circle can be chosen to be the fixed point *A*. Because of rotation symmetry, there is no loss of generality if we restrict ourselves to taking only the indicated point *A* as the fixed point. For this reason, we take as the outcome set for this experiment the possible positions on the circle of the point *P* with regard to the fixed point *A*, and assign an equal probability to each position.

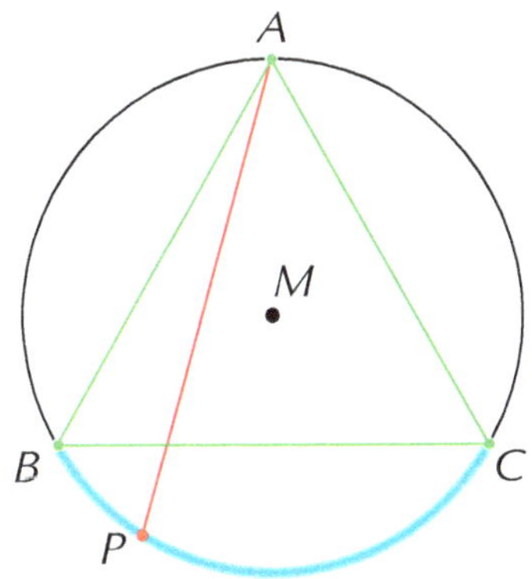

The chosen chord *AP* will have a length that is more than the side of the inscribed equilateral triangle if *P* is chosen on the arc $\widehat{BC}$ (so not on the arcs $\widehat{AB}$ and $\widehat{AC}$).

The probability for *P* being on $\widehat{BC}$ thus equals $\frac{1}{3}$, since the length of the arc $\widehat{BC}$ equals one third of the length of the full perimeter of the circle.

3.3.2 Second solution

Instead of considering chords through a fixed point of the circle, we can as well take chords with one fixed direction. Due to rotation symmetry all directions are just as probable and therefore, choosing a fixed direction creates no loss of generality. So, we now consider as possible outcomes only those chords that are mutually parallel.

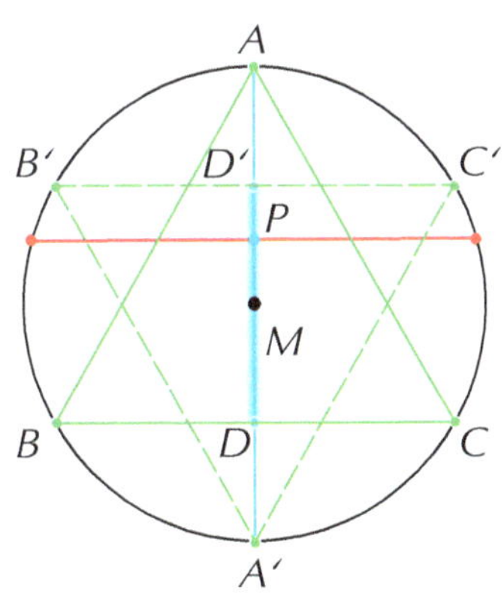

We choose as fixed direction for the chords the direction of the side *BC* of the inscribed equilateral triangle *ABC*.

A chord parallel to BC intersects the diameter AA' at a point P. The chord's position is fully determined by that point of intersection P. So it is convenient to take the points of the diameter AA' as the outcome set and to assign equal probability to each point P.

The length of the chord through P will be more than the side BC if P is in between the point of intersection D of BC and AA', and its mirror point D' respective to the centre M of the circle. Since the length of $\overline{DD'}$ is half the length of the diameter $\overline{AA'}$, the probability of the event considered equals $\frac{1}{2}$ in this sample space.

3.3.3 *Third solution*

A chord of the circle is, however, also fully determined by its midpoint, i.e. the foot P of the line perpendicular to the chord and containing the centre M of the circle. The distance between M and the side BC of the inscribed equilateral triangle ABC equals half the radius of the circle, so $MD = \frac{r}{2}$.

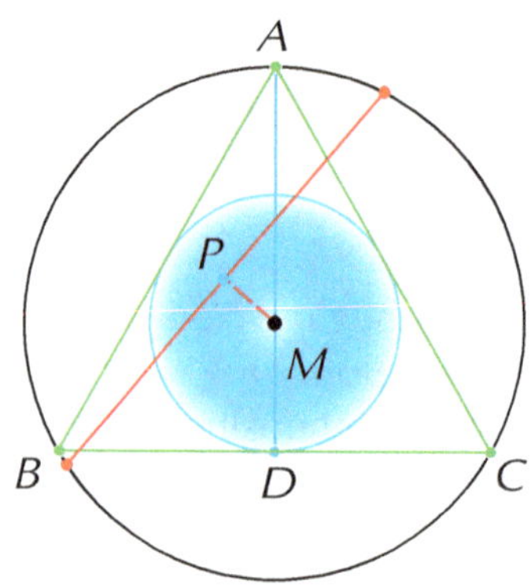

In this case, the length of the chosen chord will be more than the side of the inscribed triangle if the distance between its midpoint P and M is less than $MD = \frac{r}{2}$, this means if P is situated in the small circle with centre M and radius $\frac{r}{2}$. The area of this small circle equals $\pi\left(\frac{r}{2}\right)^2 = \frac{\pi r^2}{4}$, while the area of the full circle equals πr^2. The probability of the event considered equals $\frac{1}{4}$ in this sample space.

How is it possible that these three solutions, related to three different choices of outcome sets that seem equally convincing, nevertheless result in different probabilities? When the French mathematician Joseph Bertrand presented this example in 1989 in his *Calcul des Probabilités*, these different solutions were still considered as being paradoxes. But since the structure of the probability theory was determined in an axiomatic system, we know that all three solutions are acceptable, however relative to the chosen outcome set and its related sample space.

That is why it makes no sense trying to solve a probability problem without first defining the sample space in which one can give the problem an appropriate formalization. Be aware that it is not realistic to relate all problems to a uniform sample space. For example, when tossing a drawing pin, one can theoretically assign to the two possible figurative outcomes

an equal probability $\frac{1}{2}$ indeed, but experimental practice will not confirm this model.

In practice, one will test a theoretically designed sample space statistically by means of experimental samples, and according to trends in their results, adjust the model if necessary.

Summary

Although modern probability calculus had its origin in analysing problems in games of chance, it developed alongside mathematical statistics and became an indispensable tool for describing phenomena that are determined by chance, such as death rates of seniors in a certain region, telephone traffic on weekend days, the spreading of a contagious disease, the effectiveness of a certain medicine, etc. In spite of its theoretical basis and formal procedures, it delivers relevant information and its predictions surprise the intuition, but these are always confirmed by experiments and simulation.

4 From an evening stroll to operational networks

4.1 The seven bridges of Königsberg

The next historical tale illustrates that problems which appear at first sight to be playful and of little use, can lead to fundamentally new theories with a wide spectrum of useful applications, if presented to mathematical brains.

The town of Königsberg, present-day Kaliningrad on the Baltic Sea, lies at the confluence of two rivers: the old and the new Pregel that wind around the island of Kneiphof. In the eighteenth century the many banks were connected by seven bridges as shown below:

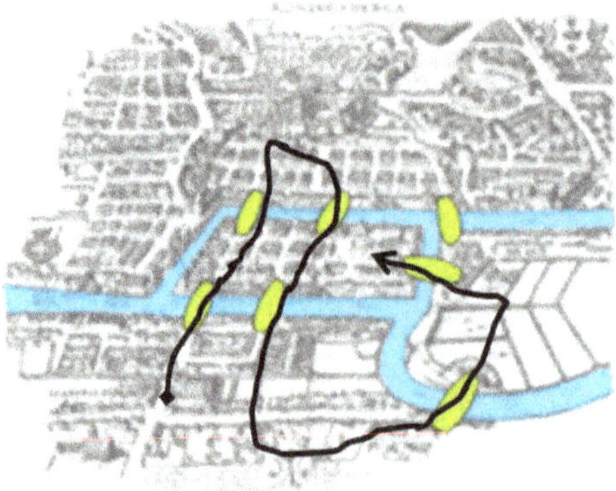

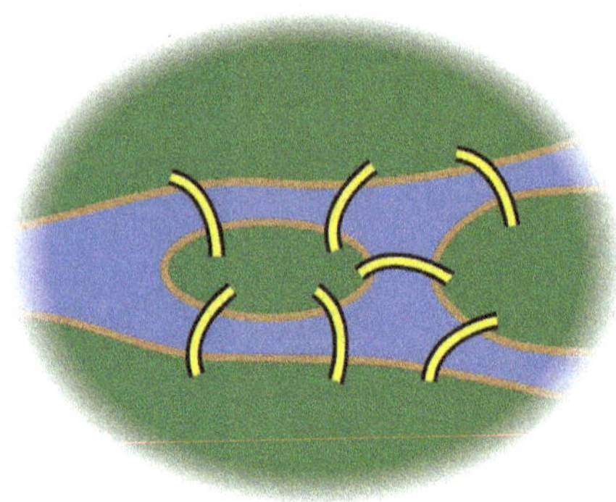

Already in those days it was a pleasure to stroll around the bridges of Königsberg. This led the inhabitants to the very playful question of whether it would be possible to make a walk in which they would cross all the bridges only once, regardless of whether they would end up on the same bank where they had started or not. From whatever place they started and in whatever order they crossed the bridges, all trials ended in failure. In the year 1736, when the problem had become a public passion, it came to the ears of Euler, 28 years old at the time, and working in Saint Petersburg. After a profound analysis and abstract approach to the problem, he could prove that such a walk in Königsberg was impossible. The way he succeeded in this became the historical starting point of the new 'geometry of situations', which has developed further into the so-called 'graph theory'. That theory has applications in many kinds of networks like sociograms (psychology), organization diagrams (economics), communication networks, transport networks, electric circuits, game strategies, chemical structure formulas, genealogical trees, decision theory, information technology and cybernetics, operational research, etc.

4.2 Eulerian paths

The basic elements of graph theory are as simple as they are efficient. It is evident that the scenery of the different banks and the architecture of the bridges don't play any fundamental role in solving the problem. The first step that Euler took was mathematizing the problem in an abstract form that allowed generalization, as in the diagram below:

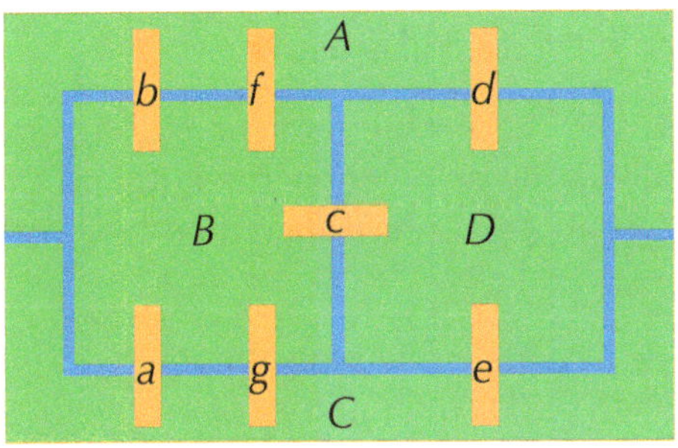

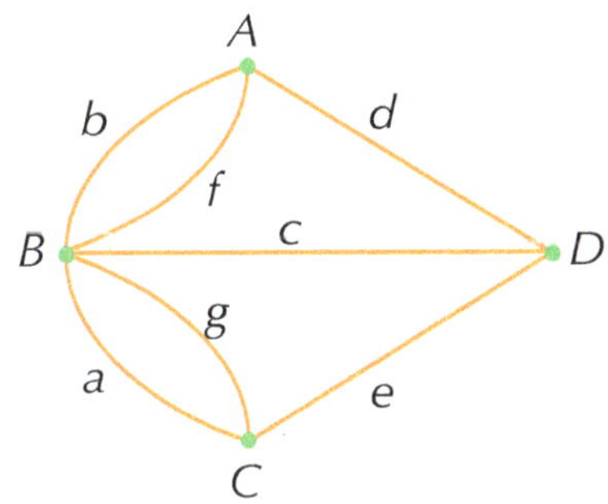

The banks are represented by points, in this context we call them vertices, and the bridges as arcs, which we call edges. The whole diagram is called a graph. The edges are not directed (there are no arrow heads) because bridges can be crossed in both directions. In 1736 one-way streets were still not an issue. To analyse the problem we restrict ourselves to a minimum of terminology and symbols. The full arsenal of graph theory is of course more subtle and more complex. In the previous graph we call a sequence of edges that starts at X and ends at Y a (simple) path connecting X to Y.

Examples

(g, f, b) is a path connecting C to B and (d, f, g, e) is a path connecting D to D, so it is a cycle.

Observe that in the graph of the bridges of Königsberg there is, for any pair of vertices X and Y, a path connecting X to Y, in other words we can get from any place to any place, which is why we call it a connected graph. Not all graphs are connected, as the following example shows:

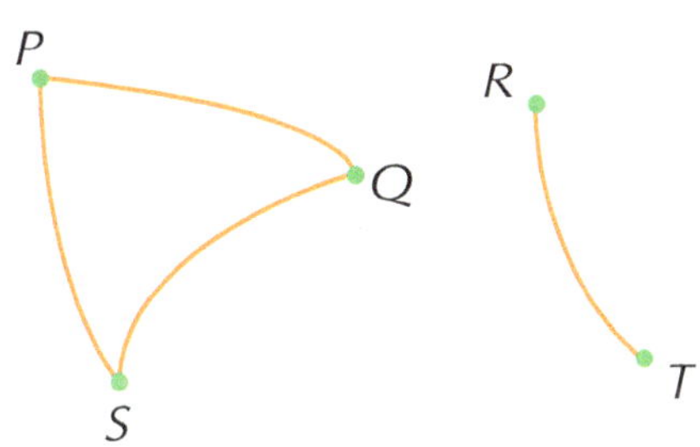

In this graph there is, for example, no path connecting Q to R. The graph is disconnected and clearly falls apart into two connected subgraphs.

The number of edges in a vertex is the order of the vertex. The order of a vertex can be even or odd.

Examples (in the graph of the Königsberg bridges)

The order of vertex B is 5 (odd) and the order of vertex C is 3 (odd).

In fact, all four vertices in this graph have an odd order.

Using this terminology we can reformulate the problem of the walk over the Königsberg bridges like this: 'Is it possible, in the graph concerned, to find a path that uses all the edges precisely once?' Such a path has since then been called an Eulerian path.

In order to solve the problem, Euler proved the following theorem: 'A graph contains an Eulerian path if it is connected and if the number of odd vertices is just 0 or 2.'

The proof is based on the argument that, except for the starting and ending vertex, one has to leave every vertex as many times as arriving at it, so the order of all the other vertices has to be even. The order of both the starting and the ending vertex may, however, be odd, since one was already in the first and stays in the last.

Because in the problem of the Königsberg bridges there are four vertices with an odd order, an Eulerian walk along the bridges is not possible.

Probably not for this reason, but later on one more bridge was built over the old Pregel, connecting banks C and D. We can now repeat the question for the eight bridges. In this new situation the graph looks like this:

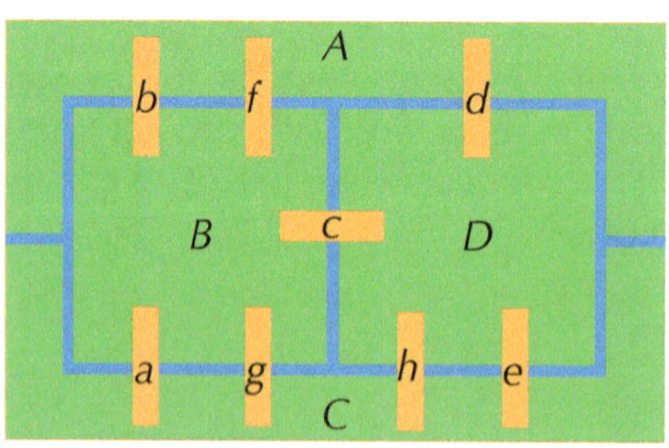

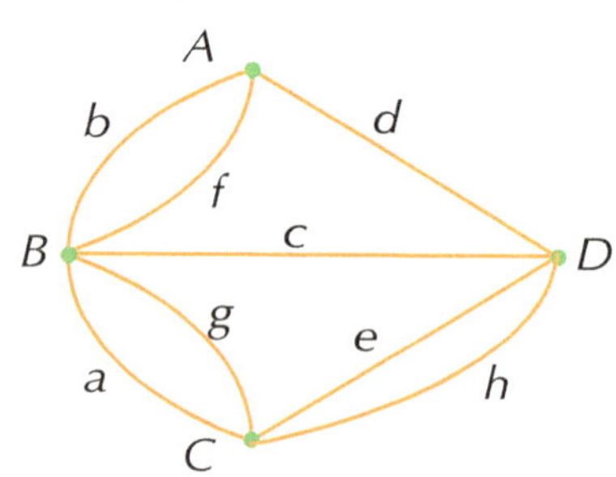

Observe that both C and D have an even order now and only A and B still have an odd order. According to the theorem this graph contains an Eulerian path that starts at A and ends at B or vice versa.

For instance, the path (*g, e, d, f, a, h, c, b*) defines one such desirable walk (from *B* to *A*). But it is not the only one possible, thus providing many evenings of delight. The joy of exploring them is yours!

With the additional condition that you should return to your starting point, in other words if you long for an Eulerian cycle, then all vertices have to be even. In this case, even the eight bridges are not enough. For this purpose one can start building again. The question of where an extra bridge has to be built makes another pleasant exercise.

Probably in your childhood you already did Eulerian experiments, without realizing it. The game that challenged you to use your pencil and follow the lines of a given figure, without lifting the pen and without going over parts of the figure more than once, is nothing more than a search for an Eulerian path in the figure. With the next two figures you can have another go, unless you learned something in the previous paragraph.

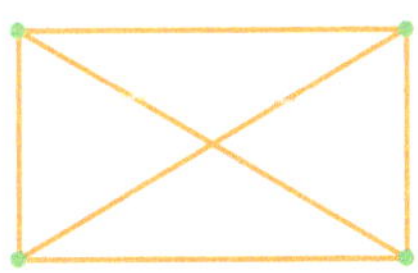

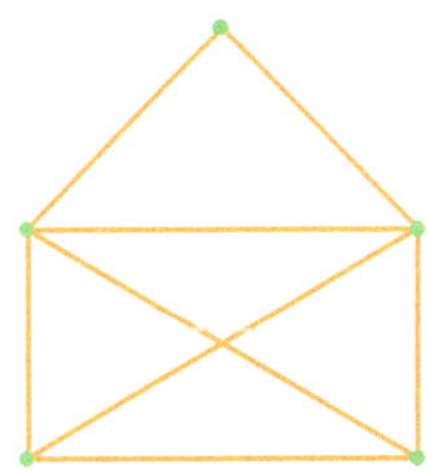

To avoid ambiguity in the number of vertices it is advisable that the edges do not intersect in a graph. In both figures we can better disentangle the diagonals and represent the graphs as follows:

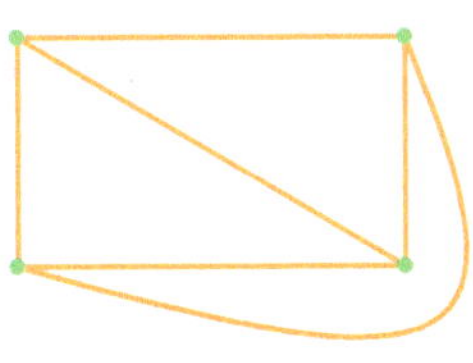

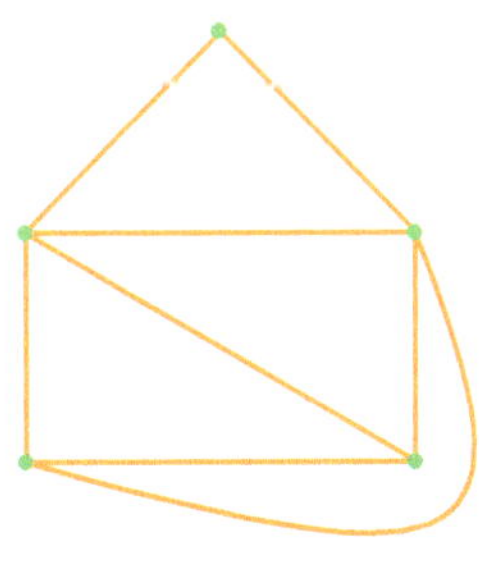

A graph in which no edges intersect (apart from vertices) is called a planar graph.

The graph of the first figure contains four vertices with odd order, so there is no Eulerian path for it. The graph of the second figure contains precisely two vertices of odd order and the order of the other three vertices is even, so you can draw an Eulerian path in it by starting at the one odd vertex and arriving at the other. Please, treat yourself to this pleasure once more.

In a (regular) octahedron (a figure consisting of two four-sided pyramids joined together) there even exists an Eulerian cycle, since in its planar graph all vertices have even order, namely order four.

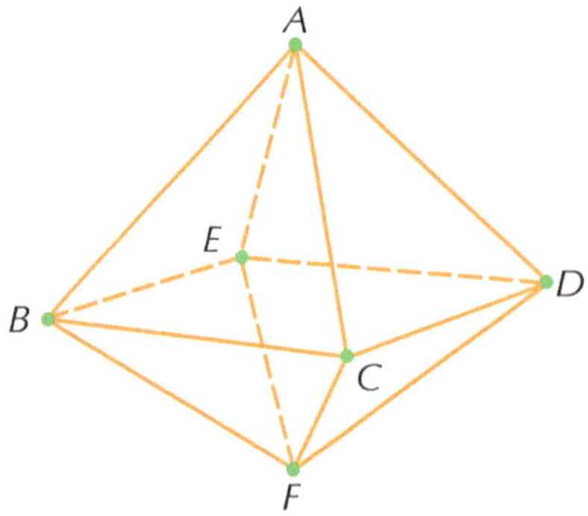

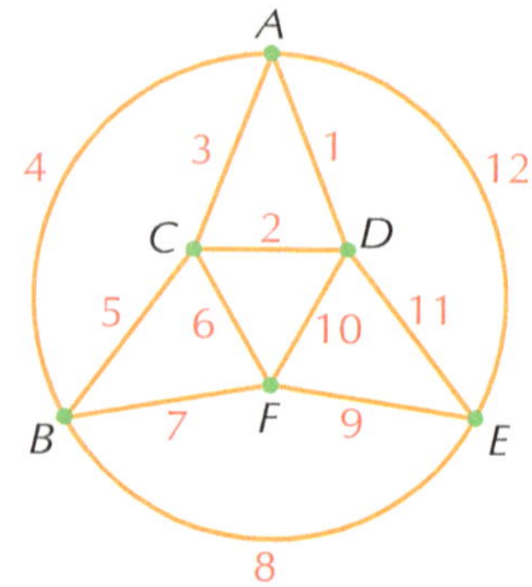

Such a cycle is, for instance: (1, 2, 3, 4, 5, 6, 7, 8, 9, 10, 11, 12). As an exercise you can examine all other regular solids, like a tetrahedron, a hexahedron (cube), a dodecahedron or an icosahedron. Since the symmetry of regular polyhedra implies that all vertices have the same order, it is sufficient to find the order of just one vertex of the solid considered (see Chapter 3, 2.4).

4.3 Problems of graph theory

As already mentioned in the introduction, graph theory can be applied in many domains. But these applications can focus on totally different goals than the Eulerian walks. We give some examples.

Example 1 Hamiltonian paths and cycles

We want to make a tour around the world visiting twenty towns, spread all over the globe, more or less like the vertices of a (regular) dodecahedron. The edges of the dodecahedron indicate flight routes, as shown in the following figure and graph:

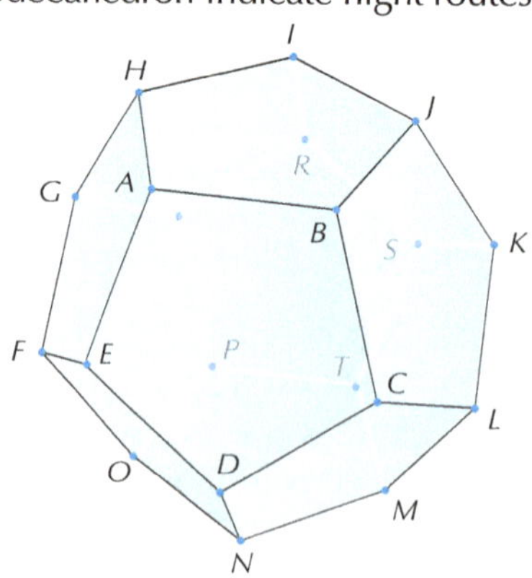

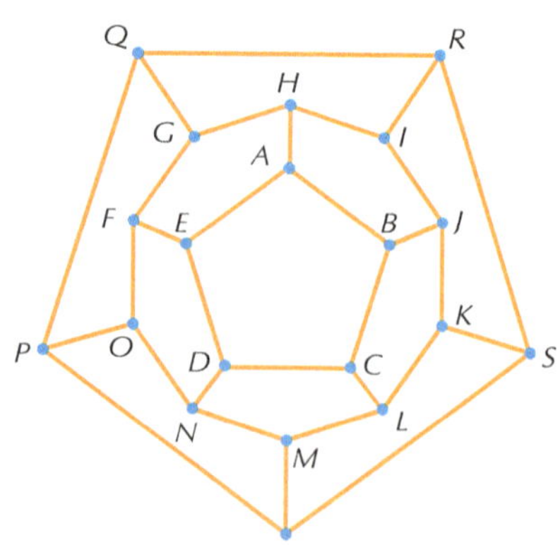

An obvious question would be: 'Does there exist a path (flight scheme) by which we visit every vertex (town) precisely once?' In such a path all vertices, instead of all edges, are visited once. Such a path we call a Hamiltonian path. If we want to end in the town where we started, which seems desirable in this example, then we speak about a Hamiltonian cycle. The conditions for such a cycle to exist are totally different from those for an Eulerian cycle. As can be seen, every vertex in the previous graph has odd order (three). So there is no Eulerian cycle, but nobody will be interested in a tour that uses all possible flights. But there exist several Hamiltonian cycles, like for instance (*P, O, F, E, A, H, G, Q, R, I, J, B, C, D, N, M, L, K, S, T, P*).

A problem that also deals with a Hamiltonian path is that of the knight on the chess-board: 'Does there exist a path in which the knight, with its characteristic jumps (edges), occupies each square (vertex) of the board precisely once?' The answer turns out to be positive. We wish you pleasant jumping!

You may also be curious about the pseudo-regular or Archimedian solids like a soccer football, in which the seams represent the edges. There are twenty hexagonal patches and twelve pentagonal ones, which can also be represented in a planar graph. Are you able to find a Hamiltonian cycle in that graph?

Example 2 Shortest ways, fastest ways, cheapest ways, etc.

In transport problems it can be interesting to find an optimal route with regard to some criterion one wants to meet. According to that criterion one can make preferences among the different paths connecting vertex *A* to vertex *B*. The simplest definition of 'length' of a path is of course the number of edges in the path. According to that definition the shortest path is the one with the least edges. But for the real practice of transport this is not a realistic criterion, since many aspects can have an influence. Does the edge represent a motorway, a waterway, an airway,...? Is there a toll to be paid? What services have to be made use of? Etc.

In the graph of such a transport network, a weight is generally assigned to every edge according to the criterion one has in mind: speed, cost, efficiency, etc.

In this way the search for the best path becomes more refined and the methods demand a deepening of the theory. In the following graph, with the weight of the edges taken into account, the best path connecting *A* to *J* is the path (*A, H, F, D, C, E, J*) with the least total value being 14. For finding such paths systematically, there exist appropriate theorems and algorithms that we cannot introduce given the restricted scope of this book.

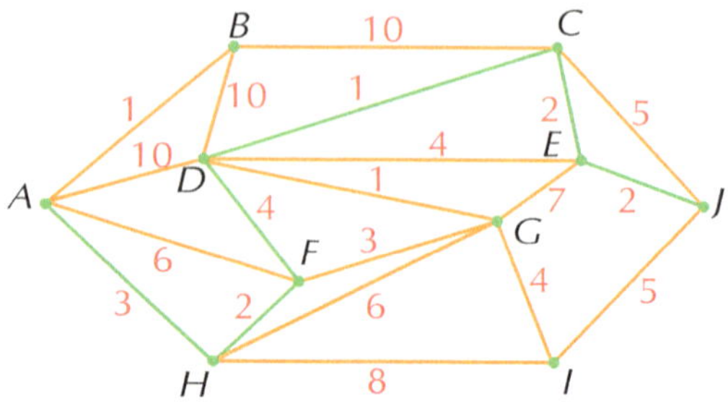

Example 3 Chromatic numbers

On a map we can consider the different countries to be the vertices of a graph and a common border of two countries to be an edge connecting the vertices. Two vertices connected by an edge (neighbouring countries) are therefore called adjacent vertices. A famous problem of graph theory is the colouring problem, i.e. 'What is the minimal number of colours needed to colour a map so that no neighbouring countries will be given the same colour?' This problem can be stated in general for any graph: 'What is the minimal number of colours needed to colour the vertices of a graph so that no adjacent vertices will be given the same colour?' This minimal number of colours is called the 'chromatic number' of the graph.

In order to solve this problem for a planar map, and taking into account all geographic rearrangements due to wars or independence movements, people had to call in both graph theory and the help of the computer to make the enormous amount of calculations necessary. Now we know the answer for a planar map, and it is: 'four'. The fact that less won't do is shown by the following abstract situation:

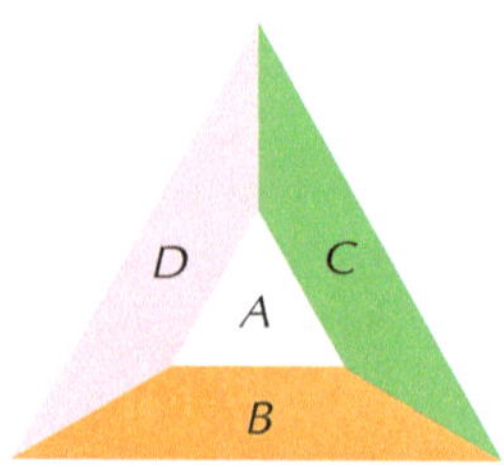

But whether the number 'four' would be sufficient in any imaginable situation, well, that's quite another cup of tea.

If all countries had straight borders in line, in other words, if the map could be drawn by only intersecting straight lines, then two colours would be sufficient. This can easily be proved by the recursive method. It is obviously true if there is only one line as a border. And if it is true for n lines, then it is also true for n+1 lines.

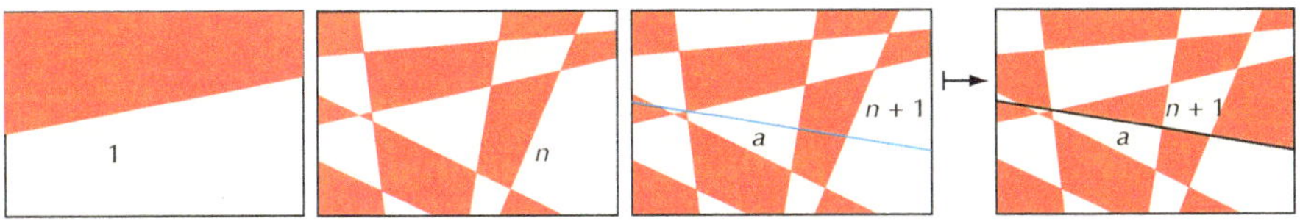

For 1 OK   If for n then for $n + 1$ because

In the situation of n lines, add one more line, for example a. a divides the plane in two half planes. Keep the colouring as it was in the one half plane, but change the colours in the other.

But the countries of the world are not situated on a plane but on a globe. What is the chromatic number of a map on a sphere? It is 'four' as well. The chromatic number of a map on a Möbius band (see left figure) is 'six' and the chromatic number of a map on a torus (doughnut) is 'seven' (see right figure).

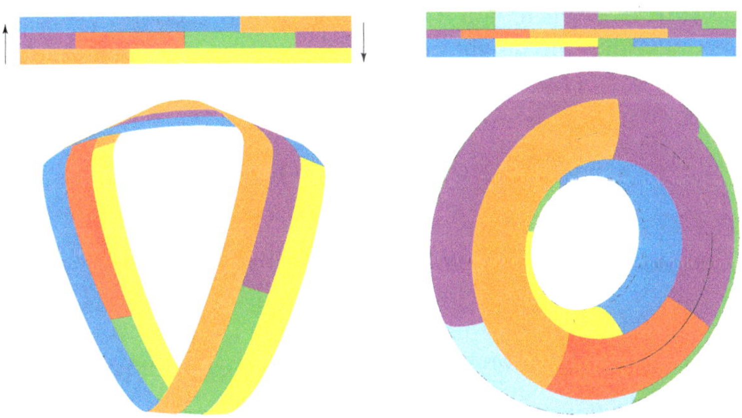

Summary

From the examples shown it is clear that the origin of a useful theory does not necessarily have to lie in the restricted domain of specific practical problems, but can just as well result from a playful idea or an intriguing puzzle. For this reason absolutely no question offends the dignity of mathematics.

5 The ideal measurements of 'Miss Tin-Can' and 'Mr Candy-Cone'

We seldom wonder why certain objects we come across every day in our role of consumer, have the particular shape and the dimensions they have. This is not merely accidental, as seen from the following two examples.

5.1 Cylinder with a given volume and minimal area

Ready-made food is often sold in cans. Why do cans usually have a cylindric shape? The manufacturer is of course interested in packaging with minimal cost. This is the case if the container requires a minimal area of packaging material for a given volume. We know that a sphere is most favourable in this, but it is obvious why we don't see piles of spheres in supermarkets: it would be quite a show when customers take one from the pile. The next frugal shape is the cylinder. This one can be stacked in a more stable way.

What are the ideal proportions of an economical cylinder, insofar as they are not already determined by the nature of the product to be packed, as it would be for asparagus when its length has to be taken into account? Well, it proves to be cylinders of which the height precisely equals the diameter. But how do we know? Well, it takes a bit of upper secondary school mathematics. Derivatives, do you remember?

The only rule needed to understand the calculations that follow is:

$$D\,(a \cdot u^n) = a \cdot n \cdot u^{n-1} \cdot Du$$

in which D stands for derivative (the slope of a curve), a stands for a constant and u for a function of an independent variable, usually written as x. Let's give it a try. It is not terrible if you miss some fragments of the film here and there, just pretend that you understand everything. A film by Ingmar Bergman is harder to get.

A cylinder has two dimensions, namely its height h and the radius of its base r.

The volume V of the cylinder is given by the formula $V = \pi\, r^2 h$.

The total area A is given by $A = 2\pi r h + 2\pi r^2$, in which the first term expresses the lateral area and the second term the area of the bottom and the top surface.

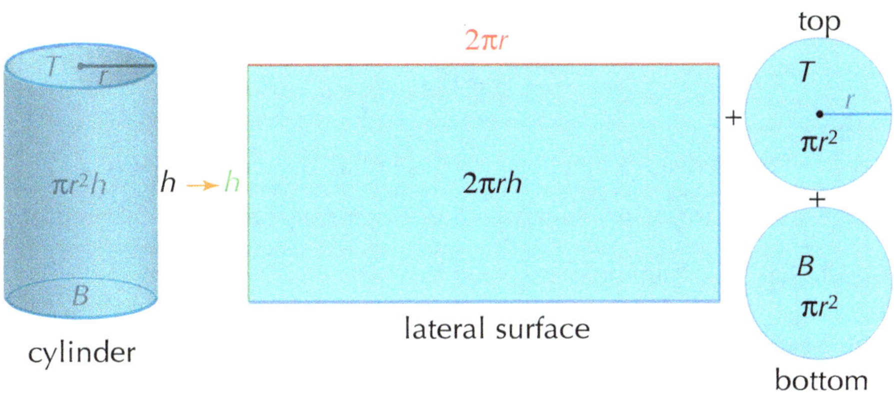

If the volume of the cylinder is fixed, then the values of *r* and *h* relate to each other and any choice for the value of *r* implies a value for *h*. By these values one can calculate the total area of the cylinder. A variation of *r* implies a variation of *A* as a function of *r*. Our aim is to find out whether the function *A* has a minimum for a certain value of *r* and what value of *h* is related to this value of *r*.

Let a certain volume *V* be stated beforehand for the cylindric can. Then, to keep the calculations simple, we can write *V* as the product of π and the third power of a positive number *k*, without loss of generality.

So $V = \pi r^2 h = \pi k^3$. which implies that: $h = \dfrac{k^3}{r^2}$

We express *A* as a function of *r*: $A = 2\pi rh + 2\pi r^2$

$$= 2\pi\left(r \cdot \frac{k^3}{r^2} + r^2\right)$$

$$= 2\pi\left(\frac{k^3}{r} + r^2\right)$$

In order to retrieve the familiar notations of function analysis, we will substitute:

$A = y$ and $r = x$, which leads to: $y = 2\pi\left(\dfrac{k^3}{x} + x^2\right)$

We have to find the possible minimum of this function. To do this, we calculate the derivative function and its zeroes:

$Dy = 2\pi\left(\dfrac{-k^3}{x^2} + 2x\right)$, because $D\left(\dfrac{1}{x}\right) = Dx^{-1} = -1 \cdot x^{-2} = \dfrac{-1}{x^2}$ and $Dx^2 = 2x$

The only zero is found by:

$$\frac{-k^3}{x^2}+2x=0 \Leftrightarrow \frac{2x^3-k^3}{x^2}=0 \Leftrightarrow x=\frac{k}{\sqrt[3]{2}}$$

The sign of the derivative function changes in this zero from – to +.

This means that for $x=\dfrac{k}{\sqrt[3]{2}}$ the function y, representing the total area of the cylinder, reaches a minimum.

And let's have a look now at the proportions of the dimensions, h and r of the cylinder having minimal area:

$$r=x=\frac{k}{\sqrt[3]{2}} \quad \text{und} \quad h=\frac{k^3}{r^2}=\frac{k^3}{\left(\frac{k}{\sqrt[3]{2}}\right)^2}=\sqrt[3]{4}\,k$$

also

$$\frac{h}{r}=\frac{\sqrt[3]{4}\,k}{\frac{k}{\sqrt[3]{2}}}=\sqrt[3]{4}\cdot\sqrt[3]{2}=\sqrt[3]{8}=2$$

As mentioned before, the height is the double of the radius, so it equals the diameter of the cylinder! 'Miss Tin-Can' is as wide as she is high, not elegant really, but the ideal measurements for a minimal total area.

5.2 Cone with a given volume and minimal lateral area.

In many countries, delicacies are sold in conical bags. It can be ice-cream or candies, but the Belgians even pretend that potato chips only taste good when eaten from a conical shaped paper bag. Cones are the most convenient shape to fit any size of hand and they remain solidly in place while the contents disappear into greedy mouths. If we decide on a certain volume for such a cone, what are its proportions, if we want to use a minimum of material?

This problem resembles the one of the tin can. Because the result is so surprisingly beautiful, we can't resist the temptation of showing its calculation.

The dimensions of a cone are: the height h, the apothem a and the radius r of the base.

They are related according to the Pythagorean theorem, namely: $r^2 + h^2 = a^2$.

The volume V of a cone is given by the formula $V=\dfrac{\pi h r^2}{3}$.

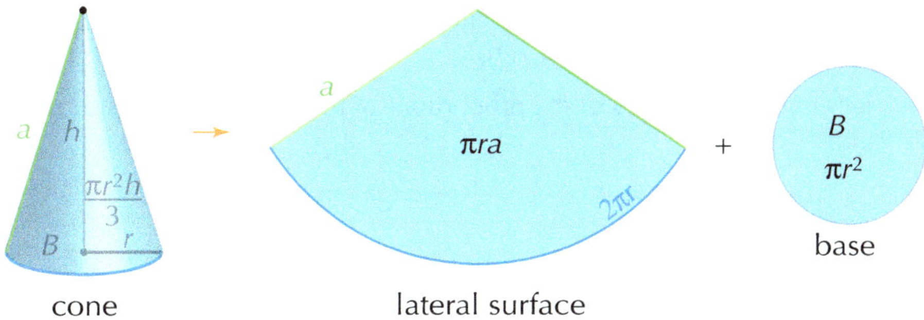

The area of the lateral surface is given by the formula $A = \pi ra$ (as indicated in the figure in the middle).

Just as we did for the cylinder, we can write the given volume V of the cone as $V = \frac{\pi k^3}{3}$ so as to keep the calculations simple and without loss of generality.

From $\quad V = \frac{\pi h r^2}{3} = \frac{\pi k^3}{3} \quad$ it follows that: $\quad h = \frac{k^3}{r^2}$

and from $r^2 + h^2 = a^2$ it follows that $\quad a^2 = r^2 + \frac{k^6}{r^4} \quad$ hence $\quad a = \sqrt{r^2 + \frac{k^6}{r^4}}$.

By means of these formulae we can express the lateral area A as a function of r:

$$\begin{aligned} A &= \pi ra \\ &= \pi r\sqrt{r^2 + \frac{k^6}{r^4}} \\ &= \pi\sqrt{r^4 + \frac{k^6}{r^2}} \end{aligned}$$

By substituting $\quad A = y \quad$ and $\quad r = x \quad$ we get: $\quad y = \pi\sqrt{x^4 + \frac{k^6}{x^2}}$.

We are trying to find out whether this function reaches a minimum for a certain value of x.

We therefore calculate the derivative function and its zeroes:

$$Dy = \pi\,\frac{1}{2}\,\frac{4x^3 - 2\frac{k^6}{x^3}}{\sqrt{x^4 + \frac{k^6}{x^2}}}$$

$$Dy = 0 \;\Leftrightarrow\; 4x^3 - 2\frac{k^6}{x^3} = 0 \;\Leftrightarrow\; x^6 = \frac{k^6}{2} \;\overset{x=r>0}{\Leftrightarrow}\; x = \frac{k}{\sqrt[6]{2}}$$

Because the sign of the derivative function changes from – to + in that zero, the function y has a minimum in that zero.

Now, let's have a close look at the proportions of the dimensions of the cone for this value of x:

$$x = r = \frac{k}{\sqrt[6]{2}}$$

$$h = \frac{k^3}{r^2} = \frac{k^3}{\left(\frac{k}{\sqrt[6]{2}}\right)^2} = \sqrt[6]{4}\,k$$

$$a = \sqrt{r^2 + h^2} = \sqrt{\frac{k^2}{\sqrt[6]{4}} + \sqrt[6]{16}\,k^2} = \sqrt{\frac{k^2 + 2k^2}{\sqrt[6]{4}}} = \frac{\sqrt{3}}{\sqrt[6]{2}}\,k$$

resulting in $\frac{h}{r} = \frac{\sqrt[6]{4}\,k}{\frac{k}{\sqrt[6]{2}}} = \sqrt[6]{8} = \sqrt{2}$ and $\frac{a}{r} = \frac{\frac{\sqrt{3}}{\sqrt[6]{2}}k}{\frac{k}{\sqrt[6]{2}}} = \sqrt{3}$.

Taking into account that $\frac{r}{r} = 1 = \sqrt{1}$, we conclude that the dimensions of 'Mister Candy-Cone' relate as $\sqrt{1}$, $\sqrt{2}$, $\sqrt{3}$ It couldn't be more elegant!

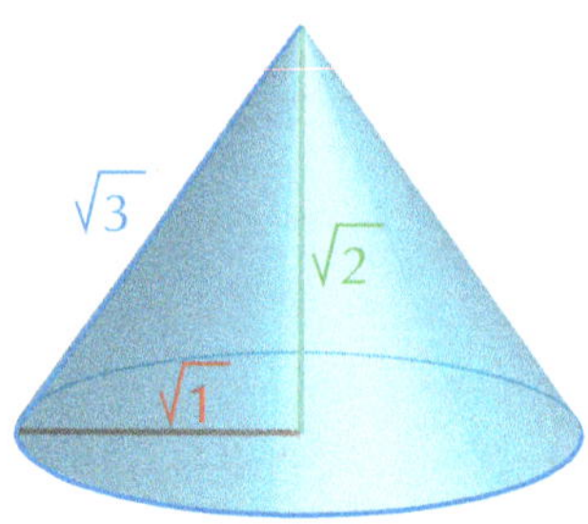

These are also the ideal proportions for an Indian tepee, or in general, for any cone having a minimal lateral area for a given volume.

Summary

Minimum and maximum problems appear in all applied sciences that make use of functions. Differential calculus is an appropriate means for calculating these extremes. Also in nature, many phenomena can be understood as a tendency towards a minimum (energy) or a maximum (rendering).

6 I can tell by your tail if you are reliable

In the field of information and communication technology as well as in the field of economic transactions and financial transfers, often specific code-numbers are used. Books and magazines have a standard number, bank cards and credit cards have an identification number, items in the supermarket have a bar code, and contracts, mobile phones, passports, airline tickets, etc. carry a reference number.

In correspondence, orders, transfers, purchases and such like, it is obviously important that the communication of these numbers must be error-free. Besides enormous loss of time, it may also have nasty consequences for all parties involved. For this reason, generally a control number is added to those numbers, enabling the detection of possible errors. Often the control number consists of a supplement of one or more digits at the end of the number. Those digits then make a number that is calculated using the digits of the original number. Maximum efficiency is aimed at in the simplest way.

If a satellite, launched to observe the planet Mars for example, transmits digital photographs, it is not only important to find out whether certain signals went astray by all kinds of interference, but also one should be able to localise the errors and to fix them. One can hardly track down the satellite and query its original communication.

The same goes for when we play a CD. When we occasionally hear some scratching noise we realize that the pits (the small indentations) and lands (space in between the indentations) on the spiral track that the laser reads from the surface of the CD are damaged, but knowing this brings us no satisfaction. What we want is a CD player that does not only detect those errors, but also repairs them, so that our musical pleasure is unaffected. It seems that already at the moment of purchase a CD can contain up to 500,000 errors, and by inattentive use it can easily run up to 1,000,000. However, refined mathematics ensure that we do not hear those errors.

The abstract algebra and number theory used were already developed a long time before and this just out of pure intellectual curiosity. It concerns a combination of the binary number system, designed by Leibniz already in the seventeenth century, modular arithmetics, considered by Gauss back in the nineteenth century, and linear algebra, which also had its cradle in the nineteenth century.

6.1 Control digits

6.1.1 Bank account numbers

An account number issued by Belgian banks is composed of twelve digits, arranged in three groups, like this: 123-1234567-12. The first group of three digits indicates the identification number of the bank, the second group of seven digits forms the individual number of the account and the third group of two digits a control number. That control number is the remainder of dividing the number represented by the first 10 digits by 97, while noting that the remainders 1 through 9 are written as 01, 02,..., 09, and the remainder 0 is substituted by 97. All possible control numbers are hence: 01, 02,..., 96, 97.

Example (of a fictitious bank account number)

000-1532845-51 In this, 51 is the remainder of the division of 0001532845 by 97.

Suppose that one or more digits is incorrect. Then the control number will no longer match unless the difference respective to the correct number is a multiple of 97, which is very improbable.

Example

Suppose that for a money transfer to the bank account number mentioned the digit 3 erroneously got replaced by 9, and thus the number 000-1592845-51 was transferred. Then the computer program checking this will report an error because the remainder of dividing 0001592845 by 97 is 08 and doesn't match the given control number 51. The transfer will not be acknowledged. The transfer order is returned for correction, since the system does not allow an unambiguous correction. But at least some trouble has been avoided.

Very often, for calculations with big numbers, the remainders of the division by a certain number n are being used. This we call calculating modulo n, modular arithmetic or also 'clock-arithmetic', because for $n = 12$ this is used in adding hours to the clock time. In this way, calculating with big numbers is reduced to calculating with their remainders, which are all less than n. In this one remainder r stands for an infinite number of numbers of the kind $kn + r$, in which kn is an integer multiple of n. For certain aspects of arithmetic this ambiguity is of no importance, as we will show further on.

Example

If it is 9 o'clock and if we add 11 hours, then the hands of an analogue clock cannot show 20 o'clock but will show 8 o'clock, and this is the remainder of the division of 20 by 12.

We say that 8 is congruent to 20 modulo 12. And we write this as:

$$8 \cong 20 \pmod{12}.$$

In a Belgian bank account number the first 10 digits make a number that is congruent to the control number modulo 97.
In our example: $0001532845 \cong 51 \pmod{97}$.

6.1.2 Credit card numbers

The numbers of credit cards, such as Visa, Mastercard, American Express, etc, have a control digit that is calculated according to the CODABAR system. The number is made up of four groups of four digits each, making sixteen digits all together. The last digit x_{16} is determined by means of a number a that is calculated from the first fifteen digits $x_1, x_2, x_3, \ldots, x_{14}, x_{15}$ and a number n that equals the number of digits which have an odd rank and are more than 4. The number a is then calculated as follows:

$$a = 2\,(x_1 + x_3 + x_5 + x_7 + x_9 + x_{11} + x_{13} + x_{15}) + n + (x_2 + x_4 + x_6 + x_8 + x_{10} + x_{12} + x_{14})$$

In other words, a is a sum in which the digits of odd rank get a double weight, and if they are more than 4, their weight is further incremented by one. In contrast, the digits of even rank keep their normal weight.

The last digit x_{16} is chosen so that the sum of a and x_{16} is a multiple of 10, in other words:

$$a + x_{16} \cong 0 \pmod{10}$$

Example (for a fictitious credit card number)

For the number 6402 7152 0323 450x_{16} we define x_{16} as follows:

$n = 3$ (since the digits 6, 7 and 5 on the odd places are more than 4)

$$a = 2\,(6 + 0 + 7 + 5 + 0 + 2 + 4 + 0) + 3 + (4 + 2 + 1 + 2 + 3 + 3 + 5) = 71$$

$$a + x_{16} = 71 + x_{16} = 71 + 9 = 80 \cong 0 \pmod{10}$$

The last digit has to be 9 in order to have a valid credit card number.

This CODABAR system can be applied to any number of digits.

Credit cards
Securing credit cards has a lot of aspects according to how they are used. When shown at the cashier's desk, a photograph is an additional means of confirming the identity of the bearer. However, for purchases over the Internet, control digits as well as security codes are necessary.

6.1.3 Book and magazine numbers

All books have an international identification number, namely the 'International Standard Book Number', the ISBN number in short. Before 2007 this number was composed of ten digits $x_1, x_2, \ldots, x_9, x_{10}$ of which the last one x_{10} is a control digit. This control digit is the remainder of the division by 11 of a number a that is the result of the weighted sum in which each of the first nine digits is multiplied by its rank (counted from left to right):

$$a = 1x_1 + 2x_2 + 3x_3 + 4x_4 + 5x_5 + 6x_6 + 7x_7 + 8x_8 + 9x_9$$

$$a \cong x_{10} \pmod{11}$$

The remainders of a division by 11 can only range from 0 to 10, and since 10 is the only remainder among them having two digits, this one is replaced by the character X. This way all remainders are represented by exactly one symbol.

Example (of an existing fine mathematics textbook!)

In the ISBN number 90-021-7115-x_{10} the last digit is calculated as follows:

$$\begin{aligned} a &= 1 \cdot 9 + 2 \cdot 0 + 3 \cdot 0 + 4 \cdot 2 + 5 \cdot 1 + 6 \cdot 7 + 7 \cdot 1 + 8 \cdot 1 + 9 \cdot 5 \\ &= 9 + 8 + 5 + 42 + 7 + 8 + 45 = 124 = 11 \cdot 11 + 3 \end{aligned}$$

x_{10} is the remainder of the division of 124 by 11, i.e. $124 \cong x_{10} \pmod{11}$

So $x_{10} = 3$.

So the full number is: 90-021-7115-3. Take care to order the right book!

Since the beginning of 2007 a prefix 978 or 979 is added to the ISBN numbers to implement them in the EAN code system explained in the next subsection.

Also magazines have their own identification number composed of eight digits. In this ISSN number (International Standard Serial Number) the last digit is a control digit too, but it is calculated in a slightly different way. The last digit in the ISSN number is chosen so that eight times the first digit plus seven times the second plus six times the third plus five times the fourth plus four times the fifth plus three times the sixth plus two times the seventh plus one time the eighth digit is divisible by 11.

Example

The ISSN number of some issue of 'National Geographic' is 1930-8116. The last digit x_8 is determined by:

$$8 \cdot 1 + 7 \cdot 9 + 6 \cdot 3 + 5 \cdot 0 + 4 \cdot 8 + 3 \cdot 1 + 2 \cdot 1 + x_8 \cong 0 \pmod{11}$$

$$126 + x_8 \cong 0 \pmod{11}$$

Since $126 + 6 = 132 = 12 \cdot 11$ is a multiple of 11, x_8 can only be 6.

In the case where x_8 equals ten, also in this system it will be substituted by X since only one symbol is allowed in the last place.

6.1.4 Bar codes

Most items in European shops are provided with a bar code and 13 digits underneath. That is the EAN code (European Article Numbering-code). In the USA and Canada the UPC code (Universal Product Code) can still be used, which can be implemented in the EAN system simply by adding a 0 in front.

Also these numbers have an internal syntax that allows control. This is done by making a weighted sum in which alternatively the weight 1, 3, 1, 3, 1,... etc. is assigned to the digits. The last digit x_{13} is chosen so that the weighted sum is divisible by 10.

Example

The number under the bar code of a certain brand of black olives is 3 560070 105939. In this, the thirteenth digit x_{13} is calculated as follows:

$$1\cdot 3+3\cdot 5+1\cdot 6+3\cdot 0+1\cdot 0+3\cdot 7+1\cdot 0+3\cdot 1+1\cdot 0+3\cdot 5+1\cdot 9+3\cdot 3+x_{13} \cong 0 \pmod{10}$$

$$81 + x_{13} \cong 0 \pmod{10}$$

$$x_{13} = 9$$

The last digit x_{13} has to be 9 in order to make the sum a multiple of ten.

So, again we have some possibilities for 'recreation'. Children can do lovely sums using all those credit cards, books, magazines and the items in the pantry and check their code number!

Bar code

The price label of the pyjamas of this well-know safari horse is revealed by its own bar code.

6.2 Error correction using Hamming codes

In the previous item we gave examples of a few systems of error detection. Errors were noticed but not corrected: therefore the systems were inadequate. In this section we explain a system that can also correct the detected errors to a certain extent. This method is used, for instance, in transmitting radio signals via satellites or for reading CD's.

When coding information signals, whether impulses or waves, they are always digitised. This means the information is changed into a sequence consisting of digits 1 (power) and digits 0 (no power). So, the coding of signals results in binary numbers of a certain length that in this context are called 'code words'. The way in which the vocabulary of code words is built is often related to geometrical models or to methods of linear algebra. These we will not examine, since they are part of the specialised field of code theory.

We restrict ourselves to the example of a vocabulary of binary numbers consisting of four digits. These are the following sixteen binary numbers, representing the numbers zero to fifteen:

```
0000  0001  0010  0011  0100  0101  0110  0111
1000  1001  1010  1011  1100  1101  1110  1111
```

The Mars Exploration Rover A (the Spirit) is one of the two robots that landed on Mars in January 2004, with the objective of doing geological and climatological research and investigating possible signs of life. This information is transmitted to the Earth as digitized radio signals.
On a CD, the laser reads symbols that are composed of eight binary digits that are physical pits (tiny indentations) and lands (space between the pits). By means of redundant digits, the system can also correct errors.

There are precisely sixteen numbers of this kind, since on any of the four positions we have the choice of the digit 0 or 1, This makes $2 \cdot 2 \cdot 2 \cdot 2 = 2^4 = 16$ possibilities.

By means of these code words we can send and receive information. But interference during transmission can make it erroneous. This, however, only in two ways: a 1 is changed to a 0 or vice versa. To avoid this we can extend the code words by a tail of, for instance, three extra digits so that they correspond to the original correct digits in a certain context. By means of this context, not only error detection, but also error correction becomes possible, however only within certain limits.

Example

We send code word 1011 as a seven digit number $1011x_5x_6x_7$.

The last three digits are calculated each respectively in relation to three of the four original digits of the code word:

x_5	so that	$x_1 + x_2 + x_3 + x_5$	$\cong 0 \pmod 2$	(1)
x_6	so that	$x_1 + x_3 + x_4 + x_6$	$\cong 0 \pmod 2$	(2)
x_7	so that	$x_2 + x_3 + x_4 + x_7$	$\doteq 0 \pmod 2$	(3)

In modulo 2 there are only two possible remainders, since all even numbers produce remainder 0 and all odd numbers produce remainder 1. So the previous sums determine the last three digits x_5, x_6 and x_7 so that the total number of terms equal to 1 in each sum is even.

This can be visualized by means of the Venn diagrams of three sets *A*, *B* and *C*, which determine precisely seven different enclosures:

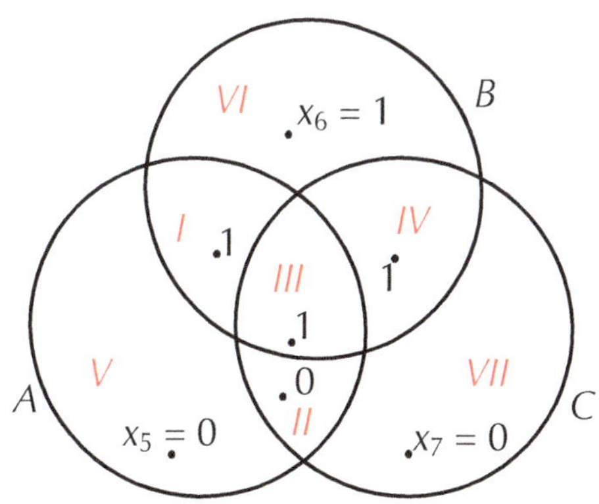

We put the first four digits of the code number respectively in the fields *I*, *II*, *III* and *IV*.

The fifth digit x_5 is put in field *V* and is chosen so that in the set *A* the total number of digits 1 is even, i.e. $x_5 = 0$.

The sixth digit x_6 is put in field *VI* and is chosen so that in the set *B* the total number of digits 1 is even, i.e. $x_6 = 1$.

The seventh digit x_7 is put in field *VII* and is chosen so that in the set *C* the total number of digits 1 is even, i.e. $x_7 = 0$.

So, instead of the proper code word `1011`, we send the extended code word `1011010`.

Suppose that due to interference one digit is received erroneously, for instance that the third digit got 0 instead of 1, in other words the number `1001010` was received. Then, by means of this tail-context, the error will be detected and repaired instantly. The false 'note' will be replaced by a correct 'note' again, because from the control equations (1), (2) and (3) it can be seen that:

$1 + 0 + 0 + 0 \cong 0$? (mod 2) wrong equation (in set *A*)

$1 + 0 + 1 + 1 \cong 0$? (mod 2) wrong equation (in set *B*)

$0 + 0 + 1 + 0 \cong 0$? (mod 2) wrong equation (in set *C*)

Obviously an error occurred in the digit that is situated in the field that is part of all sets *A*, *B* and *C*, i.e. the digit 0 in field III. So this digit should be 1 in order to make all equations correct again.

We can interpret this technique of correcting also in another way. By extending the binary code words from four digits to code words of seven digits via the equations (1), (2) and (3) we get sixteen words of length seven, namely:

```
0000000 0001011 0010111 0011100 0100101 0101110 0110010 0111001
1000110 1001101 1010001 1011010 1100011 1101000 1110100 1111111
```

These numbers are the only acceptable code words in the available 'vocabulary'. But there are all together $2^7 = 128$ binary numbers with seven digits. The number `1001010` is one of these 128 numbers, but it is not an acceptable code word. It will be detected immediately as a wrong combination of digits. But how can the original correct code word be found when receiving this wrong combination? This is done by the principle of the 'nearest code word'.

Among the 128 binary numbers having seven digits that could be received, there are seven that differ precisely in one position from the existing code word `1011001`. Together with this code word they make a class of eight numbers in which there is only one acceptable code word at a 'distance' of no more than one from the other words in this class. Whenever a binary number of this class is received that is not a code word, like the number `1001010`, the correction system will take the only acceptable code number in the class as being the correct message.

In the example we have given, the system is able to correct one error. If more digits were erroneous, then a reliable correction is no longer possible. We will have to refine the system by using even longer contextual tails. The technique of 'redundancy' of information is the basis of these error correcting code systems, which are called Hamming codes after the American mathematician Richard W. Hamming.

Corrections like these are similar to what happens in everyday language. Not all combinations of characters are meaningful words. The erroneous character combination 'patato' does not appear in the dictionary. But there is a word in the vocabulary that differs in only one character from it, namely the word 'potato'. We say that the distance of the character combination 'patato' to the word 'potato' equals 1, since both only differ by exactly one character. In other words, in the vocabulary the word 'potato' is the nearest neighbour of the character combination 'patato'. If more characters became erroneous, like for instance 'topapa', then it would become difficult to understand what is meant, unless more context is given. Some music lovers believe that there is no need to be very careful in handling a CD, because the integrated correction system is powerful enough to correct all damage. This is, of course, not true. Handle your favourite music as carefully as possible!

Richard W. Hamming (1915–1998)

A golden medal with the image of Richard W. Hamming. Since 1986 it has been awarded every year, together with the associated prize, for "exceptional contributions to information sciences, information systems and information technology". Hamming himself played an important role in the development of computer sciences, for instance with his system of error detection in codes.

Summary

The previous examples show how fundamental mathematics is used for encoding information and protecting it from errors. This is, however, just the tip of the enormous iceberg formed by contemporary cryptology and security systems. Only a few people are aware of how much mathematics is at the basis of familiar daily operations like electronic banking, the use of passwords for entering protected Internet sites, personal codes, digital signatures, etc. A lot of these protection techniques are based on the double aspect of the calculation capacity of the modern computer. On the one hand, mathematics supplies practical procedures for testing whether a large number (for instance of 200 digits) is a prime number. On the other hand, it isn't possible, not even with the fastest computers, to factorize a product of two such prime numbers within the span of a human life. This is due to the fact that the sequence of the prime numbers cannot be produced by a formula but only by means of rather slow screening methods. Each time you enter your bank code or use the Internet, the computer will make calculations with such big prime numbers to verify whether your identity actually corresponds to the additional information you supply by keying in code numbers and passwords. In the war between hackers and prime numbers, the latter win an easy victory, thank goodness!

The Enigma encoding machine

This machine was used by the Germans during the second world war for encoding secret messages. It was based on a system that applied to each character of a word a different way of substituting it by another. This was performed by scramblers (rotors) that turned one position each time a character was keyed in, with another starting position of the scramblers chosen every day. The system was cracked, however, by the allies, with the help of mathematicians like Alan Turing.

7 The fascination of fractals and deterministic chaos

Until modern computers showed up with their phenomenal arithmetic capacity, certain problems related to the complex behaviour of dynamic systems such as the atmosphere, Brownian motion, fluid dynamics, convection, turbulence and non-linear variations in general, were beyond the reach of mathematics. The aim of simplicity and the linear approach of processes that essentially are not linear, was for long the only manageable method to obtain a mathematical hold on certain physical processes. Many phenomena, whose complexity was associated with chance and randomness, necessarily remained not understood. The possibilities of the computer, however, opened up new perspectives for mathematical research that lead to the development of recent theories that carry names like 'fractals' and 'chaos theory'.

The word 'chaos' in daily conversation is generally used in the mythological meaning of a condition of total lack of order and arbitrary confusion. Chaos theory, however, shows that elementary basic rules produce weird complexity and that complex structures can have elementary causes. That is why this theory is sometimes illustrated by the metaphor of the butterfly whose feeble wing flapping later on triggers a hurricane.

'Chaos theory' studies the evolutions in time of dynamic processes, generally described by means of differential equations (these are equations in which also variables appear representing measures for the fluctuations of the process, in other words: derivatives). Because of the extreme sensitivity to the smallest perturbation in the initial conditions, caused by the approximation of measurements, many of those systems show complex behaviour that puts restrictions on the predictability of the results. 'Complex behaviour' is not to be taken as a synonym for 'arbitrary or random behaviour', but as a kind of complexity that is visualized in 'strange attractors'. These are patterns that are typical of the dynamic systems considered.

Well-known problems of chaos theory are:

- the weather forecast (If one measures three parameters of the atmosphere today, namely temperature, air pressure and windspeed, to what extent can the weather of the following days be forecasted, for instance by a model based on the differential equations of Lorenz?)
- the stability of the solar system (If one knows at the present moment the position of the sun, the planets and their satellites, is one able then to predict whether the orbits of the celestial bodies remain stable, for instance by a model based on the laws of Newton and Kepler?)

In the first place we relate the theory of fractals to the names of the French mathematician Gaston Julia and the Polish-French-American mathematician Benoît Mandelbrot, who both found fascinating prototypes of fractals, such as the 'Seahorse Valley' and the 'Gingerbread man'.

Julia fractal

with seahorse figures. This fractal is named after the French mathematician Gaston Julia (1893–1978), pioneer of research into the complexity of iterative processes.

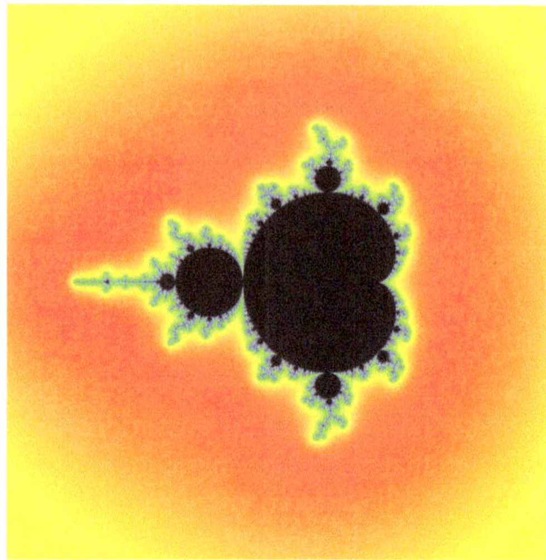

Mandelbrot fractal

also called the gingerbread man. Benoît Mandelbrot (1924–2010) came up with the name 'fractal' for these weird self-similar figures and is to be considered the father of fractal theory.

Fractals are figures that visualize iterative processes: infinitely repeated constructions or algorithms, in which the output obtained again functions as the new input. This results in self-similarity: close-ups of certain parts of the figure show the same figure as the original, to whatever scale the figure is magnified.

Popular examples of fractals are:

- a tree (each branch and every twig resembles a downsized tree)
- a cauliflower (each floret or curd resembles a downsized cauliflower).

Although the study of fractals and deterministic chaos have a different origin, these two domains are converging more and more. On the one hand because iteration plays a fundamental role in both, on the other hand because fractals can be considered as prototypes of 'strange attractors'. Many shapes and populations in nature that deal with growth can be imitated by fractals. Complex patterns like that of a fern or a palm tree seem to be the result of iteration of a simple basic figure. If the complex can be originated by multiple repetition of the simple, then the most complicated shapes and phenomena reveal their secrets. Expectations are very high that in the near future these new theories, which are barely hatched, will provide a better insight into the phenomena of hydrostatics, electrodynamics, biology, economics, climatology, etc. In the scope of this section we can only give some simple examples of how iterative constructions and algorithms lie at the basis of both fractals and chaos.

7.1 Iteration as a recipe for fractals

7.1.1 The Cantor set

We start from a line segment with a given length, for instance length 1.

We divide the segment into three parts with equal length and delete the middle part. (*)

This way we get two segments of which the length has shrunk by factor three.

We repeat procedure (*) on the remaining segments.

In the second step we get four segments of which the length is $\frac{1}{9}$ of the length of the original segment.

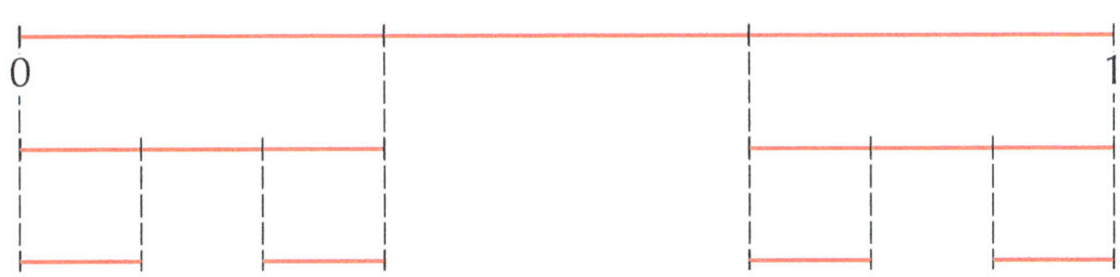

We keep on repeating this procedure on the remaining segments.

Thus a set of segments is created of which the number reaches the infinite while their length tends to zero. This set is called the 'Cantor set' and it is a fractal.

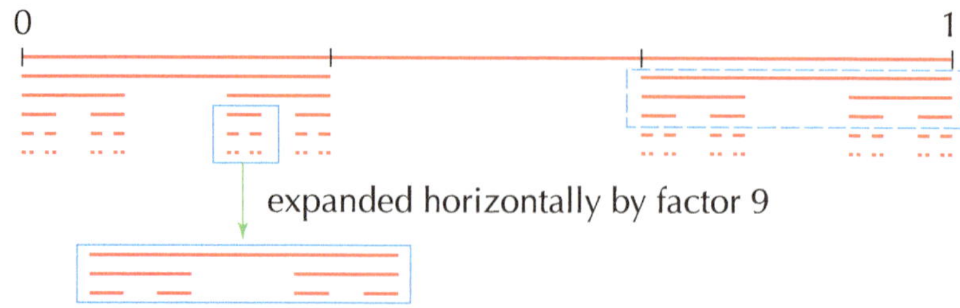

Each close-up of a part of this set shows the same figure: a set of segments and gaps in-between, having fixed proportions (see the boxes).

In the construction of the Cantor set two parameters play an essential role, namely:

- the reduction factor by which the segments are scaled (three)
- the number of remaining segments for the next step (two).

With these numbers we can calculate a number that characterizes the fractal and that is called its 'fractal dimension'. This dimension enables the comparison of a fractal with other sets of its kind.

The definition of the concept 'fractal dimension' is inspired by a characteristic of the usual geometrical dimensions, such as a line being one-dimensional, a plane being two-dimensional, and the space we live in being three-dimensional. More precisely: if one takes a cell in such a geometrical space, for instance a segment on the line, a square in the plane or a cube in the space, then the cell can be filled up by smaller cells, of which the size is a reduction by factor n of the original cell. The number m of these smaller cells can then be written as $m = n^{d,}$ and in this, d is the dimension of the space in which the cell was considered.

Examples:

line segment (dimension d = 1)	square (dimension d = 2)	cube (dimension d = 3)

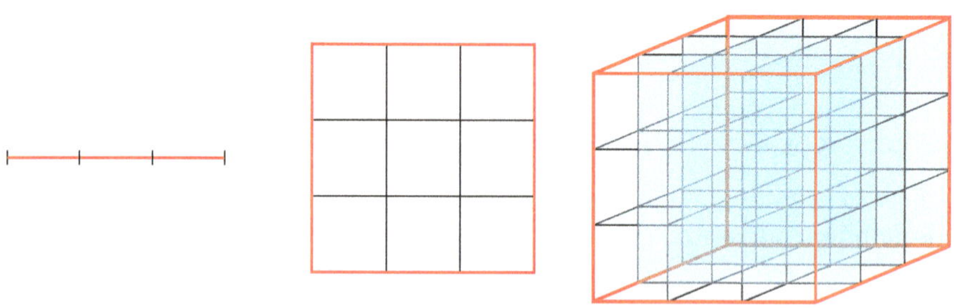

factor of reduction: 3	factor of reduction: 3	factor of reduction: 3
number of segments: 3	number of squares: 9	number of cubes: 27
formula: $3 = 3^1$	formula: $9 = 3^2$	formula: $27 = 3^3$

The general formula hence is: $m = n^d$ where n = factor of reduction, d = dimension, m = number of smaller cells.

By means of logarithms we can express d as a function of n and m as follows:

$$m = n^d \quad \Leftrightarrow \quad \log m = \log(n^d) \quad \Leftrightarrow \quad \log m = d.\log n \quad \Leftrightarrow \quad \frac{\log m}{\log n} = d$$

If we apply this formula to the Cantor set, then we get:

$$d = \frac{\log 2}{\log 3} = 0.6309...$$

Thus we get a non-integer number for the dimension of the Cantor set, since in the construction of it we did not retain all the smaller cells. It is exactly this characteristic of having a non-integer dimension that gave rise to the name fractal (fraction, broken number). The Cantor set thus has a dimension that is higher than the one of a point (0), but less than the dimension of a line (1).

7.1.2 The Koch snowflake

We start with an equilateral triangle, for instance with all sides having length 1.

We construct the Koch snowflake by repeating the following procedure (&):

- divide every side into three parts, equal in length (factor of reduction = 3)
- take the middle segment as the base of a smaller equilateral triangle, placed outside the previous cell
- omit the base of this new smaller triangle. One bigger cell (side) is thus replaced by four smaller cells (sides): m=4

The first construction step is shown below:

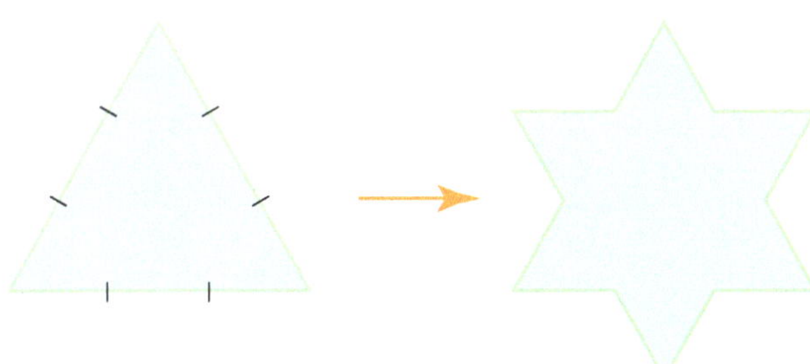

We repeat procedure (&) on the new set and keep on doing this on every new output. In this way we get a fractal of which the edge is serrated in a very subtle way and gets an infinite length, while the enclosed area stays finite.

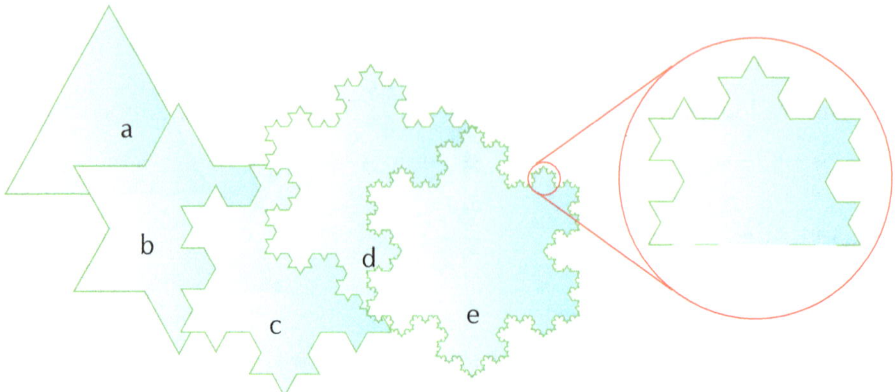

The close-up in the figure shows the self-similarity of this fractal.

Due to its resemblance to a snowflake, it is called the Koch snowflake, after the Swedish mathematician Helge Von Koch, who designed it in 1904. The dimension *d* of the Koch snowflake is found from the factor of reduction 3 of a side and the number 4 of the smaller cells replacing it in the next step. This leads to:

$$d = \frac{\log 4}{\log 3} = 1.2618...$$

This is again not an integer number but a number in between the dimension of a line (1) and the dimension of a plane (2).

The Koch snowflake as drawn above is a planar figure and not three-dimensional like a real snowflake. We can make some three-dimensional imitation by starting from a tetrahedron of which we connect the midpoints of all edges by straight segments, and so dividing all faces into four equal triangles that relate by reduction factor 2 to the original faces. On the triangle in the middle we construct a new tetrahedron with that small triangle as its base. So the number of smaller cells (triangles) is 6 all together.

Repeating this procedure results in a fractal of which the area tends to infinite but its volume stays finite. Its dimension is: $d = \frac{\log 6}{\log 3} = 2.5849...$

This is a number in between the dimension of a plane (2) and the dimension of the space (3).

It doesn't ask much of our imagination to apply the procedures, like the ones we used in the construction of the Cantor set (omitting of cells) or the Koch snowflake (adding of cells), to other basic figures. In the next figures you can easily recognise the procedures.

The Sierpinski triangle:

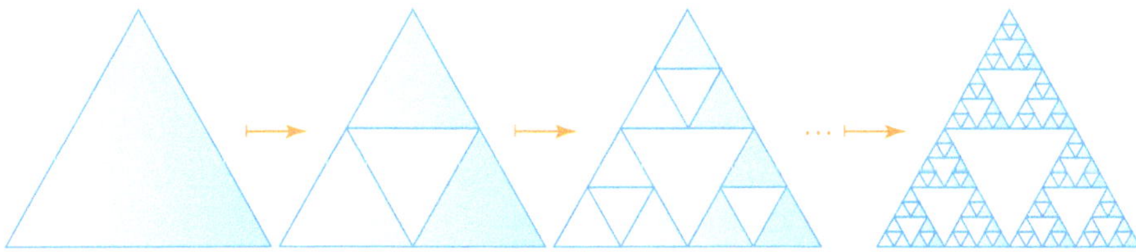

The Pythagorean tree:

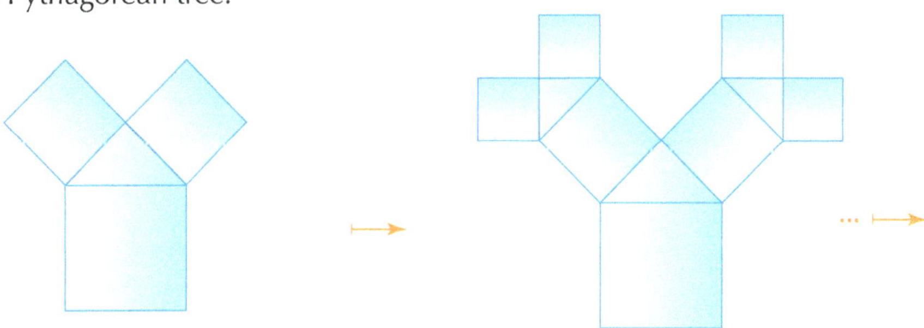

Have fun in creating your own fractal!

7.1.3 The Julia sets

Instead of geometrical constructions, also iterative algorithms can be used to construct fractals. If such an algorithm is applied to numbers, then in order to get a visualization of the process, it must be linked to some method for depicting it.

In the 1920's, while analysing the behaviour of functions of complex numbers, Gaston Julia found himself on the track of intricate patterns. His fellow Frenchman Benoît Mandelbrot augmented and completed the publications of Julia's work around 1975.

A complex number is a number of the form a+b*i*, where a and b are real numbers and where *i* is the imaginary unit complying to $i^2 = -1$. Hence i is a square root of –1. The calculation rules for complex numbers are analogous to those of real numbers.

Examples:

$$(5 + 3i) + (-2 + 4i) = 5 + 3i - 2 + 4i = 5 - 2 + 3i + 4i = 3 + (3 + 4)i = 3 + 7i$$

$$(5 + 3i) \cdot (-2 + 4i) = 5 \cdot (-2) + 3i \cdot (-2) + 5 \cdot (4i) + (3i) \cdot (4i)$$

$$= -10 - 6i + 20i + 12i^2 = -10 - 12 + (20 - 6)i = -22 + 14i$$

$$(5 + 3i)^2 = (5 + 3i) \cdot (5 + 3i) = 25 + 15i + 15i - 9 = 16 + 30i$$

A complex number $a + bi$ is unambiguously defined by a pair of real numbers (a, b), which can be considered as the coordinate of a point in the plane provided with a Cartesian reference system. In this way, every complex number can be mapped onto a unique point in the plane.

Julia considered the iterations of a quadratic complex function f of the type $f(z) = z^2 + c$, in which z and c are complex numbers. The variable z takes successive iterative values starting from an arbitrarily chosen z_0. The number c is chosen as a fixed constant that unambiguously defines the function f. The iteration of the function is expressed in the following diagram:

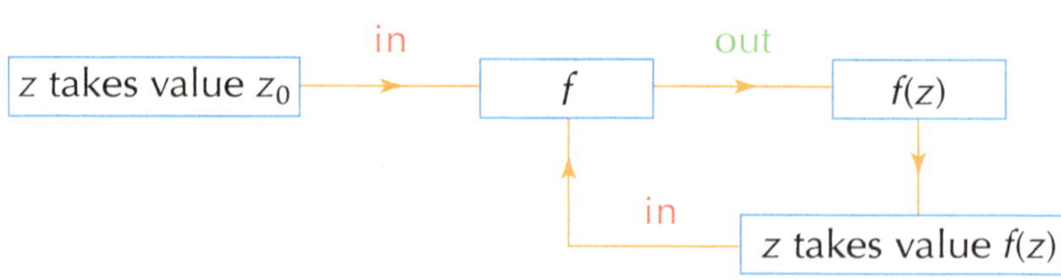

Each time the output is taken as the new input, so that $f(z_1) = z_2$ etc.

Example:

Let $f(z) = z^2 + (1 + i)$ this means that here: $c = 1 + i$.

The iterative values starting from $z_0 = 1 + 0i = 1$ are consequently:

$$z_1 = f(z_0) = 1^2 + (1+i) = 2 + i$$
$$z_2 = f(z_1) = (2+i)^2 + (1+i) = 4+5i$$
$$z_3 = f(z_2) = (4+5i)^2 + (1+i) = -8 + 41i$$

etc.

Now the big question is: 'Where do these iterations lead to?'

A different behaviour is possible, depending in the first place on the function itself (i.e. on the value of c), but also on the chosen starting value z_0.

The distance between the origin of the reference system and the image point of the complex number $a + bi$, i.e. the point with coordinate (a, b), is given by $\sqrt{a^2 + b^2}$ (according to the Pythagorean theorem).

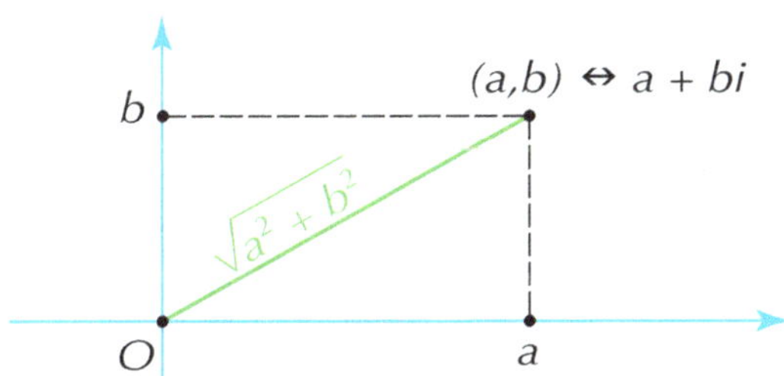

It is easy to check that the image points of the successive iterative values in the given example move further and further away from the origin. Therefore we say that the (discrete) path of this sequence of iterative values diverges to the infinite. But other examples can be given in which this doesn't occur.

Let's take as starting points, for instance, for the same function f the numbers i and $\sqrt{2}\,i$ respectively.

Consequently we then get:

$z_1 = f(z_0) = f(i) = i^2 + (1 + i) = i$	$z_1 = f(z_0) = f(\sqrt{2}i) = (\sqrt{2}i)^2 + (1 + i) = -1 + i$
$z_2 = f(z_1) = f(i) = i^2 + (1 + i) = i$	$z_2 = f(z_1) = (-1 + i)^2 + (1 + i) = 1 - i$
$z_3 = f(z_2) = f(i) = i^2 + (1 + i) = i$	$z_3 = f(z_2) = (1 - i)^2 + (1 + i) = 1 - i$
etc.	etc.

The path starting in $z_0 = i$ appears to be constantly i and the path starting in $z_0 = \sqrt{2}\,i$ remains $1 - i$ from the second step onwards.

For any function f like this we can make a distinction between the starting points for which the path diverges to the infinite and those for which this is not the case and give the respective starting points a different colour in the plane. For example, the starting points for which the paths diverge in white (invisible) and the others in black (visible) or vice versa. The edge of the final figure that arises is called the Julia set of the function *f*. Because for every point in the plane, i.e. for every complex number *z*, all iterations have to be calculated and their behaviour analysed, it is clear that without the help of a computer this would be a hopeless task. But with a program of hardly twenty lines a contemporary computer can draw the corresponding Julia set for any given function *f* in a minimum of time.

Examples:

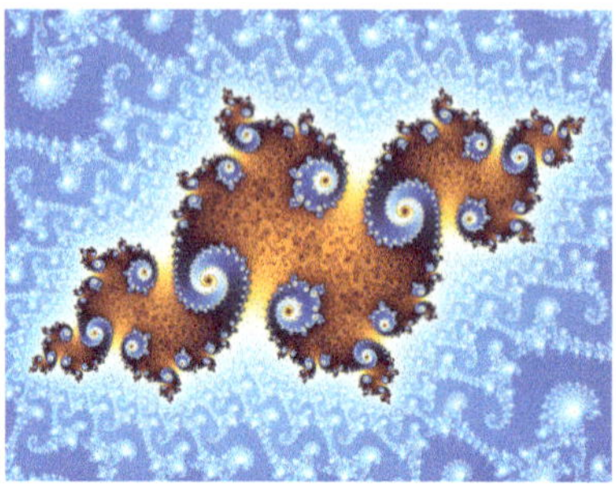

Both in the case of non-connected clouds of points, as in the first figure, and in the case of connected clouds, as in the second figure, the self-similarity can be noticed at any scale. This confirms the fractal nature of these figures.

7.1.4 Mandelbrot's 'gingerbread man'

The most famous fractal, the true emblem of fractal theory, is without doubt linked to the name of Mandelbrot, who discovered it by pursuing the ideas of Julia about complex functions of the kind $f(z) = z^2 + c$.

In contrast to Julia, Mandelbrot took the same starting point for all iterations, namely the origin of the reference system, i.e. $z_0 = 0$, but he considered *c* to be variable. In the case where the path diverges for the given value of *c*, the image point of *c* is coloured white (invisible), in the other case the image point of *c* is coloured black. One can also allow a greater variety of colours depending on the speed of convergence. Then the result is a full-coloured figure. The edge of the final figure that arises by colouring all *c*-values in this way is the Mandelbrot set, and because of its funny shape it is also called the 'gingerbread man'.

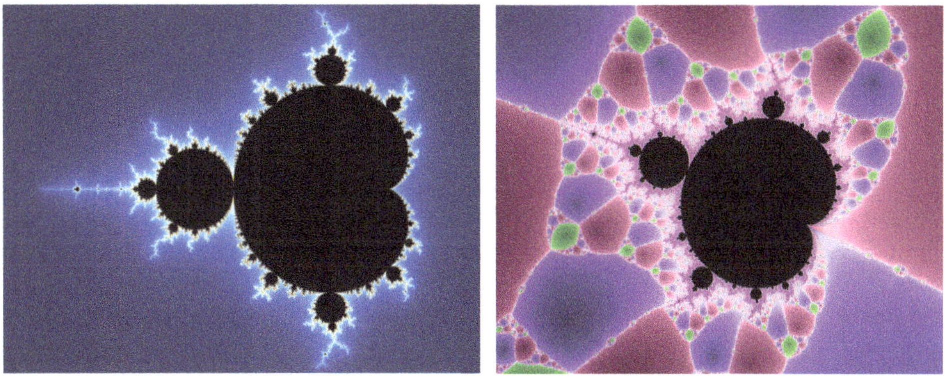

The self-similarity of the figure results in an infinite number of scaled down 'gingerbread men' again. The complicated structure of this fractal can be admired in the different close-ups of fragments of its edge: whirlpools, zippers, cactuses, seahorses etc.

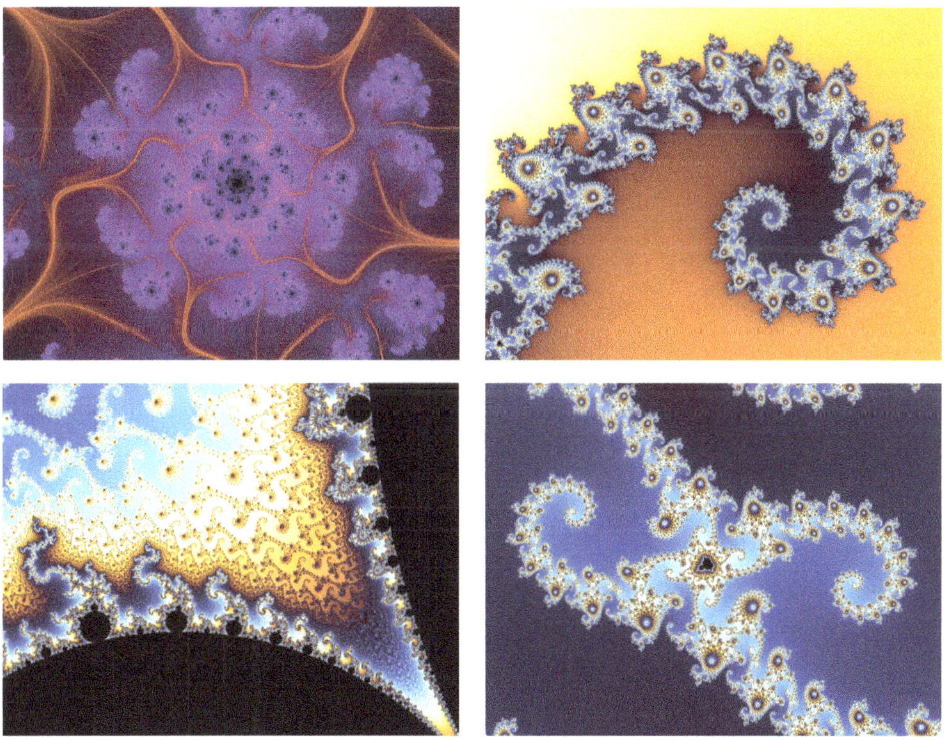

Because the 'gingerbread man' interprets all possible c-values of the function $f(z) = z^2 + c$, it's no surprise that the Mandelbrot-set contains at some place all models of the Julia sets in miniature: fascinating works of art generated by the most simple algorithms with a mysterious background. What follows will show that the ambitions of the theory of fractals aim higher than just creating fascinating beauty.

7.2 Chaotic behaviour of deterministic systems

7.2.1 Dynamic systems

A dynamic system can be described as a (continuous or discrete) succession of states that change with the value of a specific parameter, which could for instance be time. As the origin of chaos theory we can consider two historically important publications: one about the stability of orbits of planets in a planetary system and one about weather forecasting. At the beginning of the twentieth century Henri Poincaré reported in his *Science et Hypothèse* on his findings on the impossibility of predicting, by Newton's laws of gravity, the long-term behaviour of a system of no more than three bodies. In 1963 the meteorologist Edward Lorenz published in the *Journal of the Atmospheric Sciences* a remarkable article on the evolution of a weather simulation, based on a system of differential equations with no more than three variables. In this he stated that a minor change in the initial value of the variables rather quickly leads to major changes in the calculated results. This implies that there are inevitable restrictions to the time span over which reliable weather forecasts can be made. That is why farmers keep on praying for rain and tourists for sunny weather, while nobody will bother the Gods for an eclipse.

Since then mathematical methods have been developed that allow us, by means of the computer, to grasp the deeper core of these problems. This methodology of chaos theory consists in:

- the choice of the variables appropriate to the system
- the definition of the equations that specify the system
- the introduction of the initial values for the variables
- the iteration of the solutions over a variation of the parameter (e.g. time) within certain intervals
- the representation of those iterative values as points in the so-called 'phase space'
- the characterisation of the obtained paths (orbits): stable, periodical, quasi-periodical, chaotic)
- the tracing of 'strange attractors' that appear as distinctive patterns
- the analysis of the 'strange attractors' by means of sections (Poincaré sections) that show instantaneous states of the system

We illustrate this methodology by means of the example described in the above-mentioned article by Lorenz. For his research, Lorenz restricted himself to a system with only three variables, namely temperature, atmospheric pressure and wind speed. This is obviously a oversimplified model for the complicated phenomenon of atmospheric evolution, but it is indeed appropriate for the purpose of the research: namely investigating the long-term behaviour of this kind of system.

Inspired by the equations that B. Saltzman had already set up in 1962 for a simple kind of convection, he built a set of simultaneous differential equations that has become a classic in chaos theory ever since:

$$\begin{cases} \dfrac{dx}{dt} = -10x + 10y \\ \dfrac{dy}{dt} = 28x - y - xz \\ \dfrac{dz}{dt} = -\dfrac{8}{3}z + xy \end{cases}$$

In this, $\frac{dx}{dt}$, $\frac{dy}{dt}$ and $\frac{dz}{dt}$ are the derivatives of the three variables *x*, *y* and *z* as a function of the time *t*. They offer a measure for the rate of change of the variables at the moment *t*. The presence of the terms of the second degree *xz* and *xy* shows that the system is not linear and this prevents an exact general solution. By means of the equations we can, however, numerically calculate approximate solutions. We start with a set of initial values x_0, y_0 and z_0, that define a point with coordinate (x_0, y_0, z_0) in the three-dimensional phase space. The iterations of the successive states of the system are calculated for a small increment of time, e.g. $\Delta t = 1$ minute. So, after a minute, we reach the point with coordinate (x_1, y_1, z_1), i.e. the point representing the first iterative value.

You may object that for forecasting the weather of the next minute one doesn't need the whole caboodle of differential calculus. Consider that the calculation has to be repeated for each next point (every minute) and that an initially small error in the value of the variables gradually expands. Also from experience one knows that the weather after 24 x 60 minutes, i.e. the next day, is not necessarily the same as today. In this way one can follow the iterations from the chosen starting point over the period of, let's say, a whole year. This means then 365 x 24 x 60 = 525,600 iterations (Lorenz restricted himself to 3,000 iterations), of which the image points in the phase space make a (discrete) path. This procedure is checked for starting points very close to the original one, for instance by increasing the precision in terms of the number of decimals in the measurements of the variables.

Suppose that a certain path returns to its starting point after one year, then the calculations for the next year will of course run through the same iterations, and over again for the following years. We call such a path a stable orbit.

Suppose this only occurs after two years, then the cycle of iterative values will repeat every two years. Then we say that the orbit is periodical with period two.

It can also be that the path comes close to the starting point, and that this happens over again. Such a path is called a quasi-periodical orbit.

The orbits of the iterative points in models of this last kind make a complicated phase portrait, staying however within a finite part of space. Although starting from points that are very close to one another (the coordinates could, for instance, differ only in the sixth decimal), the orbits can after a while move surprisingly far away from each other. The layered structure of these almost-but-never merging orbits looks like a kind of fine-meshed puff pastry that at every close-up will show the same complicated structure, typical of its fractal nature. It was exactly to this kind of figures that the Belgian David Ruelle and the Dutchman Floris Takens later on in their study of turbulence gave the name 'strange attractor'.

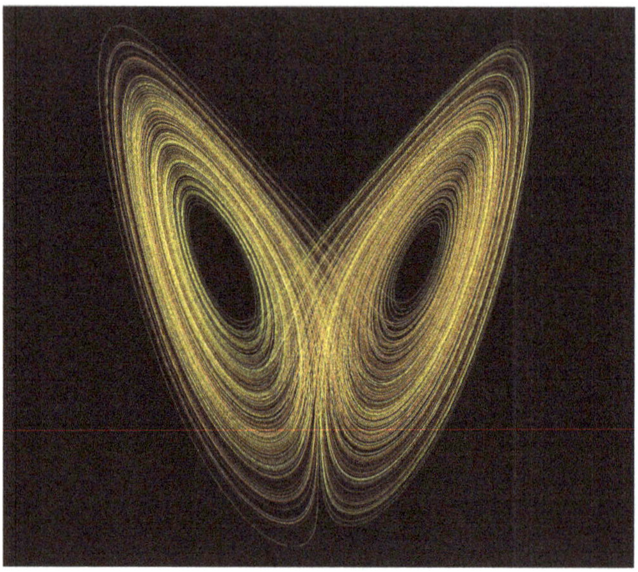

If we select in the phase portrait those points for which one of the variables has a constant value (for example the points that correspond to a pressure of 1 bar) we get a planar section of the attractor, which is called a Poincaré section. The points in the section are determined by the other two variables corresponding to the chosen constant value for the first variable. Lorenz selected the points for which the variable z was maximal. It was Lorenz himself who introduced the metaphor of the butterfly for the phenomenon of deterministic chaos. The figure shows that some order is to be found in the chaos, but because of its complicated structure it keeps its information well-hidden from us. We should not cherish the hope that this model of Lorentz could make a reliable forecast of the weather for more than a few days.

7.2.2 The growth of populations

The population growth of a certain species of animals (e.g. rabbits) in a specific ecosystem is not of the linear kind.

Even if a constant factor k were assumed for the population growth, some restraint must also be taken into account, caused by gradual setting in of food shortage and the presence of predators. In the case of rabbits these could for instance be foxes. If x is the magnitude of a population at a certain moment and if k is the factor of growth in one fertile period, then the magnitude after one such period is kx, to be tempered, however, by some correction term. This correction term is relative to the original population and its growth, namely $(kx)x = kx^2$. In this way we get a function of the form $f(x) = kx - kx^2$ or also $f(x) = kx(1 - x)$ that is called the logistic function. For positive values of k, this function has only positive values in between its zeroes, which are 0 and 1. The graphic representation relative to a Cartesian reference system in the plane of such a function is a parabola of which the vertex has coordinate $\left(\frac{1}{2}, \frac{k}{4}\right)$.

The derivative of this function is $f'(x) = k - 2kx$ or also $f'(x) = k(1 - 2x)$. This expresses for every value of x the velocity of increment of the function f and also determines the slope of the tangent line to the graph of f at the point with coordinate $(x, f(x))$.

Henri Poincaré (1854–1912)

French mathematician, engineer and science philosopher. He is considered to be one of the last universal spirits and the godfather of chaos theory. His findings concerning the three-body problem initiated the notion of deterministic chaos. Also in the field of special relativity he had already achieved results before Einstein's publications.

Example

For $k = 2.5$ we get $f(x) = 2.5x(1 - x)$ and $f'(x) = 2.5(1 - 2x)$.

For $x = 0.2$ we get $f(x) = 0.4$ and $f'(x) = 1.5$.

x	0	0.2	0.5	0.6	1
$f'(x)$	2.5	1.5	0	–0.5	–2.5
$f(x)$	0	0.4	0.625	0.6	0

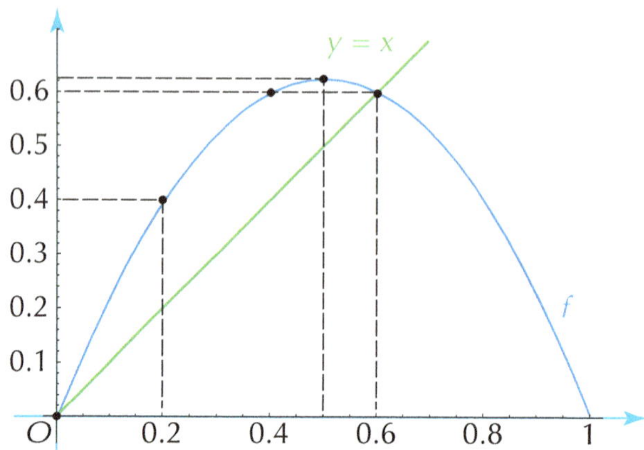

In the iteration of the function f, starting from a value x_0 between 0 and 1, a lot of peculiarities show up that are related to both deterministic chaos and fractals. Although some of those aspects were already studied in 1844 by the Belgian François Verhulst (1804-1849), only in 1980 were the spotlights turned on the behaviour of this function by the discoveries of the American physicist Mitchell Feigenbaum. The diagram of the typical bifurcation (the 'doubling waterfall'), that appears in the iterative behaviour of the function f has in his honour also been called 'fig tree' since then. Most remarkable is that the iteration of the function f behaves totally differently for different values of the growth factor k between 0 and 4, and this at very precise transitions.

Example 1

Let $k = 2.5$, then $f(x) = 2.5x(1 - x)$ and $f'(x) = 2.5(1 - 2x)$.

Regardless of the chosen starting point x_0 between 0 and 1, all iteration paths lead to the point of intersection of the graph of f and the bisector of the axes, i.e. the line having the equation $y = x$. The coordinate of this point is (0.6 , 0.6). So the sequence of the iterative values for x converges to the value 0.6 and becomes stable there since $f(0.6) = 0.6$. The iterative paths for $x_0 = 0.2$ and $x_0 = 0.7$ respectively can be constructed in the figure as follows:

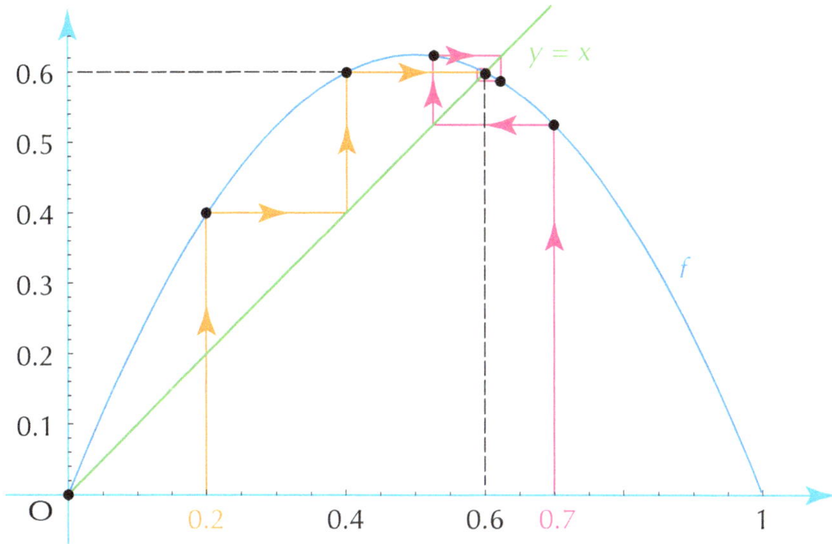

Since $f(x_0) = f(0.2) = 0.4 = x_1$, we can find the new input x_1 starting from the value of the function $f(0.2)$ by first drawing a horizontal line through the point having coordinate (0.2 , 0.4) and next a vertical line through the point of intersection of this horizontal line and the bisector. The intersection of this vertical line and the x-axis gives the value $x_1 = 0.4$ (because the points on the bisector have equal coordinate numbers). For the following iterative values the construction is analogous. This construction very well visualizes the convergence onto the intersection point of f and the bisector. The actual sequence of iterative values of x can be seen on the x-axis as the intersections with the successive vertical lines. These values clearly converge on the stable value 0.6 . Hence this value appears to be an attractor for all paths starting in whatever x_0 between 0 and 1. Such a 'fixpoint' (fixed point) makes the most simple model for an attractor.

For each value of k less than 3, more or less the same happens, of course with adapted fixpoints depending on k. However, if k is given a value between 3 and 3.449... suddenly a different behaviour appears.

Example

Let $k = 3.2$ then $f(x) = 3.2\,x\,(1 - x)$ and $f'(x) = 3.2\,(1 - 2x)$.

We draw the iterations for the starting values $x_0 = 0.2$ and $x_0 = 0.7$:

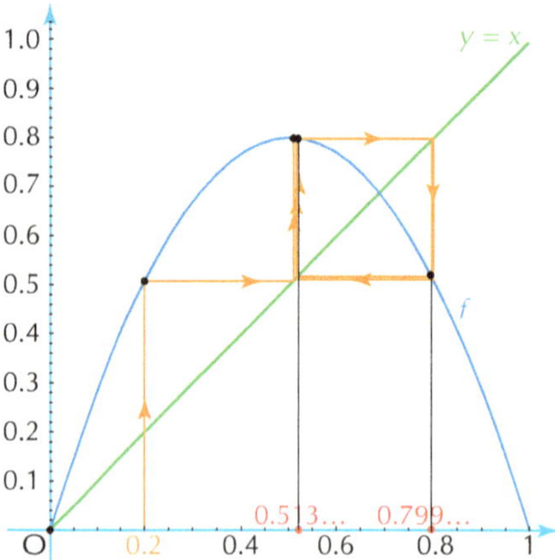

$x_0 = 0.2$

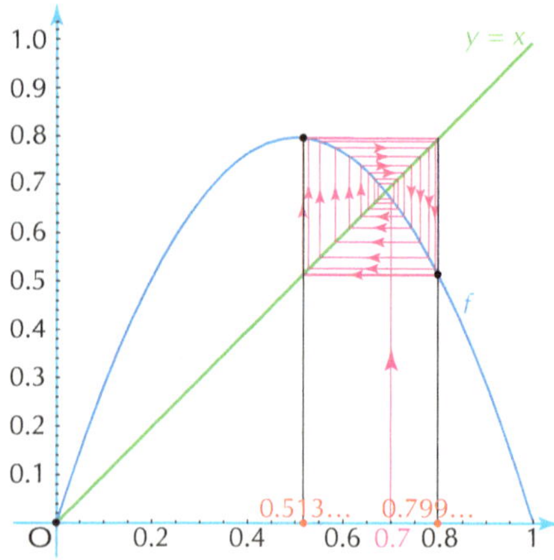

$x_0 = 0.7$

Now we notice that the sequence of iterative values finally alternates between two values, namely 0.513... and 0.799... , and this for any starting value x_0 between 0 and 1. This means that now the attractor changed into a periodical cycle with period 2. Indeed, we notice that $f(0.513...) = 0.799...$ and $f(0.799...) = 0.513...$. So, when the value of k exceeds 3, a bifurcation (a splitting, a doubling) appears in the attractor. Analogous behaviour occurs for increasing values of k until 3.449... .

If we give k a value between 3.449... and 3.545... , then the attractor doubles once more to a cycle period 4.

Example

Let $k = 3.5$ then $f(x) = 3.5x\,(1 - x)$ and $f'(x) = 3.5\,(1 - 2x)$.

We draw the iterations for the starting values $x_0 = 0.2$ und $x_0 = 0.7$:

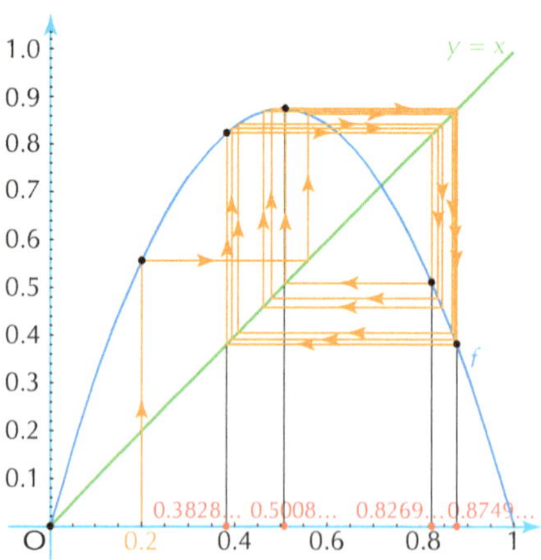

$x_0 = 0.2$

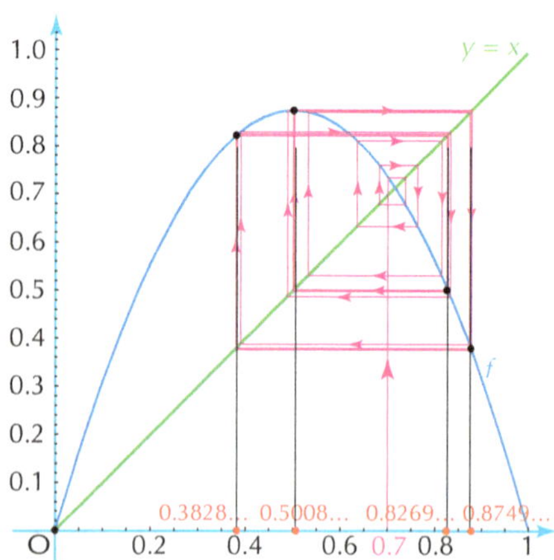

$x_0 = 0.7$

A cycle with period 4 has appeared. The period doubled a second time. The periodic values are 0.3828... , 0.8269... , 0.5008... and 0.8749... .

This phenomenon of period doubling continues for increasing values of k in successive intervals of which the length shrinks with factor 4.6692016... . This constant was discovered in 1980 by the American physicist Mitchell Feigenbaum. It has since been called the Feigenbaum constant.

For $k = 3.56$ the period of the iterations equals 8, for $k = 3.567$ the period equals 16, and so on for successively shrinking k-intervals. Beyond the value 3.58 for k, the period has already doubled so many times that the iterative behaviour has become chaotic. The 'spider web' of the constructions of the iterations gets so dense that it seems to fill up the whole available area. The iterative values themselves jump around between 0 and 1 in an unpredictable way and gradually fill the whole interval between 0 and 1. The attractor has become the entire interval.

Example

Let $k = 3.9$ then $f(x) = 3.9x(1 - x)$ and $f'(x) = 3.9(1 - 2x)$.

We draw the iterations for the starting values $x_0 = 0.2$ und $x_0 = 0.7$:

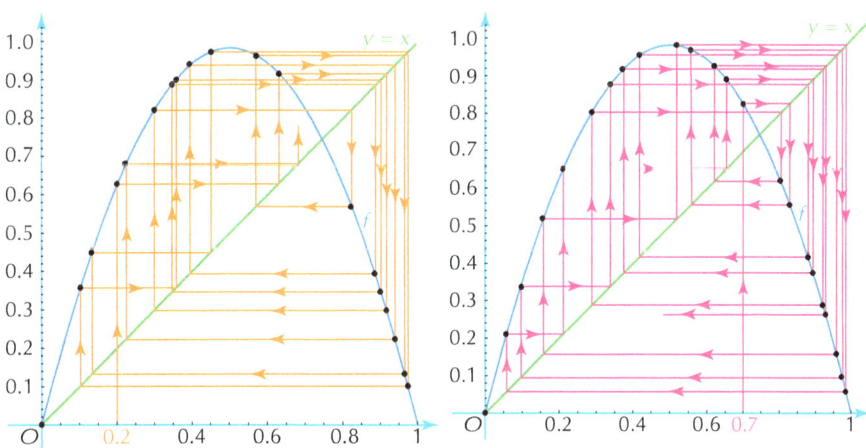

$x_0 = 0.2$ $x_0 = 0.7$

This chaotic behaviour appears in this way for values of k in the interval between 3.58 and 4. Strangely enough there are particular values in this interval for which the behaviour of the iterations becomes periodical again. For $k = 3.739$ we get period 5 and for $k = 3.835$ period 3. In the monstrous pool that this chaotic area is, there apparently exist spots where one can nevertheless bring normal fish to the surface.

This weird story could be explained in a more extensive algebraic environment and is linked to the increasing degree of the iterative functions $f(x)$, $f(f(x))$,

$f(f(f(x)))$,..., and to the behaviour of tangents in the intersections of the graphs of these functions and the bisector, and also to the discerning relations between the different periods that occur. These backgrounds go beyond the scope of this section.

The behaviour of the logistic function is thus very sensitive to the value of k, of which we could smoothly turn up the value with something like a volume control to watch the spectacle of the doubling waterfall changing into complete chaos with local windows of restored order. This process is represented in the figure of the 'fig tree' below.

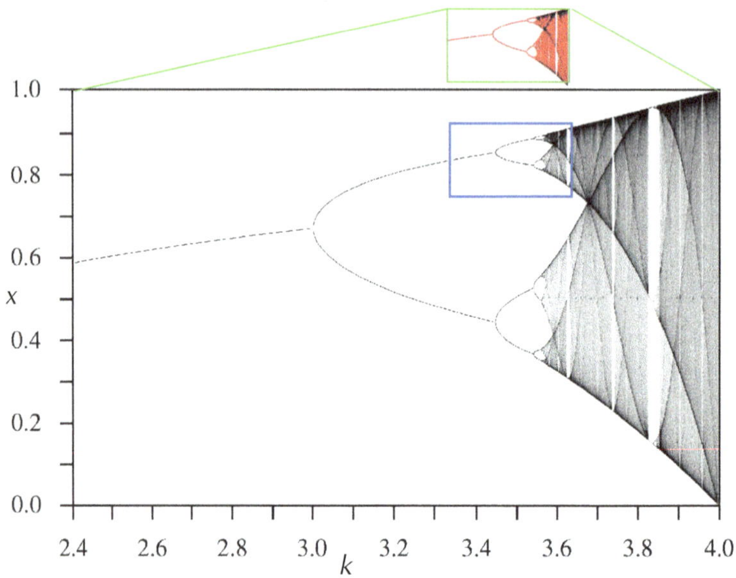

In this bifurcation diagram the value of k increases from left to right. The dark spots show where chaos occurs. The white areas indicate where some (periodical) order recurs locally. The diagram has a fractal structure, as is shown from the suggested self-similarity.

It didn't take long to find out that this weird behaviour did not only appear in the logistic function but in every function that had exactly one maximum in the interval [0, 1], in other words a function of which the graph is a 'hill' or a 'peak'. For the uttermost astonishment notice that the same Feigenbaum constant reappears as the key for the bifurcations. This proves that the phenomenon of the 'doubling waterfall' was universal for the phenomenon of deterministic chaos. Consequently a certain correspondence was found between the fig-tree diagram and Mandelbrot's gingerbread man, especially between the bifurcation pattern and the regularity with which baby-gingerbread-men appear in the self-similarity of the Mandelbrot fractal.

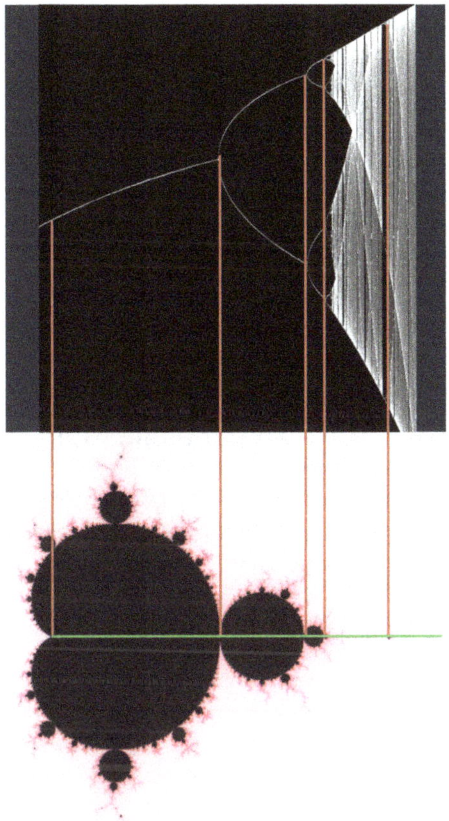

The bifurcation diagram illustrates the intimate embrace of chaos and fractals from which many beautiful offspring are to be expected.

Summary

In this section a glimpse was given of new theories in mathematics that could only be developed since the computer has been available. The possibilities to also examine the behaviour of non-linear systems and to visualize patterns of remarkable order in the complexity of their structures, has raised the importance of these theories since the 1980's. Before the introduction of fractals and chaos theory, who would have taken it seriously that there is a relationship between so very different phenomena as our heart rate, the functioning of our brain, the flight of a flock of birds, the stability of planet orbits, the presence of holes in the asteroid zone, the occurrence of cyclones, the antics of the stock exchange, the growth of plant and animal populations, the fluctuations in chemical reactions and even something as banal as the successive folding from the left and the right of a sheet of paper?

8 Recursive algorithms for the calculation of roots

I want to end this chapter reporting on a research study that I did long ago in the year 1981 in preparation for a presentation at a congress of mathematicians and that, much to my own surprise but also to my great pleasure, ended successfully. The discourse sketched below may prove that anyone can make discoveries in Alice's Wonderland if only they let themselves be guided by their curiosity, whether or not under some pressure of time. The story takes place in the domains of planar geometry and algebra.

8.1 Sections of an interval with a parallelograph

A parallelograph is an instrument by which exclusively parallel lines can be drawn. So it is the appropriate tool to interpret the properties and the invariants of the affine plane (see Chapter 2, 1.2).

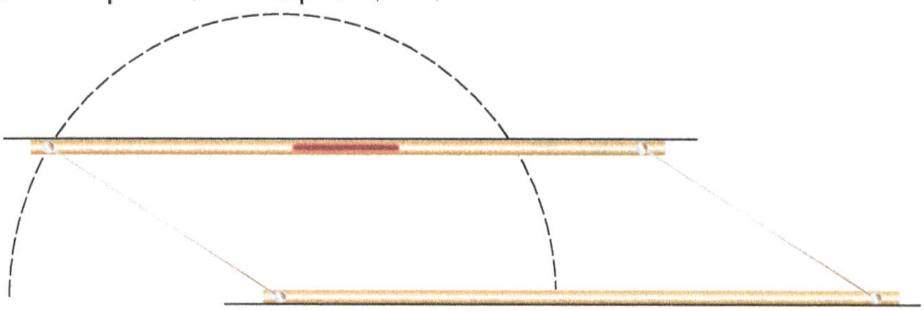

The basis of our discourse is a geometrical problem in which the theory of real numbers is involved:

'What can we learn from the analogy between the following constructions, especially about the situations on the right?'

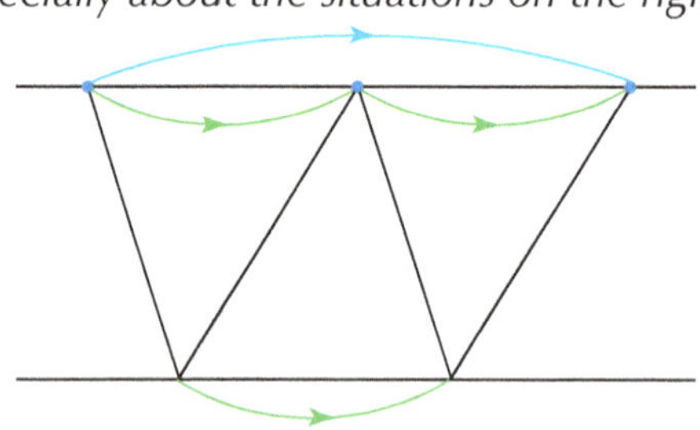

A composition of a translation with itself

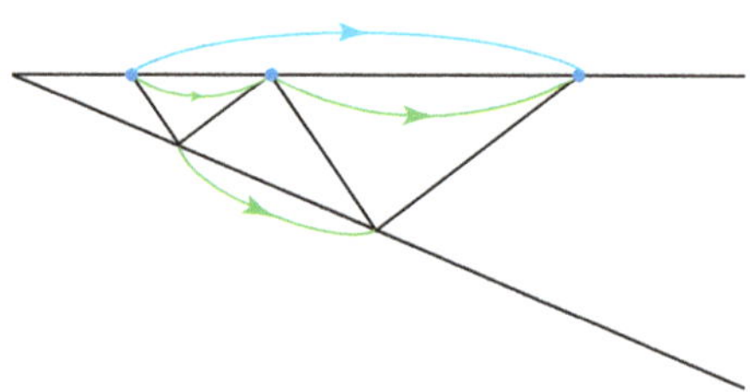

A composition of a homothety with itself

From the projective point of view both constructions are the same, but in the affine plane they do anyhow have different properties, as will be shown further on.

If we calibrate the line L, then the points of L correspond to real numbers. For any given positive numbers a and b we can calculate the value of the number x in the following diagrams:

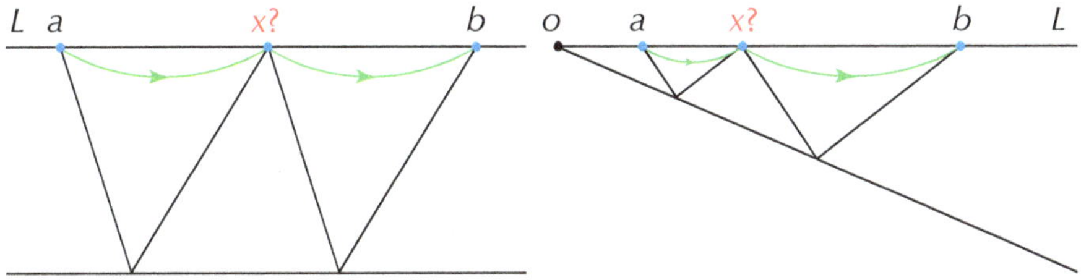

$$a + 2(x - a) = b$$
$$2x = a + b$$
$$x = \frac{(a + b)}{2}$$

x is the arithmetic mean of a and b

$$a \cdot \left(\frac{x}{a}\right)^2 = b$$
$$x^2 = ab$$
$$x = \sqrt{ab}$$

x is the geometric mean of a and b

Can we invert the construction in both situations? In other words: can we find the point that represents the number x, by means of only parallel lines and starting from the points that represent the numbers a and b? For the arithmetic mean the answer is positive (as can be seen from the next construction), but for the geometric mean the answer is negative.

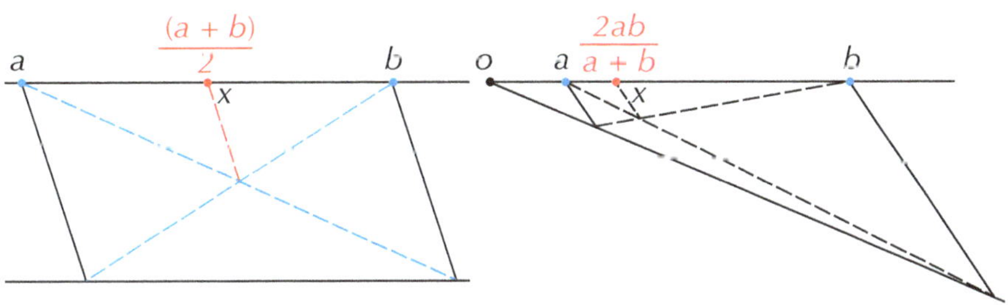

If we make a construction in the situation on the right analogous to the situation on the left, then we find x to be the harmonic mean of a and b, namely $\frac{2ab}{a+b}$. This is seen from the similarity of triangles: the parallel sides of the trapezium relate like a and b; the point of intersection of the diagonals divides them in parts that relate like the parallel sides, thus like a and b; the ratio of $x - a$ to $b - x$ is also the same, so:

$$\frac{x-a}{b-x} = \frac{a}{b} \Leftrightarrow bx - ab = ab - ax \Leftrightarrow (a+b)x = 2ab \Leftrightarrow x = \frac{2ab}{a+b}$$

There exist certain order relations between the harmonic, the geometric and the arithmetic mean of two positive numbers *a* and *b*, which are easy to prove algebraically, but which also can be seen from the geometrical figure:

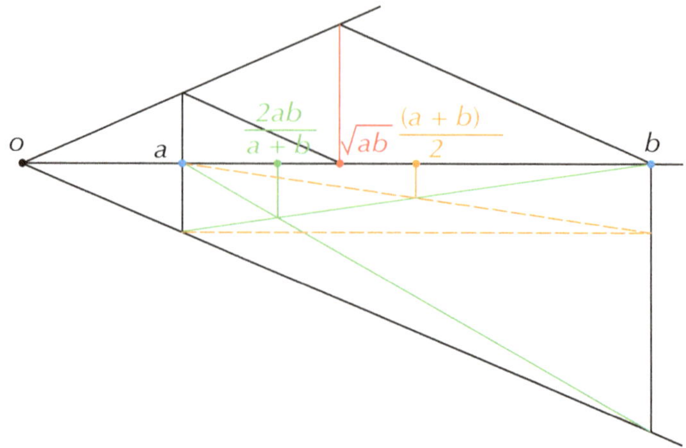

We notice that: $\frac{2ab}{a+b} < \sqrt{ab} < \frac{a+b}{2}$ (for $a \neq b$)

There is a beautiful coincidence: the geometric mean $\sqrt{ab}$ of *a* and *b* is also the geometric mean of $\frac{2ab}{a+b}$ and $\frac{a+b}{2}$, since:

$$\sqrt{\frac{2ab}{a+b} \cdot \frac{a+b}{2}} = \sqrt{ab}$$

For the calculation of the geometric mean of two positive numbers *a* and *b*, we can thus iterate on their harmonic and arithmetic mean.

Moreover, we can construct the image points of all these numbers by using only parallel lines. In each successive iteration the length of the interval between the arithmetic and the harmonic mean of *a* and *b* shrinks to less than half of the length of the original interval set up by *a* and *b*, since:

$$\frac{a+b}{2} - \frac{2ab}{a+b} = \frac{(a+b)^2 - 4ab}{2(a+b)} = \frac{(b-a)^2}{2(a+b)} < \frac{(b-a)\cdot(b+a)}{2(a+b)} = \frac{b-a}{2}$$

This means that the cluster of intervals set up by the successive harmonic and corresponding arithmetic means converges to the geometric mean. We can approximate the last to any required precision by the other two, in both an algebraical and a geometrical way.

For the purpose of generalization we now introduce symbols for the different means of two positive numbers a and b:

$H_{1/2}$ = harmonic mean

$G_{1/2}$ = geometric mean

$A_{1/2}$ = arithmetic mean

The above shows that $H_{1/2} \cdot A_{1/2} = a \cdot b$, and that is why the $G_{1/2}$ of a and b also is the $G_{1/2}$ of $H_{1/2}$ and $A_{1/2}$.

In an analogous way we now construct the division of an interval in respectively three, four,... and n parts.

8.1.1 Arithmetic, geometric and harmonic thirds of an interval:

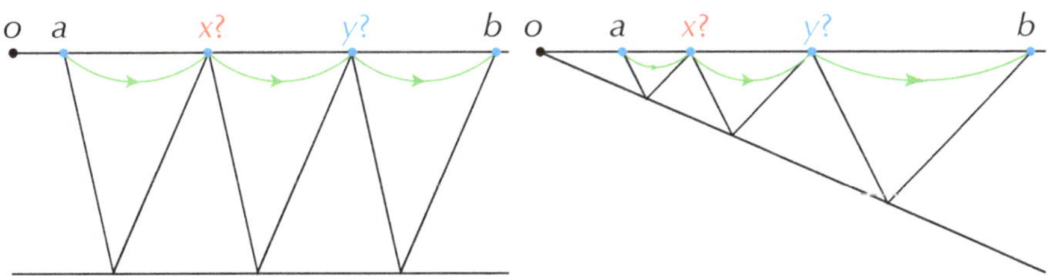

$$x = \frac{2a+b}{3} = A_{1/3} \qquad x = \sqrt[3]{a^2\, b} = G_{1/3}$$

$$y = \frac{a+2b}{3} = A_{2/3} \qquad y = \sqrt[3]{a\, b^2} = G_{2/3}$$

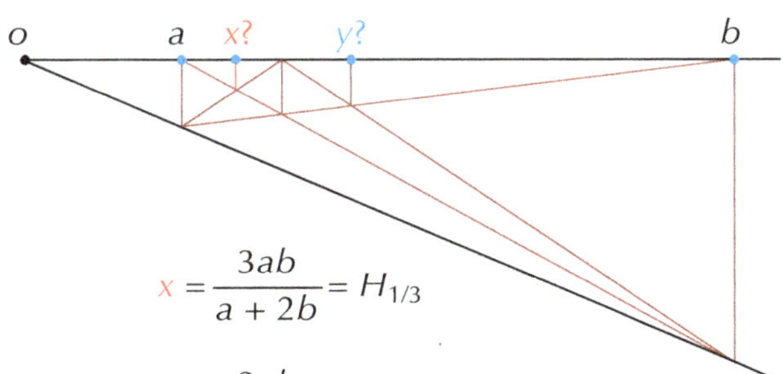

$$x = \frac{3ab}{a+2b} = H_{1/3}$$

$$y = \frac{3ab}{2a+b} = H_{2/3}$$

Observe that: $A_{1/3} \cdot H_{2/3} = ab$, $A_{2/3} \cdot H_{1/3} = ab$ and $G_{1/3} \cdot G_{2/3} = ab$.

8.1.2 Arithmetic, geometric and harmonic fourths of an interval:

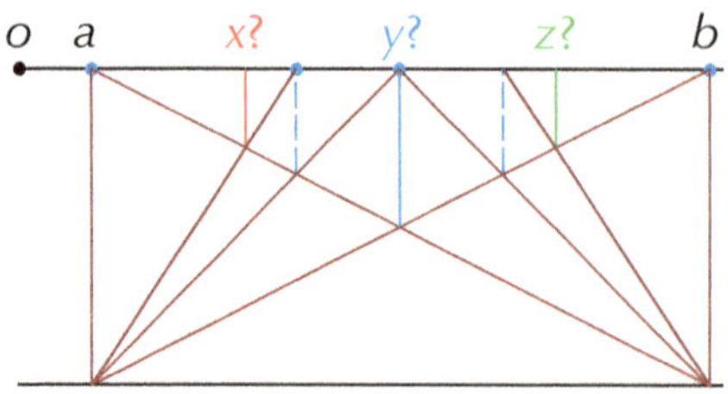

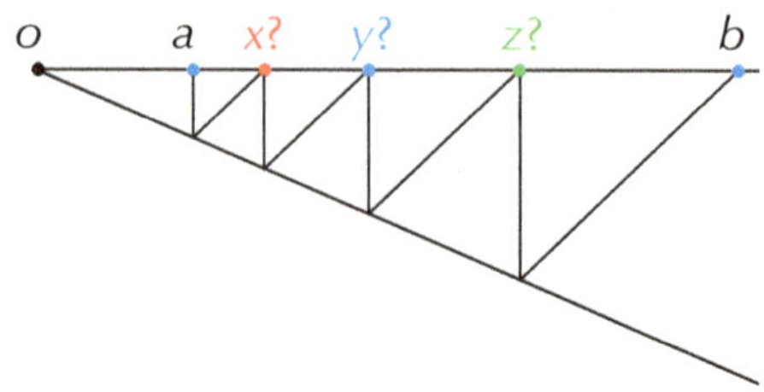

$$x = \frac{3a + b}{4} = A_{1/4} \qquad x = \sqrt[4]{a^3 b} = G_{1/4}$$

$$y = \frac{2a + 2b}{4} = A_{2/4} = A_{1/2} \qquad y = \sqrt[4]{a^2 b^2} = G_{2/4} = G_{1/2}$$

$$z = \frac{a + 3b}{4} = A_{3/4} \qquad z = \sqrt[4]{a\, b^3} = G_{3/4}$$

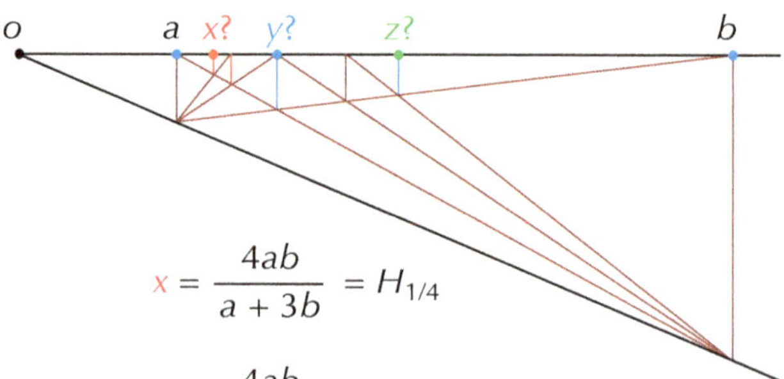

$$x = \frac{4ab}{a + 3b} = H_{1/4}$$

$$y = \frac{4ab}{2a + 2b} = H_{2/4} = H_{1/2}$$

$$z = \frac{4ab}{3a + b} = H_{3/4}$$

Observe that: $A_{1/4} \cdot H_{3/4} = ab$, $A_{2/4} \cdot H_{2/4} = ab$, $A_{3/4} \cdot H_{1/4} = ab$,
$G_{2/4} \cdot G_{2/4} = ab$ and $G_{1/4} \cdot G_{3/4} = ab$.

8.1.3 Arithmetic, geometric and harmonic *n*-th parts of an interval:

The procedure applied in 8.1.1 and 8.1.2 can be generalized to the arithmetic, geometric and harmonic *n*-th parts of an interval. The following formulas apply to the *i*-th and the $(n-i)$-th parts, where $0 < i < n$:

$$A_{i/n} = \frac{(n-i)a+ib}{n} \qquad G_{i/n} = \sqrt[n]{a^{n-i}b^{i}} \qquad H_{i/n} = \frac{nab}{ia+(n-i)b}$$

$$A_{(n-i)/n} = \frac{ia+(n-i)b}{n} \qquad G_{(n-i)/n} = \sqrt[n]{a^{i}b^{n-i}} \qquad H_{(n-i)/n} = \frac{nab}{(n-i)a+ib}$$

Observe that: $A_{i/n} \cdot H_{(n-i)/n} = ab$, $A_{(n-i)/n} \cdot H_{i/n} = ab$ and $G_{i/n} \cdot G_{(n-i)/n} = ab$.

All these results can be proved by similarity of triangles (for the more adept among us, also by means of cross-ratios in the projective plane).

Although this is in itself already a beautiful bit of geometry, entwined with algebra, the most spectacular is still to come when we link this to the calculation of roots.

8.2 Recursive formulas and nomograms for the calculation of roots

8.2.1 Calculation of the square root

The square root of a positive number c can be considered to be the geometric mean $G_{1/2}$ of the numbers 1 and c, since $\sqrt{1 \cdot c} = \sqrt{c}$.

Because of the property $\sqrt{\frac{1}{c}} = \frac{1}{\sqrt{c}}$, we can restrict ourselves to the case where $1 < c$.

If x is too low an approximate value for $\sqrt{c}$, then $\frac{c}{x}$ is too high an approximate value for $\sqrt{c}$, since:

$$x < \sqrt{c} \;\Rightarrow\; \frac{c}{x} > \frac{c}{\sqrt{c}} \;\Rightarrow\; \frac{c}{x} > \sqrt{c}$$

So, the interval $\left[x, \frac{c}{x}\right]$ contains $\sqrt{c}$.

Now $\sqrt{c}$ also is the geometric mean $G_{1/2}$ of x and $\frac{c}{x}$,since $\sqrt{x \cdot \frac{c}{x}} = \sqrt{c}$.

So we can start iterations from the harmonic mean $H_{1/2}$ and the arithmetic mean $A_{1/2}$ of the interval $\left[x, \frac{c}{x}\right]$, of which we know (from the formulas in 8.1) that they include $G_{1/2} = \sqrt{c}$ and converge to $\sqrt{c}$.

So we have:
$$H_{1/2} = \frac{2x\cdot\frac{c}{x}}{x+\frac{c}{x}} = \frac{2c}{\frac{x^2+c}{x}} = \frac{2cx}{x^2+c}$$

$$A_{1/2} = \frac{x+\frac{c}{x}}{2} = \frac{\frac{x^2+c}{x}}{2} = \frac{x^2+c}{2x}$$

where
$$H_{1/2} = \frac{2cx}{x^2+c} < \sqrt{c} < \frac{x^2+c}{2x} = A_{1/2} \quad \text{and} \quad H_{1/2}\cdot A_{1/2} = c$$

Moreover, this algorithm, which one can proceed via the lower bound $H_{1/2}$ as well as via the upper bound $A_{1/2}$, can be expressed beautifully with the successive constructions of $H_{1/2}$ and $A_{1/2}$, as shown in the following figure. The geometric representation of this algorithm is called a nomogram.

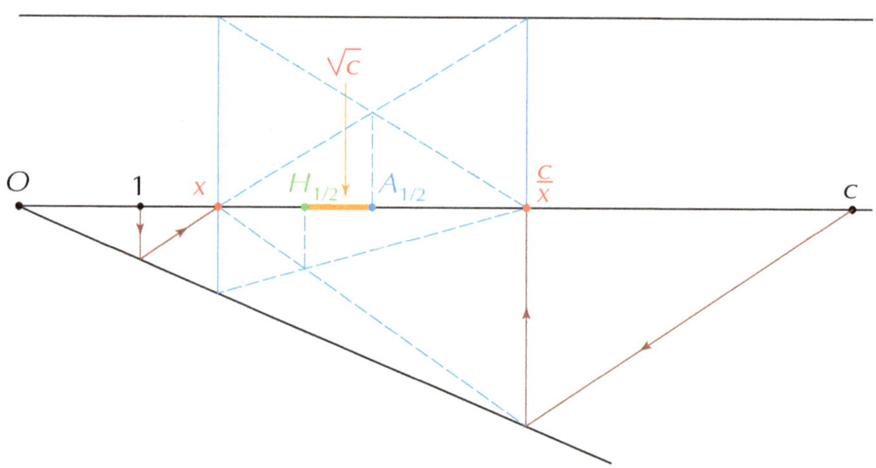

Example

For the calculation of the square root of $c = 4711$ we take as starting value the too low approximation $x_0 = 68$ ($68^2 = 4624 < 4711$), then $\frac{c}{x_0} = \frac{4711}{68} = 69.279...$

$$H_{1/2} = \frac{2cx_0}{x_0^2+c} = \frac{2\cdot 4711\cdot 68}{68^2+4711} = 68.633... = x_1$$

$$A_{1/2} = \frac{x_0^2+c}{2x_0} = \frac{68^2+4711}{2\cdot 68} = 68.639...$$

(Remember that $H_{1/2} \cdot A_{1/2} = c$ and thus also $A_{1/2} = \dfrac{c}{H_{1/2}} = \dfrac{4711}{68.633...} = 68.639...$)

After the first iteration it can already be seen that:

$$68.633... < \sqrt{4711} < 68.639... \text{, also } \sqrt{4711} = 68.63...$$

Now we iterate by replacing x_0 by $x_1 = 68.633....$ This results in:

$$H_{1/2} = \frac{2cx_1}{x_1^2 + c} = 68.6367238... = x_2$$

$$A_{1/2} = \frac{x_1^2 + c}{2x_1} = 68.636724...$$

Which results in: $\sqrt{4711} = 68.63672...$

There is one more interesting coincidence in the formula for the upper bound in the algorithm, namely the expression $\dfrac{x^2 + c}{2x}$ for $A_{1/2}$. Notice that $\sqrt{c}$ also is a zero of the function $f(x) = x^2 - c$.

Newton's method for the approximation of the zeroes of real functions starts from a first estimate x_0 and uses a tangent line in the point with coordinate $(x_0, f(x_0))$ to construct and calculate the next better approximation:

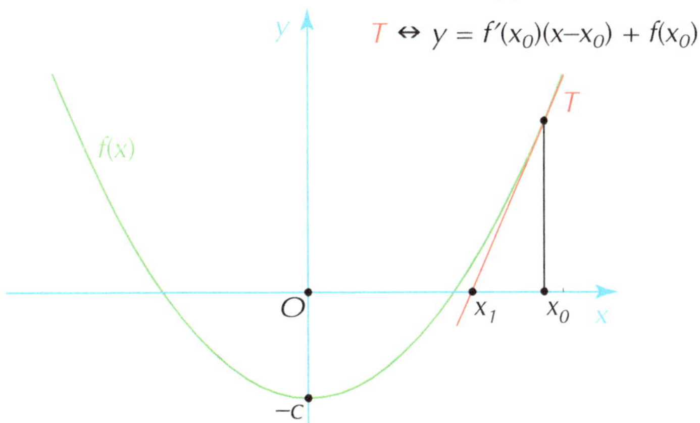

The derivative function of f is $f'(x) = 2x$. This function gives the slope of the tangents to the graph of f. The point x_1 of intersection of the tangent T with the x-axis can be calculated as follows:

$$x_1 = x_0 - \frac{f(x_0)}{f'(x_0)} = x_0 - \frac{x_0^2 - c}{2x_0} = \frac{x_0^2 + c}{2x_0}$$

This expression is exactly the upper bound $A_{1/2}$ in the given algorithm for the square root.

So, is there nothing new under the sun? There truly is! This expression has a specific geometrical meaning in the nomogram, which has nothing to do with tangents, derivatives and other sophisticated concepts of analysis. Moreover, to this upper bound there belongs a related lower bound that can also be used as an approximate value and that has a nice geometrical meaning as well. But the real reward comes, as always in mathematics, from the possibilities for generalization.

8.2.2 Calculation of the cubic root

The cubic root of a positive real number c can be considered to be the geometric (first) third $G_{1/3}$ of 1 and c, since $\sqrt[3]{1^2 \cdot c} = \sqrt[3]{c}$. Because $\sqrt[3]{\frac{1}{c}} = \frac{1}{\sqrt[3]{c}}$, we can restrict ourselves to the case where: $1 < c$.

If x is too low an approximate value for $\sqrt[3]{c}$ then $\frac{c}{x^2}$ is too high an approximate value for $\sqrt[3]{c}$, since:

$$x < \sqrt[3]{c} \;\Rightarrow\; x^2 < \sqrt[3]{c^2} \;\Rightarrow\; \frac{c}{x^2} > \frac{c}{\sqrt[3]{c^2}} \;\Rightarrow\; \frac{c}{x^2} > \sqrt[3]{c}$$

So, the interval $\left[x, \frac{c}{x^2}\right]$ contains $\sqrt[3]{c}$.

But from $x < \sqrt[3]{c}$ follows that also $x^2 < \sqrt[3]{c^2}$ and $\frac{c}{x} > \frac{c}{\sqrt[3]{c}} = \sqrt[3]{c^2}$.

So, the interval $\left[x^2, \frac{c}{x}\right]$ contains $\sqrt[3]{c^2}$.

Now $\sqrt[3]{c}$ is also the geometric first third $G_{1/3}$ of the interval $\left[x, \frac{c}{x^2}\right]$, since $\sqrt[3]{x^2 \cdot \frac{c}{x^2}} = \sqrt[3]{c}$.

The arithmetic first third $A_{1/3}$ of the interval $\left[x, \frac{c}{x^2}\right]$ is $\frac{2x + \frac{c}{x^2}}{3} = \frac{2x^3 + c}{3x^2}$ and is a new upper bound for $G_{1/3} = \sqrt[3]{c}$. This can be proved algebraically, but we confine ourselves to referring to the constructions in the following figure.

The harmonic second third $H_{2/3}$ of the interval $\left[x^2, \frac{c}{x}\right]$ is $\dfrac{3x^2 \cdot \frac{c}{x}}{2x^2 + \frac{c}{x}} = \dfrac{3cx^2}{2x^3 + c}$ and is a new lower bound for $\sqrt[3]{c^2}$ (see figure), and hence:

$\sqrt{H_{2/3}} = \sqrt{\dfrac{3cx^2}{2x^3+c}}$ is a new lower bound for $\sqrt[3]{c}$

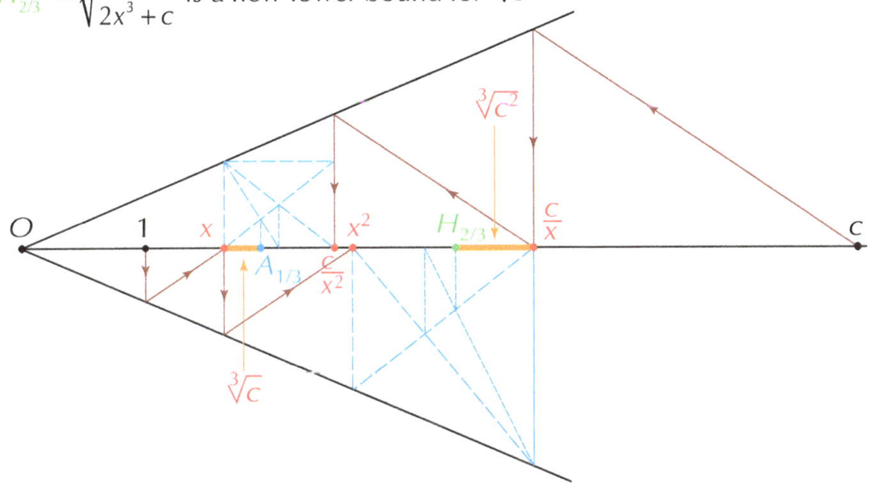

So we have the following inequalities:

$$\sqrt{H_{2/3}} = \sqrt{\frac{3cx^2}{2x^3+c}} < \sqrt[3]{c} < \frac{2x^3+c}{3x^2} = A_{1/3} \qquad (\S)$$

Moreover, $\sqrt[3]{c}$ is again the geometric first third of the interval set up by the left and the right members of these inequalities, since:

$$G_{1/3} = \sqrt[3]{\sqrt{H_{2/3}}^{\,2} \cdot A_{1/3}} = \sqrt[3]{\sqrt{\frac{3cx^2}{2x^3+c}}^{\,2} \cdot \frac{2x^3+c}{3x^2}} = \sqrt[3]{\frac{3cx^2}{2x^3+c} \cdot \frac{2x^3+c}{3x^2}} = \sqrt[3]{c}$$

So, we can iterate on these formulas. Since the length of every successive new interval shrinks to less than one-third of the previous interval, the algorithm converges in $\sqrt[3]{c}$.

Just like in the calculation of the square root, one also gets the formula in the right member of (§) by applying Newton's tangent method for approximation of the zero of the function $f(x) = x^3 - c$.

The derivative of this function is $f'(x) = 3x^2$, so

$$x - \frac{f(x)}{f'(x)} = x - \frac{x^3 - c}{3x^2} = \frac{2x^3 + c}{3x^2}$$

But the algorithm based on the iterations of the left member of (§) converges faster than the algorithm based on the right member of (§). This algorithm calculates the cubic root by means of the square root, which makes it a recursive method. Moreover, both algorithms can be expressed as nomograms.

In the previous we introduced two test intervals for the calculations of cubic roots, the first test interval being $\left[x, \frac{c}{x^2}\right]$ that contains $\sqrt[3]{c}$ and the second the test interval being $\left[x^2, \frac{c}{x}\right]$ that contains $\sqrt[3]{c^2}$ i.e. the square of $\sqrt[3]{c}$.

In the first test interval we took $A_{1/3}$ as an upper bound for $\sqrt[3]{c}$,to which we linked $H_{2/3}$ from the second test interval as a lower bound for iterating on $\sqrt[3]{c^2}$ We could have chosen other linked arithmetic and harmonic divisions in those test intervals to be the lower and upper bounds for $\sqrt[3]{c}$. Which means that there are multiple algorithms available for the calculation of the same root. So as not to make the story too long-winded, we leave this as an exercise and confine ourselves to giving the general formulas.

Omar Al-Khayyam (ca. 1044–ca. 1123)

Persian mathematician, astronomer, philosopher and poet. In mathematics he is particularly known for his original contributions to the solution of cubic equations, in which the roots are geometrically constructed. He is also famous as the author of the *Ruba'iyat*, quatrains with profound mystical and philosophical ideas. It was translated into English in 1839 by Edward Fitzgerald.

8.2.3 Calculation of the *n*-th root

Starting from a too low estimate x, the available test boxes for the n-th root $\sqrt[n]{c}$ of a positive real number c are:

(1)	$= \left[x, \frac{c}{x^{n-1}}\right]$	that contains	$\sqrt[n]{c}$
(2)	$= \left[x^2, \frac{c}{x^{n-2}}\right]$	that contains	$\sqrt[n]{c^2}$
⋮ ⋮	⋮ ⋮		
(i)	$= \left[x^i, \frac{c}{x^{n-i}}\right]$	that contains	$\sqrt[n]{c^i}$
⋮ ⋮	⋮ ⋮		
($n - i$)	$= \left[x^{n-i}, \frac{c}{x^i}\right]$	that contains	$\sqrt[n]{c^{n-i}}$
⋮ ⋮	⋮ ⋮		
($n - 2$)	$= \left[x^{n-2}, \frac{c}{x^2}\right]$	that contains	$\sqrt[n]{c^{n-2}}$
($n - 1$)	$= \left[x^{n-1}, \frac{c}{x}\right]$	that contains	$\sqrt[n]{c^{n-1}}$

All together there are $n - 1$ test intervals.

If, for instance, we take in the interval (i) the i-th arithmetic n-th part $A_{i/n}$ as (too high) an approximation for $\sqrt[n]{c^i}$, then we must link this to the harmonic ($n - i$)-th part $H_{(n-i)/n}$ in the interval ($n - i$) as (too low) an approximation for $\sqrt[n]{c^{n-i}}$ in order to make the iteration. We now have

$$A_{i/n} = \frac{(n-i)\cdot x^i + i\cdot \frac{c}{x^{n-i}}}{n} = \frac{(n-i)\cdot x^n + ic}{nx^{n-i}}$$

$$H_{(n-i)/n} = \frac{n\cdot x^{n-i}\cdot \frac{c}{x^i}}{(n-i)\cdot x^{n-i} + i\cdot \frac{c}{x^i}} = \frac{ncx^{n-i}}{(n-i)\cdot x^n + ic}$$

For $\sqrt[n]{c}$ the following inequalities apply:

$$\sqrt[n-i]{H_{(n-i)/n}} = \sqrt[n-i]{\frac{ncx^{n-i}}{(n-i)\cdot x^n + ic}} < \sqrt[n]{c} < \sqrt[i]{\frac{(n-i)\cdot x^n + ic}{nx^{n-i}}} = \sqrt[i]{A_{i/n}}$$

These formulas thus give iterative convergent algorithms for calculating the n-th root by means of the i-th or the $(n-i)$-th root, and this for any $i < n$.

This solves the problem of reductivity for the calculation of roots.

The higher the value of the root exponent in the algorithm concerned, the faster the convergence.

If we take in particular the case that $i = 1$ in the previous formulas, then we get:

$$\sqrt[n-1]{H_{(n-1)/n}} = \sqrt[n-1]{\frac{ncx^{n-1}}{(n-1)\cdot x^n + c}} < \sqrt[n]{c} < \frac{(n-1)\cdot x^n + c}{nx^{n-1}} = A_{1/n}$$

In the right member this once more produces Newton's formula for the zero $\sqrt[n]{c}$ of the function $f(x) = x^n - c$.

In this, the derivative function is $f'(x) = nx^{n-1}$, and so:

$$x - \frac{f(x)}{f'(x)} = x - \frac{x^n - c}{nx^{n-1}} = \frac{(n-1)x^n + c}{nx^{n-1}}$$

This is, however, the slowest converging of all algorithms of this type. The algorithm based on the formula in the left member is the fastest because its root exponent is the highest, namely $n - 1$.

All these algorithms do have nice nomograms. A wonderful and close link between algebra and geometry that offers a feeling of well-being to the grateful human mind.

Summary

If you had the courage to struggle through this section you will have understood that around every corner there are simple situations one can wonder about, and where beauty can be discovered, if only one persists in research. Always remember: "Mathematics is a mistress that never deceives you".

Conclusion

In this chapter we have given some examples of how mathematics can develop in various domains of its sphere of interest. Absolutely no problem is beneath its dignity: whether it is evening strolls across bridges, or mixing up hats in a cloakroom, it doesn't matter. Each of these examples shows that mathematical methodology does not content itself with providing a solution to the problem stated, but that it tries to lift its analysis to a degree of abstraction that reveals a panorama of analogies, generalizations and modifications. Whatever expedition is undertaken, the result is always a theory whose scope far exceeds the original goal. It may deal with non-Euclidean geometry, which unexpectedly also seems to have applications, or it might concern transfinite numbers, which, like a dizzily expanding universe, exceed the bounds of our imagination to a seemingly unnecessary degree. But regardless of its focus the abstraction process always rewards the effort. Often the eagerness emerges from simple propositions which then suddenly can no longer be restrained.

On top of all of this comes the astonishing observation that, despite seeming esoteric at first sight, these new spin-offs of the creative mathematical mind, nevertheless also find a practical application somewhere. The Greek Eratosthenes himself would not have expected his 'sieve' to remain the only access to the sequence of prime numbers, whose complexity twenty-three centuries later underlies the protection of codes that cannot be cracked, even by the fastest and most powerful computers. In his popular and often quoted witticism "The unreasonable effectiveness of mathematics", the respected Hungarian physicist and Nobel Prize winner, Eugene P. Wigner, expressed his amazement at the unimaginable possibilities of applying in the sciences theoretical mathematics, which originated as a mere mind game. Undoubtedly, the rationality of the human mind has to be intimately interwoven with the profound features of a universe to which its history has been linked for millions of years already. This cosy homely feeling can only put us in a happy mood.

"The pursuit of an idea is just as exciting as the pursuit of a whale"

Henry Norris Russell

نمودند نزدیکی آن مقر
درو پانزده اهل علم گزین

Chapter 3

Mathematics and civilization

Descartes influenced the future more from his chair than Napoleon from his throne

Oliver Wendell Holmes

Introduction

In this chapter we give some examples of the connection between the great changes in human society over the course of history and the specific innovations in mathematics that promoted or accelerated these developments.

It is an ironic feature of the history of civilization that the brilliant ideas of a few exceptional individuals, sometimes in very esoteric domains and far removed from everyday realities, have changed society and the living conditions of people more fundamentally than the many successive wars and political revolutions. Also, some outstanding mathematical personalities have made a not insignificant contribution to these changes.

It is not our intention to present a cavalcade of mathematical dates, facts and individuals. There are excellent standard works on the history of mathematics and you can find them listed in the bibliography at the end of this book. We shall restrict ourselves to four remarkable periods in the history of (western) civilization that are generally recognized as revolutionary changes.

First, there is the transition from a wandering nomadic culture to a sedentary agrarian culture, which also marks the transition from the occasional collection of food by hunting, fishing or berry picking, towards planned activities such as agriculture, livestock breeding, urbanization and barter trade. This is an era that begins already in the Neolithic Age and leads up to the impressive ancient cultures along major rivers like the Tigris and the Euphrates, the Nile, the Indus and the Ganges, the Hoang-ho and the Yang-tse.

R. Verhulst, *Under the Spell of Mathematics*, https://doi.org/10.1007/978-3-031-63871-8_3

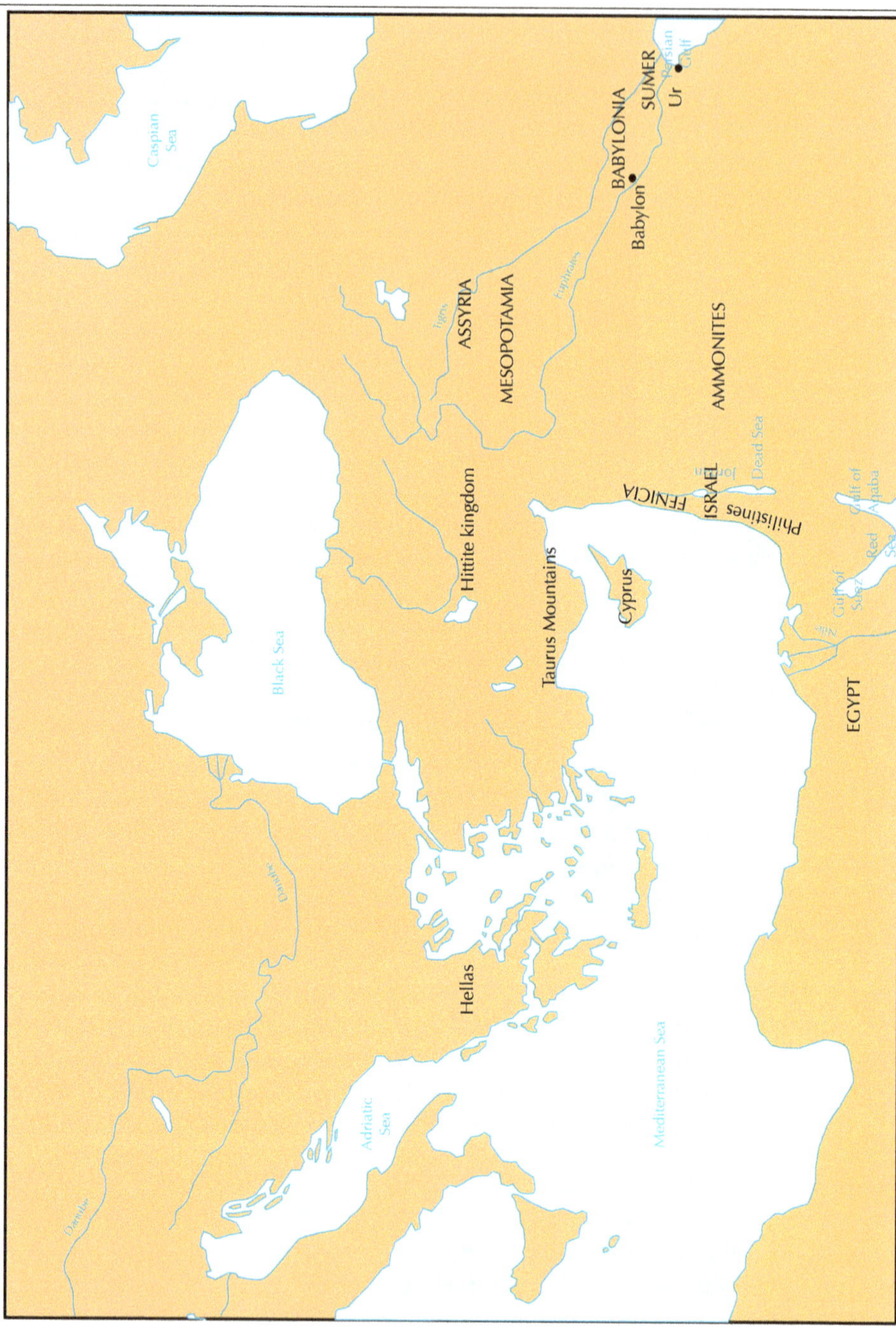

The territory of the **Middle East** was rich in ancient civilizations. It was the cradle of the Greek and the Jewish cultures, both of which are indebted to Egyptian and Babylonian culture. It can thus be said that European civilization came into being in this region.

Next, there was the fundamental revolution unleashed by the Greek spirit: a striving for rational insight into the mysteries of being, prioritizing pure intellectual understanding over everyday utility and material benefit. It was the simultaneous birth of dialectic philosophy, formal logic and the mathematical deductive method, resulting in a shift in western civilization that marked it for ever after. This rational style blossomed again during the Renaissance and through a fertile marriage to inductive experimentation it set into motion the astounding adventure of western science.

Thirdly, there was the unstoppable wave of advancing industrialization, stimulated by the successful development of natural sciences. It was based on the emerging ideal of the Age of Enlightenment coupled to a process of secularization and humanization, and accompanied by the triumphant trumpet call of logical positivism and an optimistic belief in progress.

Finally, there came our entry into modern times. The gradual advance of information technology and automation, with everybody having computers and many other sophisticated means of communication at their disposal, resulted in a network that spans more or less the entire globe. The result is a satisfying virtual life in an easy chair. After centuries of hard labour, we have finally regained the lost paradise of nomadic man.

According to one biblical story, the tree of knowledge already had a fatal attraction for the first humans. Since the loss of paradise, we have travelled a long way and ended up in close proximity to another tempting apple: the unlimited flow of information. Nobody can predict the extent of the possibilities offered by the Internet.

1 From a nomadic to an agrarian culture

When the collection of food by hunting, fishing and gathering of plants and fruits from nature made way for the production of food by agriculture and cattle breeding, also a more efficient development was needed for the primitive representations of numbers and shapes as they already appeared in the Old Stone Age (Palaeolithic).

This transition from a passive to an active attitude towards nature happened in the New Stone Age (Neolithic), about 5,000 BCE. Permanent settlements arose, mostly along the fertile banks of big rivers. Gradually villages, towns and barter trade developed. These activities could only be successful if humans had a practical arithmetic and geometry at their disposal.

In order to supervise the cattle, they had to be counted. The primitive concepts of amount and quantity, which initially were restricted to one, two and many, and which were mostly represented by actual objects, were refined to practical and well-organized numerical systems in a more abstract form. It was the combining of lower units into higher entities, denoted by particular symbols, that led to the development of numerical systems.

Moreover, the harvesting of crops required calculations of the yield, the amount of seed for sowing needed, how to distribute the resulting supply of comsumables and the tax due. Ingenious methods for multiplication and division were developed.

One of the most important problems that all agrarian cultures had to find a solution to was time measurement. The cyclic changes, related to the motion of the celestial bodies, such as day and night, the phases of the moon and the succession of seasons, were particularly suited for this. The recording in tables, based on repeated observations, marked the start of practical astronomy.

The changing level of rivers, which left fertile tillage on their banks after floods, were used for inventive irrigation systems. The techniques of land surveying and area calculation were therefore indispensable. In this way an arsenal of rules of thumb came into being, which started the development of geometry.

Barter trade required a currency and a system of weights and measures. Problems of fair barter, interest calculations, profit determination, loan calculations, etc. advanced the further development of arithmetic.

The itinerary of caravans and sea navigation were the incentive for drawing land and sea maps, based on methods for determining position and distance.

This way an arsenal of practical mathematics grew, needed for solving concrete everyday problems. Initially the required knowledge was not passed on to the next generations as a theory of formulas and properties, but rather as a compendium of rules and skills, mostly illustrated by striking examples. We will mention some remarkable facts from the most prominent ancient cultures.

1.1 Primeval numbers

The roots of the number concept of humanity are difficult to trace. Even primitive humans must have had a notion of distinctive quantities when judging the amount of wildlife stock, menacing enemies, children and other 'members' of the family.

Even in animals a limited ability for distinguishing quantities has been noticed. For instance, if there are four eggs in a bird's nest and one is taken away, then the parents will keep on taking care of the nest. But no longer if two eggs are taken.

Originally, numbers were related to concrete objects. In several traditional tribes different terms are used to indicate 'one' or 'two' depending on whether they deal with monkeys, trees or other specific items. The gradual shift to the abstract concepts 'one' and 'two' , independent of a concrete content, probably lasted many centuries. Also the symbolism by which numbers were indicated has gone through a long evolution. From raising fingers, pointing at parts of the body, laying cobblestones or notching a bone or a stick, to the writing of symbols.

The oldest known artefact referring to numbers was found in Swaziland and dates from about 35,000 BCE. It is the fibula of a baboon with 29 notches. Possibly it was used as a calendar stick.

Another old artefact is the Ishango bone, from about 20,000 BCE, found near Lake Edward on the border of Uganda and Congo. The notches could be related to the phases of the moon.

1.2 Egyptian land surveyors

We owe our knowledge of Egyptian mathematics mainly to the discovery of some ancient papyrus scrolls, such as the Rhind Papyrus (about 1650 BCE) and the Moscow Papyrus (about 1850 BCE). These very vulnerable documents were surprisingly well kept. This was due to the very dry climate in Egypt and also to the extraordinary cult of the dead Pharaohs. This is why many inscriptions, wall paintings and objects that were found in graves, temples and pyramids remained reasonably intact till present time. Also the hieroglyphics, this special symbolic script that the Egyptians prefered to use while writing on stone, revealed some facts about their mathematical knowledge.

The Egyptians wrote numbers according to a decimal system in which the successive higher ranks (powers of 10) were indicated by different signs. By adding the values of these signs they represented numbers as their sum. Such a system is called an additive numerical system. The Greek and the Romans also used an additive system.

Examples

1 = I, 10 = 𓎆, 100 = 𓍢, 1000 = 𓆼, 10000 = 𓂭 etc...

which makes 𓂭𓆼𓆼𓆼𓎆𓎆IIII = 10000 + 3000 + 20 + 4 = 13024

Usually the Egyptians wrote the ranks in the opposite order to the one above, but that is of no importance in an additive system. In such a system neither is there a need for the numeral null.

The **Rhind Papyrus**, which was found in 1856 in Luxor on the Nile, was acquired by the Scottish banker and antique dealer A. Henry Rhind. It is now in the British Museum in London. It was written about 1650 BCE by a certain Ahmes, who copied material that probably was many centuries older. The Papyrus contains 84 mathematical problems and together with the Moscow Papyrus, which contains 25 problems, it is the main source for our knowledge of Egyptian mathematics.

The operations of multiplication and division are not easy to do in an additive system. The doubling or halving of the signs was, however, no problem for the Egyptians. This method, combined with their knowledge that every number can be expressed as a sum of powers of two (which in fact refers to the binary numerical system), allowed them to do arbitrary multiplications and divisions in an ingenious way.

Example: multiplication by successive doubling

	17 x 13 = ?	
	13 = 8 + 4 + (0 · 2) + 1	(13 as a sum of powers of 2)
and	17 x 1 = 17	
		(17 x 2 = 34)
	17 x 4 = 68	(34 x 2 = 68)
	17 x 8 =136	(68 x 2 = 136)
which makes:	17 x13 =221	(making the sum)

Remarkable in Egyptian arithmetic is how they calculated with fractions. They wrote an arbitrary fraction as the sum of fractions with numerator one (unit fractions, i.e. reciprocals of integers). The unit fraction with denominator 7 (i.e. 1/7) was written as the symbol for 7 with a sign on top of it, namely: $\dot{7}$.

The Rhind papyrus contains a table that converts fractions, with numerator 2 and an odd denominator from 5 till 101, into a sum of unit fractions. In this, 2/5 is mentioned as 5: . Because arbitrary numerators were converted into a sum of powers of two (like in the binary system), also arbitrary fractions could be written as a sum of unit fractions by means of this table.

Example

The table shows: $\frac{2}{5}=\frac{1}{3}+\frac{1}{15}$ $\quad$ $(5:=\dot{3}+\dot{1}5)$

this also makes:

$$\frac{7}{5}=\frac{4}{5}+\frac{2}{5}+\frac{1}{5}$$
$$=2\cdot\frac{2}{5}+\frac{2}{5}+\frac{1}{5}$$
$$=2\cdot\left(\frac{1}{3}+\frac{1}{15}\right)+\left(\frac{1}{3}+\frac{1}{15}\right)+\frac{1}{5}$$
$$=\frac{1}{3}+\frac{1}{3}+\frac{1}{3}+\frac{1}{5}+\frac{1}{15}+\frac{1}{15}+\frac{1}{15}$$

According to historiographer Herodotos, geometry was invented by the Egyptians because they needed to mark out the farm lands every year again after the River Nile flooded its banks, and they had to make various area calculations. For constructing right angles, they would have made use of a rope that was divided into twelve (3 + 4 + 5) equal segments by means of knots. According to a property of rectangular triangles, the sum of the squares of the catheti (the smaller sides) equals the square of the hypotenuse (this is known as the Pythagorean theorem). And since $3^2 + 4^2 = 5^2$, the Egyptians assumed that in a triangle with sides of length 3,4 and 5 the angle opposite of the side with length 5 had to be a right angle (as seen in the next figure).

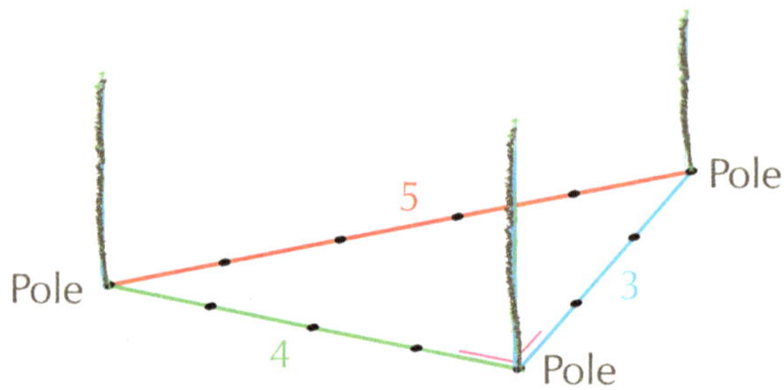

By means of three poles (or three men) and by stretching out the rope they could construct a right angle. That is why the Egyptian land surveyors were also called rope stretchers (harpedonaptai).

The King's Chamber in the Great Pyramid of Giza has dimensions that relate as 3, 4 and 5. So, the Egyptians knew the relationship between the sides of a right triangle for more than 2000 years before Pythagoras would provide a proof of it.

It is no surprise that the builders of pyramids knew formulas for the area of a triangle and the volume of a regular quadrilateral pyramid.

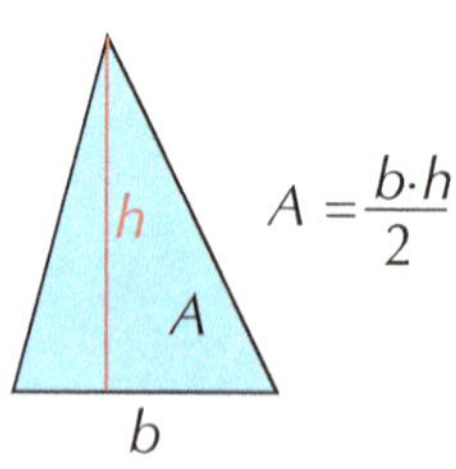

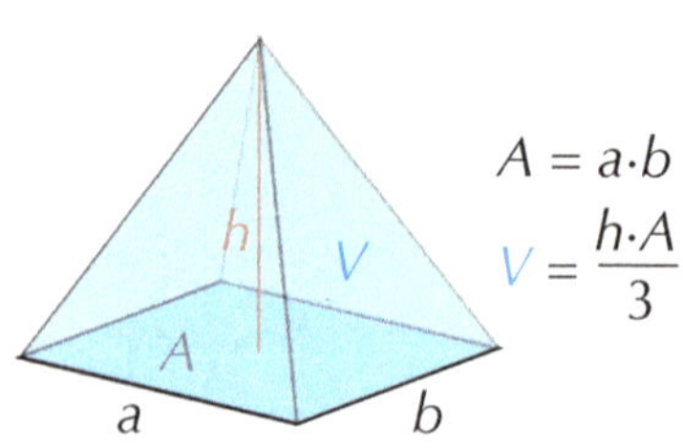

But how did they find the remarkable formulas for the area of a disc and the volume of a regular quadrilateral truncate pyramid (frustum)?

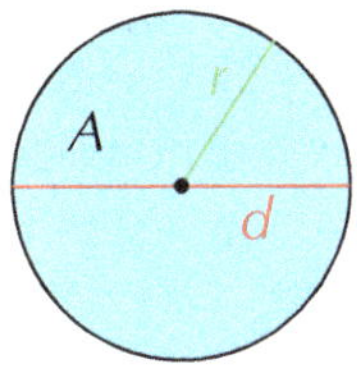

$$A = (d - \frac{d}{9})^2$$

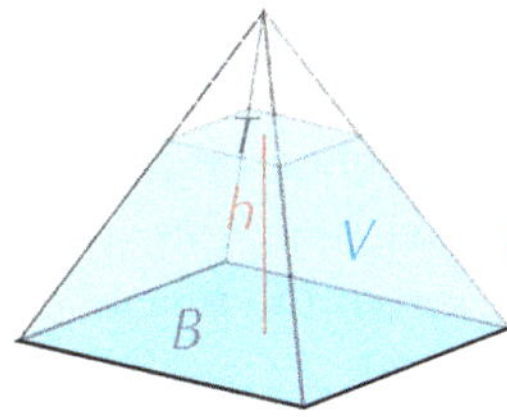

$$V = \frac{h(B + T + \sqrt{BT})}{3}$$

This last formula bewildered the mathematical historian E.T. Bell so much that he considered this result as "the biggest of all Egyptian pyramids". Moreover it shows from the given formula for the area of a disc that the Egyptians used 256/81 = 3.16… as an approximation of the number π (the ratio of the perimeter of a circle to its diameter), which is a fairly good result for that time. This approximate value is shown from the following calculation and its comparison with the correct formula for the area A of a circle, namely:

$$A = \pi r^2 \qquad (\text{where } r = \text{radius} = \frac{d}{2}).$$

And:

$$\left[d - \frac{d}{9}\right]^2 = \left[\frac{8}{9} d\right]^2 = \left[\frac{16}{9} r\right]^2 = \frac{256}{81} r^2$$

which results in an approximation of π

$$\frac{256}{81} = 3.16049.$$

The correct value of π up to five decimals is 3.14159…

1.3 Babylonian astronomers

Mesopotamia, the land of the two rivers Tigris and Euphrates, is the territory where the Sumerians, the Accadians, the Assyrians, the Chaldeans, the Babylonians and later the Persians established one of the wealthiest civilizations of ancient times. Our knowledge of the Babylonian culture is mainly based on the clay tablets, of which at present more than half a million have been excavated. On them is the so-called wedge or cuneiform script. As long as the clay was wet, the inscriptions could be wiped out and the tablet could be inscribed again. But once the clay dried, the tablets were stored or thrown away, and the valuable inscriptions were preserved for future generations. About 300 of those clay tablets contain mathematical calculations or tables. Although the cuneiform was already deciphered in 1847 by Rawlinson, it took untill 1935 before the true meaning of the mathematical tablets was understood.

The Babylonians are the fathers of the sexagesimal numerical system, which used only two symbols, namely 𒁹 (one) and 𒌋 (ten). By juxtaposition they made groups that represented the numbers from 1 till 59. Then, taking sixty as a base, they went to numbers of higher rank. Moreover, they wrote the numbers of higher rank in a *positional system*, this means that the value of a group of signs is determined by its position in the whole of the notation.

Example

𒌋𒌋𒁹𒁹𒁹 = 10 + 10 + 1 + 1 + 1 = 23

𒌋𒁹 𒌋𒌋𒌋𒁹𒁹 = 11 x 60 + 32 = 660 + 32 = 692

$(a/b)^2$	side a	hypot.		name
1;59; ;15	1;59	2;49		1
1;56;56;58;14;50;6;15	56;7	1;20;25		2
1;55;7;41;15;33;45	1;16;41	1;50;49		3
1;53;10;29;32;52;16	3;31;49	5; ;9; ;1		4
1;48;54; ;1;40	1; ;5	1;37		5
1;47; 6;41;40	5;19	8;1		6
1;43;11;56;28;26;40	38;11	59;1		7
1;41;33;45;14; 3;45	13;19	20;49		8
1;38;33;36;36	8; 1	12;49		9
1;35;10; 2;28;27;24;26;40	1;22;41	2;16;1		10
1;33;45	45	1;15		11
1;29;21;54; 2;15	27;59	48;49		12
1;27; ;3;45	2;41	4;49		13
1;25;48;51;35; 6;40	29;31	53;49		14
1;23;13;46;40	56	1;46		15

One of the most remarkable clay tablets is number **322** of the **G.A. Plimpton collection**, which is kept in the Columbia University in New York. It dates from some time between 1900 and 1600 BCE. and contains fifteen rows of numbers in four colums. Two colums are clearly related to the so-called Pythagorean triples. These are positive integers a, b and c such that $a^2 + b^2 = c^2$.

The first column contains $(a/b)^2$, which proves calculation skills and the knowledge of this relation. So also the Babylonians had knowledge of this property of rectangular triangles and its interpretation in numerical relations long before Pythagoras.

In a positional system a numeral null is indispensable to mark the possible absence of digits of a certain rank. For instance, in our decimal system the meaning of 18, 108 or 180 is totally different. Although the Babylonians were indeed aware that some ambiguity in the interpretation of numbers was inevitable without a sign for the empty places, initially they made do without. They hoped that the context would make clear what the rank of the actual groups of sign would be.

At first, they noted empty spaces in between, but later they introduced a symbol to mark the empty ranks. However, at the right-hand of a number this was never written. So they never introduced a real full sign for null. The interpretation of the numeral null as the number zero was none of their concern. For that contribution we thank the Hindus.

Why exactly sixty was chosen as the base isn't clear so far. There exist divergent theories about this.

A first theory is based on the way they were supposed to measure time, in which the annual cycle was considered to consist of 360 days, and hence a full circle was divided into 360 parts (degrees). When measuring the time in a day by the shadow of a stick placed in the middle of a hexagonal field, divided into six equal sectors being equilateral triangles, one inevitably gets the number 60 (360 / 6 = 60).

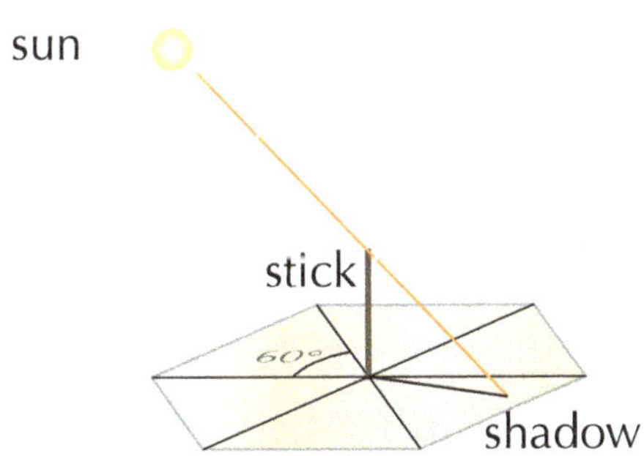

A second explanation for the number 360 has been found in a measure for long distances that was frequently used at that time in connection with the time needed to walk it. For instance, twelve of those distances in one day, while one distance was divided into thirty parts for practical reasons (e.g. many dividers). This also makes 12 x 30 = 360.

A third theory is based on the evolution of the script and the aim of bringing unity into initially different numerical systems, used as measures for quantities such as length, area and volume. Thereby they would have taken a number that was the least common multiple of 6, 10 and 12, namely 60.

Whatever may be true, the sexagesimal numerical system has proved to be so practical for angle and time measurements that we still use it for these quantities.

Examples

1 minute = 60 seconds,

1 hour = 60 minutes = 60 x 60 seconds = 3 600 seconds

3 h 13 m 29 s = (3 x 3 600 + 13 x 60 + 29) s = 11 609 s

1 degree = 1° = 60 minutes = 60′

1 minute = 1′ = 60 seconds = 60″

25° 13′ 22″ = (25 x 60 x 60 + 13 x 60 + 22)″ = 90 802″.

The Babylonians were fascinated by the constantly clear and star-filled sky above their heads. It's no surprise that in such an environment a tradition of special interest in celestial bodies has always existed. They are not only the fathers of astronomy, the pure observation of the stars with its tables and predictive calculations, but also of astrology, the more horoscope-like approach with the belief that the signs of the zodiac had an influence on personal destiny.

Their devotion to the number seven as a holy number and the arrangement of the week in seven days is ascribed to the fact that the Babylonians knew seven visible planets, among which also the sun and the moon were taken into account.

Traces of geometrical knowledge have also been found on the clay tablets. Typical of Babylonian geometry is its algebraic nature. Starting from problems concerning measurements and constructions they came to equations and sets of simultaneous equations. Equations of a higher degree were solved by means of calculations of roots.

Besides all this they were experienced accountants. Being at the crossroads of important caravan routes, they were well acquainted with all aspects of a busy barter trade: sales contracts, orders, invoices, loans (seemingly at exorbitant compound interest rates of 20 or 30 percent), etc.

Example

How long does a certain amount have to be put out at a compound annual interest rate of twenty percent in order to get double maturity value?

This problem leads to an exponential equation of the form:

$$\left(1+\frac{20}{100}\right)^{x}=2$$

This was solved by interpolation (inclusion between successive better approximations). For logarithms it was still 3000 years too early. The answer found was 3.8... years. Apparently some Babylonians got rich very quickly.

1.4 The decimal system of the Hindus

Already in prehistoric times the Hindus lived in the territory that is now called India. It is difficult to get a clear view on this old civilization, in which probably also mathematical science reached a high level. The first known written documents date from not earlier than the third century BCE and the most interesting mathematical documents from even later than 400 CE. The reason for this results, among others, from their tradition of keeping knowledge within the (higher) castes by means of principally oral communication. Also the scarcity of durable writing materials like parchment and the vulnerability of other materials like bamboo or bark certainly can be considered as significant factors.

It is not clear whether at a later date Indian mathematics was influenced by Egyptian, Babylonian or even Greek science. The street network, the irrigation canals and the structure of the buildings found in the 5000-year-old ruins of the excavated town Mohenjo Daro undeniably bear the signature of antique mathematics of an already advanced stage. Some results also indicate an indisputable originality: the use of the numeral null and the number zero; negative and irrational numbers as solutions for equations; the use of sinus tables in problems of trigonometry and astronomy.

Although differently situated in time, the Indian mathematics also unmistakably bear the characteristics of a practically oriented knowledge and do not show the Greek theoretical mind. That is why we discuss the contributions of the Hindus in this paragraph, and for the same reason also those of the Chinese and the Maya.

To the Hindus we owe our decimal numerical system and most of all the numeral null. That system was passed down to us by the Arabs, but it took a long time. In the eighth century a Sanskrit work on astronomy by an Indian scientist was translated into Arabic by al-Fazari. It explained the principles of Indian arithmetic. Also other Arab authors like Al-Khowarizmi started writing books about this. The original terms and symbols started to live their own life in the Arab world. Moreover, this happened in a different way in the eastern Bagdad and the western Andalusia. Through medieval Latin translations of these Arab works, the Indo-Arab numerals and the related methods of calculation spread across Europe.

The numerals as we write them now developed from the so-called *gobar digits* used in the western Arab world. 'Gobar' meant 'dust'. At that time the abacus (counting frame) consisted of nothing more than a sanded board in which the signs were written by means of a rod. The oldest Hindu version of the numeral null was probably nothing more than just an empty space. This is shown from the term 'sunya' (emptiness) that was used for it in Sanskrit. Later this empty space was filled up by a big dot or a small circle. The Arabs translated 'sunya' as 'sifr', which in turn was translated into Latin in two different ways: as 'cifra', as well as 'zeferium'.

From 'cifra' the French 'chiffre', the English 'cipher', the German 'Ziffer' and the Dutch 'cijfer' were derived. 'Zeferium' changed through 'zefiro' into the more generally known 'zero'.

The word 'cipher' was initially only used for null. In the book *Liber Abaci* (1202) by Leonardo Fibonacci (from Pisa) null was still considered as an outsider. The other nine numerals he called 'figura'. Our name 'null' comes exactly from the phrase: 'nulla figura' (no number).

Initially there was strife between the addicts of the abacus (the abacists) and the protagonists of the new arithmetic method with Hindu-Arab numerals (the algorists). In the middle of the sixteenth century, there was general agreement that the decimal positional system was the more practical.

Lady Aritmeticea
watching a competition between Boethius (on the left) and Pythagoras (on the right, with an abacus). The winner is clearly shown by the direction the lady is looking.
Representation with an allegorical female figure carrying two books and the banner *Typus arithmeticae*. (Gregor Reisch, *Margarita philosophica*, 1508)

1.5 Chinese calculation sticks

It is beyond doubt that Chinese mathematics goes back to a centuries old tradition. Several archeological finds provide the evidence: 7000-year-old pottery with vertical stripes; turtle shells and inscriptions in oracle bone script from the time of the Shang dynasty; bamboo or bone calculation sticks from the time of the Han dynasty; etc.

Because of the Oriental tradition of knowledge and values continuing by oral tradition over the generations certain results and customs cannot be dated precisely. So, calculating with small bamboo sticks (chou suan) was probably already practised many centuries BCE. Using these sticks the Chinese set up number configurations in the decimal positional system. The units were arranged vertically, but starting from six a horizontal stick represented five. The tens were arranged horizontally, the hundreds again vertically, etc. An empty place was left to express the null. This method is explained in the books *Sun Zi suanjing* (The Mathematical Classic by Sun Zi, about fifth century CE) and the *Xiahou Yang suanjing* (The Mathematical Classic by Xiahou Yang, about eighth century CE). In these one can respectively read the following formation rules:

> 'Units are vertical, tens horizontal, hundreds stand, thousands lie...'
>
> 'Once more than six the five is crosswise, six should not pile up, but five may not stand alone,...'

Conforming to these rules, the digits 'one' to 'nine' are represented by the following configurations, respectively in vertical and horizontal versions:

	1	2	3	4	5	6	7	8	9
vertically	𝍠	𝍡	𝍢	𝍣	𝍤	𝍥	𝍦	𝍧	𝍨
horizontally	𝍩	𝍪	𝍫	𝍬	𝍭	𝍮	𝍯	𝍰	𝍱
	10	20	30	40	50	60	70	80	90

Using these, numbers were represented as follows:

𝍧 𝍬 𝍥 $= 8 \cdot 100 + 4 \cdot 10 + 6 = 846$

𝍪   𝍮 𝍨 $= 2 \cdot 1000 + 0 \cdot 100 + 6 \cdot 10 + 9 = 2069$

In one of the oldest surviving Chinese textbooks on mathematics, the *Jiuzhāng Suànshù* (*The Nine Chapters on the Mathematical Art*, dating from the Han-dynasty (202 BCE–220 CE), we encounter problems interpreted in the form of equations and of sets of simultaneous equations.

Among other things, the following set of simultaneous equations:

$$\begin{cases} 3x + 2y + z = 39 \\ 2x + 3y + z = 34 \\ x + 2y + 3x = 26 \end{cases}$$

is solved via a modern looking schematic form, which we express nowadays by the terms 'matrix' and 'matrix transformation'. Also negative numbers were accepted as being solutions.

The Chinese found in mathematics not only an instrument to solve practical problems, but also a field of magical symbolism and religious references. The construction of altars, of which the size, the form and the placement were important for the effectiveness of ceremonies, demanded geometrical techniques in which the Pythagorean theorem was also used.

In the famous *Book of Changes*, the *I-Ching*, one finds a remarkable magic square, called *Lo-Shu*:

4	9	2
3	5	7
8	1	6

It consists of the first nine numbers. All sums of the numbers on every row and every column, as well as on the diagonals, equal 15. The Lo-Shu is the oldest example known of a magic square and was probably used in fortune-telling. According to a legend, the emperor Yu noticed the scheme for the first time about 2200 BCE. It was in a pictographic form on the shell of a turtle on the banks of the Yellow River:

Pictographic representations of the numbers in the magic square called **Lo - Shu**. In the middle, a turtle with its shell and the configuration as emperor Yu would have perceived it. On the right, a modern artistic interpretation.

1.6 The calendars of the Maya

Undoubtedly the ancient Indian cultures of South America, in the pre-Colombian era, developed independently from those on the other continents. Several million inhabitants of Guatemala, south Mexico and Belize still speak one of the thirty different Maya languages. It is remarkable that all these languages have similar numerical systems. They are related to the ancient calendars of the Maya, the Zapotec and the Olmec.

The Maya were fascinated by the cycles of the astronomical systems because they believed that through these they could foretell the future of their religious leaders and kings. The complex calculations, needed to make the different calendars match, acted as a springboard to the domain of abstract arithmetic that would go far beyond the original goal, just as it happened in Egyptian, Babylonian, Hindu and Chinese civilizations.

Coloured **Maya calendar**, in which the cycles of the sun and the moon are brought together.

These barefoot Indians probably got their inspiration not only from their fingers but also from their toes when choosing twenty as the base of their numerical system. In the Quiché language the word for twenty, namely 'huvinak', literally means 'the whole person'. The usual solar year, having 365 days, was divided into eighteen periods of twenty days and five extra days. That is why the higher rank following on 20 is not 20 x 20 but 18 x 20. In the higher ranks the factor 18 is carried along in a product with powers of twenty, like this: 18×20^2, 18×20^3, etc.

On the other hand, the holy calendar of the fortune-tellers was based on a period of 13 x 20 = 260 days. That is the average time a future mother carries a baby before birth. This period was used to foretell the future and the destiny of priests and kings.

The Maya combined both calendars in what they called 'the great cycle'. So they ended up with a period being the least common multiple of 365 and 260, namely 18980. A great cycle equalled 18980 : 365 = 52 solar years and 18980 : 260 = 73 holy years.

For representing numbers, the Maya used hieroglyphic-like symbols consisting of dots (for one) and bars (for five). They also had a special symbol for null, resembling an eye or a shell.

Examples

 $= 5 + 5 + 1 + 1 = 12$

 $= 6 \cdot 20 + 10 = 130$

 $= 3 \cdot (18 \cdot 20) + 0 \cdot 20 + 6 = 1086$

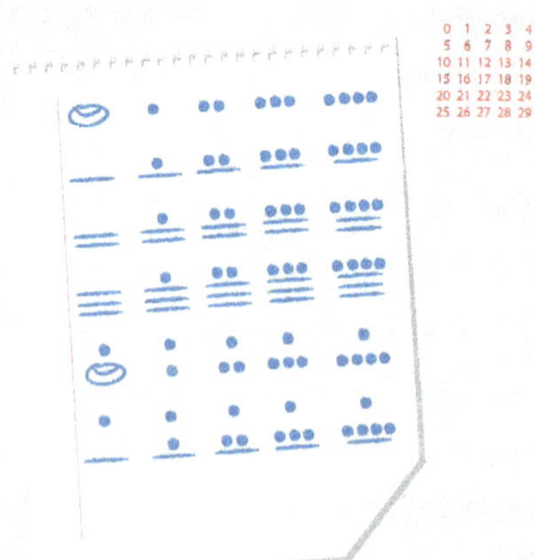

Summary

The examples mentioned show that in all civilizations similar practical problems are the cause for developing numerical systems and constructing all kinds of measuring devices. The inspiration comes from the observation of the immediate environment: your own body with fingers and toes, the sky with its regular or irregular phenomena, etc. The arithmetic and geometry that were developed served primarily the efficiency of agrarian and economic activities. They also supported constantly increasing barter trade and the complex administration of the growing villages and towns. Real industrialization is still far away: at this stage of history, knowledge is too limited and production methods are too primitive.

The perception of reality in these agrarian bureaucratic communities is in most cases still ruled by myths and religions. An overall vision of the world and the cosmos is lacking or is of touching naivety. The earth is considered as being a flat disc and the sky as a dimly lit cupola, which protects as well as limits existence.

Naturally, it is impossible to make a complete inventory or a final judgement on the knowledge of old civilizations about which rather few sources have been found. Most probably a lot got lost in the course of time or has not yet been found. However, it shows that a certain degree of abstract mathematical knowledge always goes hand in hand with its concrete applications, and both unmistakably characterize the level of civilization. The brilliant discovery of something like a symbol for the null and the use of it in a positional numeral system determines the supremacy of one arithmetic over another. The different fundamental concepts by which reality is approached, can give a totally different character to civilizations. Already, the different Hindu and the Chinese interpretations of the essence of the universe as respectively 'Brahma' and 'Tao' have led them to different attitudes to life and have determined the course of their culture. 'Brahma', which fills the whole cosmos and in which everything finds its origin but also to which it returns, and is thus perpetually invariable, leads in itself to an attitude of resignation, submission and rather inactive contemplation. On the other hand, the 'Tao' as being that 'which is not yet completed' and that remains open to formation and creativity, necessitates active collaboration.

However meritorious and remarkable, in none of these formerly cited ancient cultures can one find a trace of something like reasoning or proofs. To take care of this kind of rational approach of coherent knowledge, there would be another people and another cultural context.

2 The revolution of the Greek theoretical mind

We will probably never be able to explain thoroughly why exactly at that certain place and at that certain time some development occurred in humanity's vision of the world that proved to be unique in our history. However, some factors related to the geographical situation, the social circumstances and political structures can be put forward as being favorable for the development of a universal culture that became the synthesis and the exponent of all that had already been realized in several old civilizations.

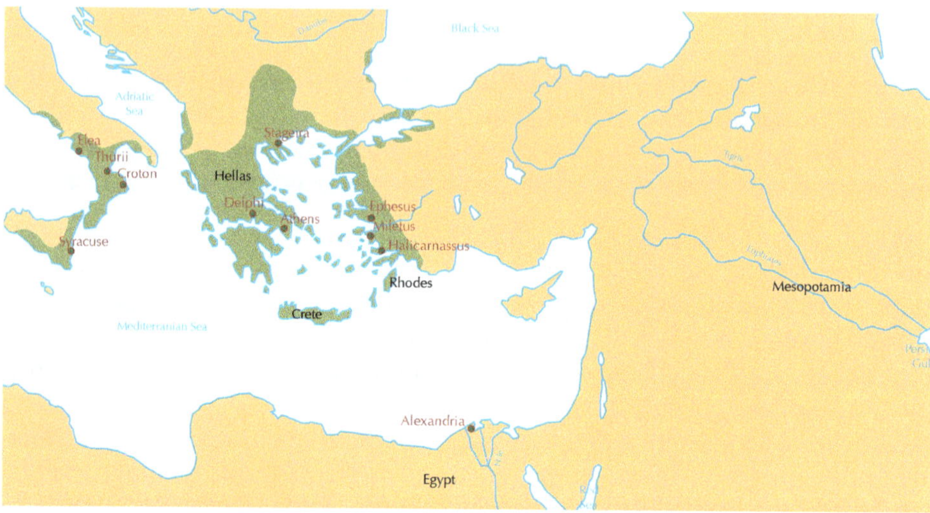

The great Hellenistic realm, spread out around the Mediterranean Sea, had points of contact with three continents: Europe, Asia and Africa. In these, several cities were no longer just administrative centres of an irrigation-economy along large rivers, as it was earlier in Egypt and Mesopotamia, but these cities were rather international trade centres, connected by a network of travel routes. In these independent city-states the old-style feudal country lords had to cede power to the new class of learned merchant-travellers and specialised artisans.

Thus the Greek 'polis' arose, with its particular types of administration such as democracy, oligarchy or tyranny. The slaves did the material work and the free citizens shared in the power and the wealth. In these circumstances a small group of privileged people could afford to reflect on the laws of nature and existential mysteries. This is how free speculative thinking was born, averse to any practical usefulness. In the time span of hardly a hundred years, dialectical philosophy, as well as formal logic and deductive mathematics came into being, and this within a community of only a few thousand people. To these achievements are linked the names of great minds, to which reference will be made until the end of times: Thales, Pythagoras, Socrates, Plato, Aristotle, Euclid, Archimedes, and many others. Since then, western civilization could never again put aside this rational attitude, to which it owes among other things its scientific successes.

How could this change from a magical mythological belief into the fearless questioning of the 'why' of things have taken place? We outline the answer to this question by briefly looking at some remarkable facts and persons.

2.1 Thales of Miletus

Thales (624–546 BCE) lived in Miletus on the coast of Asia Minor (it is nowadays an ancient Greek site in Turkey). He is considered one of the Seven Sages of antiquity and he was at the same time mathematician, astronomer, engineer, businessman, philosopher and statesman. He travelled widely, including to Mesopotamia and Egypt. In Egypt, he astonished the priest-scientists by measuring the exact height of the pyramids without any instrument. For this, he only used the length of his body and the property of equal proportions in similar triangles. Firstly, he laid himself on the ground to measure his body length, then standing up he waited until his shadow had the same length as he had measured. He decided that at that moment also the shadow of the pyramid equalled its height. This already shows a different attitude concerning the use of mathematics, i.e. starting from a general property he derived a particular application, instead of the other way round.

Thales is considered to be the founder of the Ionian school of nature philosophers, who tried to explain that the world originated from one primal substance. Thales himself considered water to be that primal substance. This attitude of explaining complicated phenomena starting from simple basic principles also left its mark on mathematics. In this way, efforts were made to derive general properties from 'evident' basic principles by a chain of logical reasoning. Also the first proofs of the following five propositions of geometry are attributed to Thales :

'any diameter bisects the circle'

'the opposite angles, created by intersecting lines, are equal'

'two triangles that have one equal side and two equal angles are congruent'

'the base angles of an isosceles triangle are equal'

'the inscribed angles pointing at a diameter in a circle are right angles'

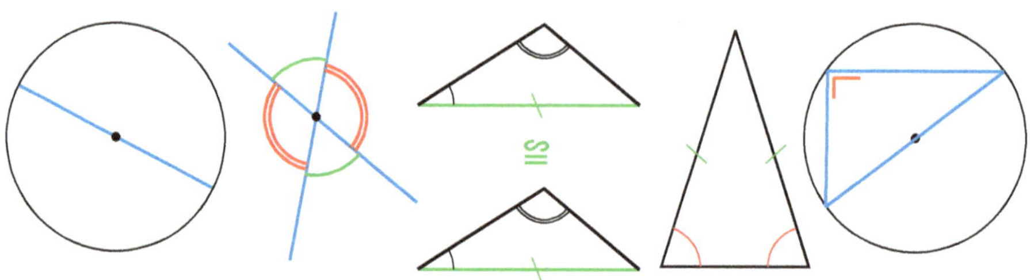

Some of these properties were already known in earlier ancient cultures, but only by measurement or observation of particular cases. Now a proof was given for the general case, regardless of the dimension of the figures observed, however big or small. This was like the starting shot for the deductive construction of geometry starting from basic principles, generally based on simple constructions of lines and circles.

There are some amusing anecdotes about Thales, even if they may not be historically correct. He seems to have been a shrewd olive merchant who, in times of abundance, bought all the presses so as to have a monopoly. There is another story about a stubborn donkey that was loaded with heavy sacks of salt and used to run in a river to lighten its burden. It is said that, one fine day, Thales knocked this bad habit out of the donkey by replacing the salt with sponges!

2.2 The school of Pythagoras at Croton

Pythagoras (about 580–500 BCE) was born on the island of Samos. After travelling through Egypt and Mesopotamia he settled in Croton, on the sole of the Italian 'boot'. Here he founded the famous school that bears his name. Probably inspired by the closed Egyptian priest caste and the Babylonian Magicians, this school was also a rather secret society, with strange rites and rules, to which a candidate was only allowed under strict conditions.The Pythagoreans believed in transmigration (metempsychosis, or the reincarnation of the soul) and had to respect several vegetarian regulations, such as the prohibition of eating lentils. They also had to observe a rule of silence. Nothing of what had been discovered could be taken out and everything had to be attributed to the master. The authority of the master was absolute and his word beyond all doubt ('autos efa' = 'ipse dixit' = 'he himself said').

The main characteristic of the Pythagorean school was its mystical devotion to the integer numbers as the basis for the whole cosmos. Pythagoras is said to have discovered the numerical relations of musical harmony when he was walking past the workshop of a coppersmith who was beating a piece of metal with hammers of various weight. He recognized the octave, generated by hammers of six and twelve pounds in a relative proportion of 1/2, the quint by hammers of eight and twelve pounds in the proportion 2/3 and the quarter by hammers of nine and twelve pounds in the proportion 3/4. He was fascinated by the fact that these proportions 1/2, 2/3 and 3/4, were exactly made up of the first four numbers 1, 2, 3 and 4 of the pyramidal tetractys figure, in which furthermore the number 10 is retrieved as its sum. Both the tetractys and the number 10 were important mythical symbols of the school.

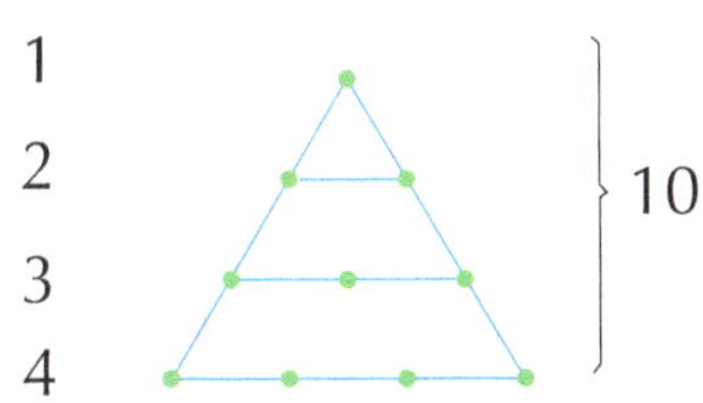

At home he successfully repeated these experiments. Firstly with intestine strings of equal length, on which he suspended weights in the above-mentioned proportions (the Pythacora). Then on a monochord of his own design, i.e. a tightened intestine string and a movable bridge by which the length of the string could be divided into the mentioned proportions.

A string that produces a C will, by halving in this way, produce the same tone again but one octave higher.
Between these tones , the other tones D, E, F, G, A and B will be produced by bringing the original length of the string to the respective proportions:

$$\frac{9}{16},\ \frac{5}{8},\ \frac{2}{3},\ \frac{3}{4},\ \frac{5}{6}\ \text{and}\ \frac{15}{16}.$$

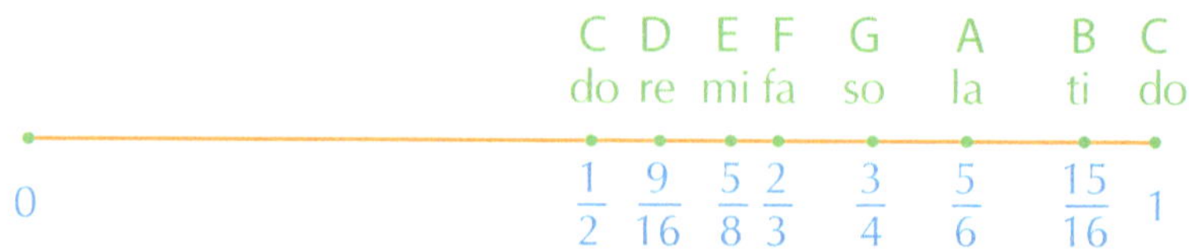

The discovery that there is a simple arithmetic relation between the abstract world of music and these numbers made the Pythagoreans believe in a general harmony in the cosmos, of which the integer numbers and mathematics formed the basis. They even went so far as to believe that the planets in their orbits in the universe produced a 'heavenly music', determined by the proportions of their orbits and their speed.

They held practical plain reckoning in low esteem, calling it 'logistic' and leaving it to tradesmen and artisans. They considered the natural numbers and their relations, called 'aritmetica', as an independent theory that was to be studied solely for its beauty. So they defined all kinds of remarkable numbers, such as the 'perfect numbers', which had the property of equalling the sum of their proper divisors, such as $6 = 1 + 2 + 3$ and $28 = 1 + 2 + 4 + 7 + 14$.

They were also interested in pairs of 'amicable numbers' that had the property of equalling the sum of each other's proper divisors, like 220 and 284:

$$220 = 1 + 2 + 4 + 71 + 142 \quad \text{and}$$
$$284 = 1 + 2 + 4 + 5 + 10 + 11 + 20 + 22 + 44 + 55 + 110.$$

Even now, twenty-six centuries later, not all problems concerning the general form of these numbers have been solved. By means of computers some very big examples have been detected. Also, it is known that every even 'perfect number' is of the form $2^{n-1}(2^n - 1)$ provided that $2^n - 1$ is prime. For $n = 2$ this formula delivers 6 and for $n = 3$ we get 28; $n = 4$ is not a valid value since $2^4 - 1 = 15$, and 15 is not prime; on the other hand $n = 4423$ is valid and the corresponding 'perfect number' consists of 2663 digits! But it still is unknown whether odd 'perfect numbers' exist.

The Pythagoreans also constructed all kinds of sequences of figurative numbers, such as triangle, square and pentagonal numbers and found their general properties in a geometrical way.

Examples:

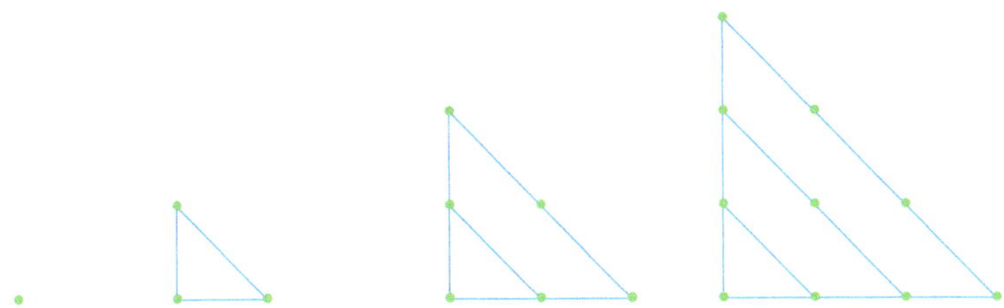

1 3 6 10

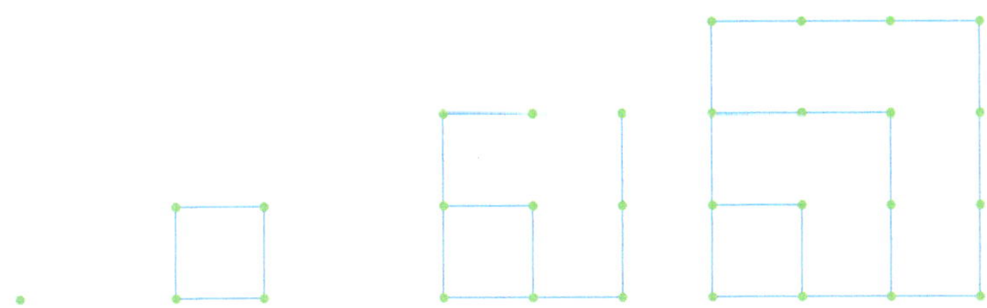

1 4 9 16

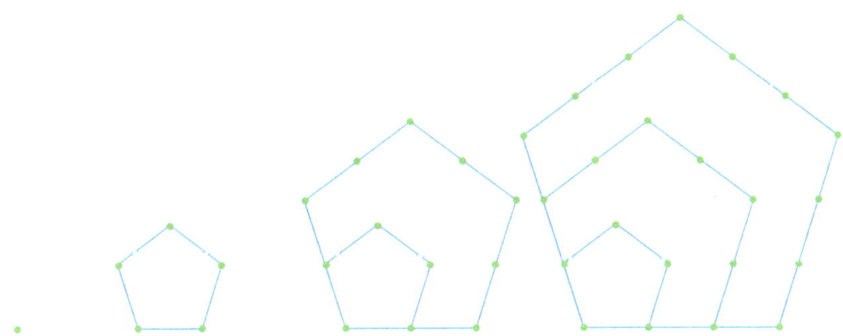

1 5 12 22

The sum of the first n natural numbers apparently equals the n-th triangle number, e.g. 1 + 2 + 3 + 4 = 10 is the fourth triangle number.

The sum of two successive triangle numbers always makes a square number and reciprocally, every square number is the sum of two successive triangle numbers. This is shown from the following construction:

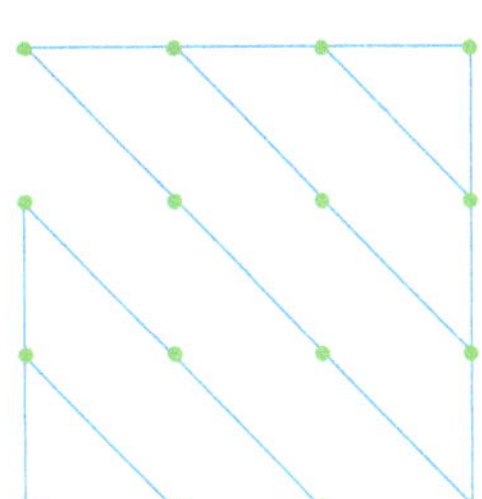

The third triangle number is 6,

the fourth triangle number is 10,

6 + 10 = 16

16 is the fourth square number.

Obviously, this school also paid a lot of attention to pure geometry, separate from number theory. For instance, the first proof ever of the well-known Pythagorean theorem, i.e. in a right-angled triangle the area of the square on the hypotenuse equals the sum of the areas of the squares of the catheti, is attributed to Pythagoras.

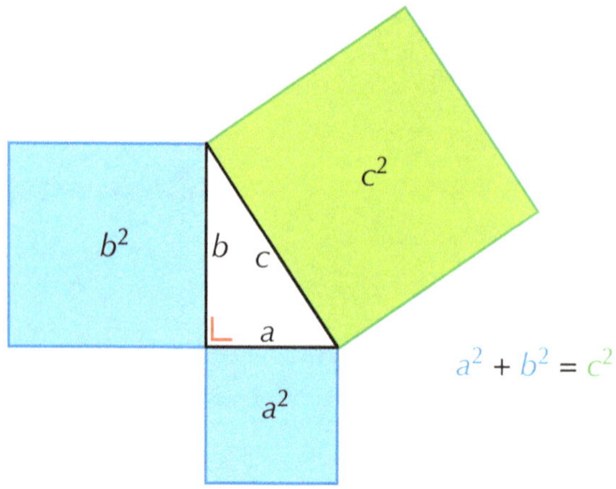

That is why this theorem, even though it was already known by the Egyptians, the Babylonians and the Hindus, nevertheless carries the name of Pythagoras. Many beautiful proofs have been found since then. A simple proof that could have been given by Pythagoras himself is based on the comparison of the next two figures:

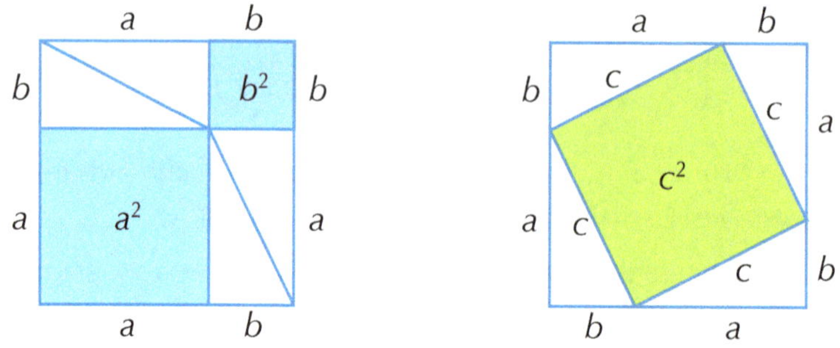

The area of the big square with side $a + b$ can be written as a sum as follows: firstly as the sum of the areas of the squares with sides respectively a and b and four right-angled triangles with catheti a and b; secondly as the sum of a square with side c and the same four right-angled triangles with catheti a and b. From this it follows that:

$$a^2 + b^2 + 4(ab \div 2) = c^2 + 4(ab \div 2)$$

$$a^2 + b^2 = c^2$$

It is a cynical caprice of the history of mathematics that exactly this theorem, the jewel in the crown of the Pythagoreans, burdened them with the astonishing conclusion that there exist line segments of which the length, referring to an arbitrary unit, cannot be expressed as a proportion of integer numbers. For instance, according to the Pythagorean theorem, the length of the diagonal of a square with side 1 has to be a number c of which the square equals $1^2 + 1^2 = 2$.

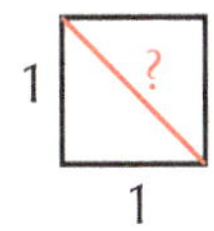

Since $c^2 = 2$ it is impossible that c equals a ratio of integer numbers! And they themselves had found an ingenious geometrical proof for this. Aristotle, however, gave a simple indirect proof in the algebraic way.

It goes like this:

let $c = \frac{a}{b}$ be a reduced fraction.

Then a and b are coprime integers (i.e. a and b have no common divisors)

and since $c^2 = \frac{a^2}{b^2} = 2$, we get $a^2 = 2b^2$, which means that a^2 is even, and so must be a.

In other words, a is a multiple of 2; let $a = 2d$,

this results in $(2d)^2 = 2b^2$ and hence in $4d^2 = 2b^2$, which makes $2d^2 = b^2$.

So also b is a multiple of 2, and thus 2 is a common divisor of a and b. But this is a contradiction of the fact that a and b are coprime.

So, the assumption of c being a ratio of two integers Is wrong.

The insight that not all lengths are rational, i.e. that there exist irrational numbers like $c = \sqrt{2}$, did not only undermine the Pythagorean belief in the integer numbers being the base for the universe, but it also subverted the whole of their geometrical measure theory, which was based precisely on the axiom that all lengths were commensurable.

It certainly can't be called a fine feature of the Pythagoreans that they initially tried to keep this 'logical scandal' a secret. According to a legend, Hippasus, who is said to have revealed the secret, was thrown into the sea from a boat. It is certain that because of paradoxes of this kind, the Greeks preferred to keep geometry separate from the apparently problematic number theory. For centuries, this has led to a purely synthetic treatment of geometry.

Later Eudoxus (about 408–355 BCE) would give a beautiful solution to the problem of irrational quantities, by making a new definition of equal proportions of quantities.

In this way, the deductive construction of geometry could regain its consistency. But it would still take till the nineteenth century before a number theoretical definition of irrational numbers by means of rational numbers would be finalized.

With their immaterial vision of reality, the Pythagoreans introduced a pure theoretical attitude into the entire Greek thinking, in which insight was the highest goal and mathematics was raised to the level of a religion. Their view that precisely the mathematical entities made up true reality, would inspire Plato to the philosophical notion of 'idealism'. It is no surprise that they saw the study of arithmetic, geometry, astronomy and music as a whole. This *quadrivium* together with *the trivium (logic, rhetoric and grammar)*, as they were called in the Middle Ages, would make up the most important study program of the later universities.

2.3 The paradoxes of Zeno of Elea

Not so far from Croton, in the sole of the Italian 'boot', a philosophical school was flourishing in Elea, of which the most famous members were undoubtedly Parmenides and Zeno. Parmenides (about 515–445 BCE) believed that all existence is essentially unchangeable (panta stei = all remains), in contradiction to his opponent Heracleitos, who stated that all is perpetually changing and moving (panta rei = all flows). To prove his viewpoint, Zeno (about 490–430 BCE), a disciple of Parmenides, thought of some paradoxes that should prove that motion is an illusion. These paradoxes deal with subtle mathematical concepts, such as infinite series and the continuum. Everybody certainly knows the paradox of Achilles who apparently could never overtake the tortoise. Zeno was arguing: in the same time that Achilles covers the tortoise's initial lead, the tortoise has built up a new lead, however smaller. This keeps on repeating, so that the tortoise remains in the lead in all the following phases!

Similar is the fate of the athlete who has to run a lap in a stadium. Firstly he will run half of it, next half of the remaining part, and so on. He will never reach the finishing line.

Obviously, just like you and me, the Greek contemporaries of Zeno found these allegations pure nonsense. However, they could not find weaknesses in his argumentation. It deeply bothered them, just as it still bothered many mathematicians and philosophers for many centuries. It took till the nineteenth century before people got a clear view of convergent series. These are sums of an infinite number of terms that, however, determine a definite number, such as

$$\frac{1}{2}+\frac{1}{4}+\frac{1}{8}+\frac{1}{16}+\ldots=1.$$

It took as long as the other problem of getting a clear view of the continuum, which cannot be identified with the countably infinite, like the sequence of natural numbers 1, 2, 3, ... Continuous time cannot be subdivided into a sequence of discrete moments. Science at that time had not yet arrived at that insight.

Also the other paradoxes of Zeno — known as 'the arrow' (that at any moment of its flight is stationary somewhere and hence doesn't fly at all) and the 'dichotomy' (so as to move from point *A* to point *B*, one has to pass through the middle point *C*, but first through the point *D* halfway to *C*, etc., which means that one doesn't even get started) — all deal with essentially the same problem.

These paradoxes, just like so many others of later times, have the benefit that they challenge thinking. Although they were meant to prove that movement is an illusion and that all is unchangeable, they at least provoked movement in thinking and changes in science.

2.4 The five Platonic solids

Plato (429–348 BCE), the 'broad shouldered', the meaning of his name, is mainly known as the father of philosophy. Being a pupil of Socrates he was an addict of his method of dialogue and maieutics (midwifery). By this method, pupils were brought to the true insight by critical reflection on judiciously put questions. In his famous *Dialogues* he described this 'dragging' method of his master in detail. Plato is also known as the founder of the Academy, a philosophical school in Athens, where his doctrine of 'idealism' was taught. Opposing the relativism and the subjectivism of the sophists, Plato was looking for a reliable way of reaching absolutely objective knowledge, which consisted of 'ideas', i.e. purely immaterial concepts that existed a priori and stood above all real things.

The concepts 'justice', 'horse', 'circle' etc... already existed even before the first just deed, the first neighing horse or the first round shape were noticed. The concrete objects only refer to pure 'ideas' and they are just means to recall knowledge lost at the moment of birth. According to Plato, the 'Good' as the highest human pursuit, arises by itself from true insight, which can only be achieved by the critical method of mathematics.

Although he was not a mathematician in the pure sense, Plato held the mathematical discipline in high esteem and, inspired by the Pythagoreans, he gave it a particular place in his thinking and in his attitude to life. Above the entrance of the Academy there was the inscription: 'Let nobody who is ignorant of geometry enter here'.

Several important mathematicians were associated with the school, such as Archytas, Theaetetus and Eudoxus.

In his dialogue *Timaeus*, Plato gives a description of the five regular polyhedra explaining them as symbols of the origin of the world and the essence of the cosmos. A regular polyhedron is a three-dimensional figure of which all faces are congruent regular polygons and of which all dihedral angles (inner angle between two faces) are equal. Moreover, all of them can be inscribed in a sphere, i.e. all their vertices are on one sphere. The Greek succeeded in proving that there are only five regular polyhedra (see below in 2.6).

	number of faces	number of vertices	number of edges
tetrahedron	4	4	6
hexahedron (cube)	6	8	12
octahedron	8	6	12
dodecahedron	12	20	30
icosahedron	20	12	30

These solids have some particular properties: there is a kind of duality between the hexahedron and the octahedron and between the dodecahedron and the icosahedron. The tetrahedron is dual to itself. The midpoints of the faces of one polyhedron set up vertices for a dual polyhedron.

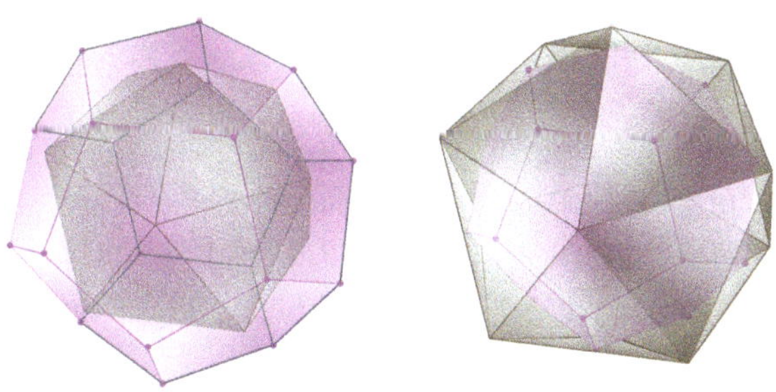

The faces of a regular dodecahedron are regular pentagons, those of the cube are squares, of all the other regular polyhedra they are equilateral triangles.

There is a fixed relation between the number of faces, vertices and edges, expressed by *Euler's formula*:

$$\text{number of faces} + \text{number of vertices} = \text{number of edges} + 2$$

This formula also applies to all irregular convex polyhedra, i.e. solids with flat faces but with no cavities. Even an everyday thing like a football must recognize that it can't avoid this formula.

A new type of football that was presented as an alternative to the usual 'team spirit' model in the 2006 World Cup, consists of twenty hexagons, twelve pentagons and sixty triangles and is closer to a sphere than the normal one with twelve pentagons and twenty hexagons. The latter is a semi-regular polyhedron that appears when truncating the twelve vertices of an icosahedron, so that the new edges have an equal length of one third of the edges of the original icosahedron. Because in a vertex of an icosahedron five edges converge, truncating makes a pentagon of every original vertex. The original triangular faces become hexagons.

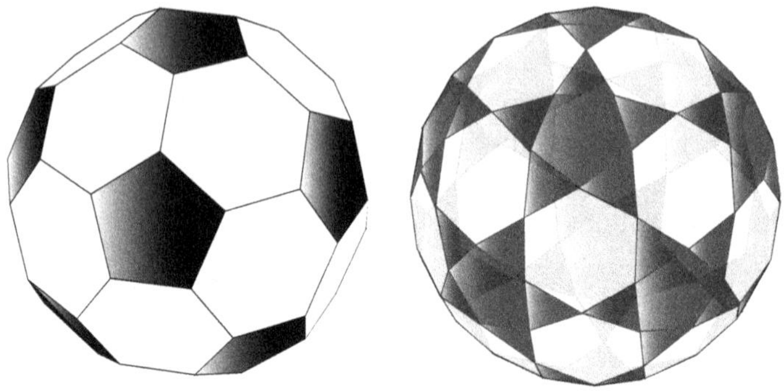

usual type alternative type

The usual type of football has 32 faces, 60 vertices and 90 edges, to which Euler's formula applies: $32 + 60 = 90 + 2$.
The alternative type has 92 faces, 90 vertices and 180 edges, to which Euler's formula also applies: $92 + 90 = 180 + 2$.

Some crystals have the shape of a regular tetra-, hexa- or octahedron. Also in the skeletons of microscopic marine animals called radiolara, the shapes of a regular dodecahedron and icosahedron have been found. Also other shapes combining hexagons and pentagons (like in the football) appear. An amusing anecdote tells about a biologist informing his mathematician-friend that he had found a skeleton that only consisted of hexagons. The mathematician immediately figured that this was impossible because of Euler's formula, to which the biologist answered that God is almighty and doesn't have to lose out to some mathematical theorems. But later he had to admit that not all the faces of his specimen were hexagons!

The regular tetrahedron and dodecahedron, like the cube as a regular hexahedron, were already known by the Pythagoreans. The regular octahedron and icosahedron seem to have been discovered by Theaetetus. However, these five solids are nowadays usually named 'Platonic solids'. Plato associated these five solids with the four basic elements, which according to Empedocles were: fire (tetrahedron), earth (cube), air (octahedron) and water (icosahedron). The ethereal cosmos itself was represented by the dodecahedron.

In the book *Mysterium Cosmographicum* by Johannes Kepler (1596) we re-encounter these representations.

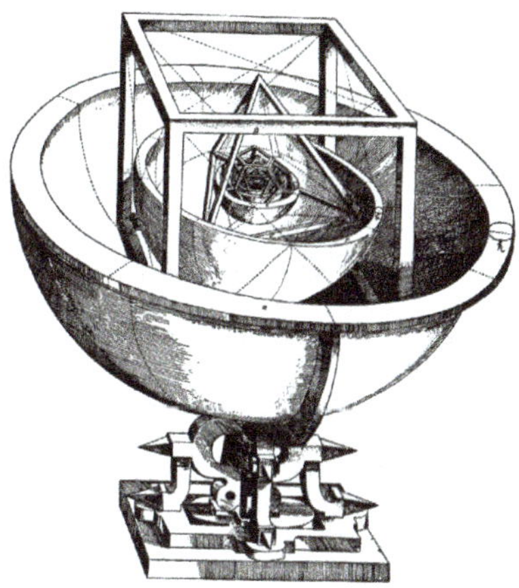

In his *Mysterium Cosmographicum* Johannes Kepler expresses Plato's opinion that the world has an harmonic basis in the idealized numerical proportions and in the shapes of the regular solids. Therefore he designed the model shown on the left in which the distances of the planets from the sun are represented by a nesting of Platonic solids.

It also seems that Plato introduced the strict requirement of performing all geometric constructions by compass and ruler only, excluding all other instruments. This led to the famous 'classical' problems, i.e. the square of the circle, the tripartition of an angle and the doubling of the cube (i.e. the problem of the Apollo altar in Delos).

The square of the circle is the problem of constructing, by compass and ruler, a line segment of which the length equals the side of a square having the same area as the given circle.

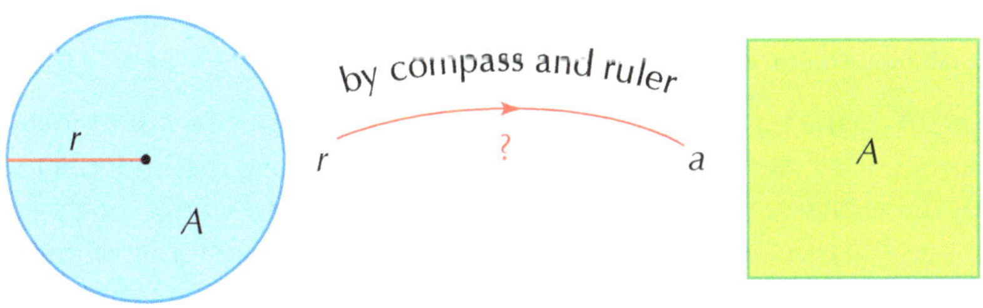

The tripartition of an angle is the problem of constructing two half lines that divide an angle into three equal parts, just like a bisector divides an angle into two equal parts.

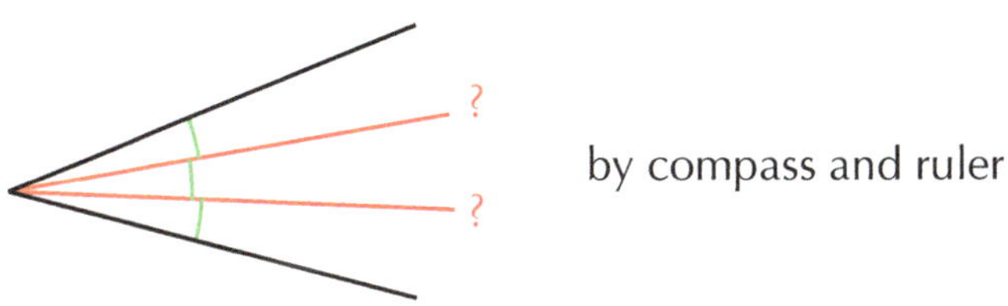

The doubling of the cube means the construction of the edge of a cube that has precisely double the volume of the original cube. The altar of Apollo at Delos had the shape of a cube. During the plague they hoped to propitiate Apollo by doubling the volume of his altar.

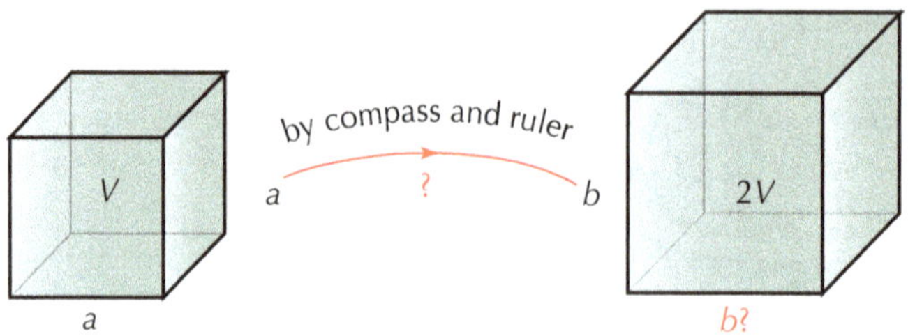

Only much later, in the nineteenth century, did these three problems prove to be insoluble. They all deal with the possibility of constructing, by compass and ruler, line segments of which the length is one of the following particular numbers:

$\sqrt{\pi}$	(square of the circle),
cosinus 20°	(tripartition of an angle of 60°),
$\sqrt[3]{2}$	(doubling of the cube).

In analytic geometry, a line is interpreted algebraically by an equation of the first degree and a circle by an equation of the second degree. A sequential construction by compass and ruler can hence be interpreted as a solution of a set of simultaneous equations of only the first and second degree. Because $\sqrt{\pi}$, cos20° and $\sqrt[3]{2}$ are not solutions to such sets of simultaneous equations, it is clear now that these problems are insoluble. No wonder that 'squaring the circle' has become a metaphor for a 'mission impossible'.

Plato has set up the basis for the entire Western philosophy by creating the image of an eternal, idealistic and objective reality, rising above the imperfect material world of deceptive phenomena. His belief that the knowledge of these higher realities can only be acquired by pure spiritual contemplation and thinking has certainly, not only within Greek culture but for a long time afterwards as well, pushed aside practical applications of this knowledge, which were considered not interesting and undesirable. Even the brilliant contemporary Archimedes was ashamed of his experimental methods, in spite of them leading to excellent results. He always tried to give his discoveries a formal appearance, rather than reporting on the 'method' he actually practised. We will come to this later.

Moral standards, based on the coincidence of the truth, the aesthetic and the good, as stated by Plato, became the strongest supports of Christian theology and philosophy in the Middle Ages.

2.5 Aristotelian logic

Aristotle (384–322 BCE) originally came from Stageira, an Ionic colony in Macedonia, hence his surname 'The Stageirean'. He was a pupil of Plato and the personal teacher of Alexander the Great. He is principally known as the father of (formal) logic. Obsessed by the urge to explain the coherence of natural and biological phenomena, he set up a whole system of premises and consequences. He considered reason to be essential for gaining an insight into these causal connections. He presumed that there exists a logical order in the world that can be discovered by the active intellect.

In contrast to his teacher Plato, Aristotle was convinced that the universal concepts were to be found in (and not above) sensory perceptions. The human mind is structured in such way that it is predestinated to discover them. In this way Plato's dualistic world of 'ideas' and 'material reality' was unified. For acquiring the general concepts and truth, experience was needed. By induction and abstraction universal premises were found, from which new truths could be derived by deduction. In his *Organon*, he set up a whole arsenal of structural rules for reasoning that were applied over more than 23 centuries, and this without any criticism or logical suspicion. Only in the second half of the nineteenth century did the contributions of mathematicians complete and finish this ingenious logical construction.

One of those famous argument forms of Aristotle is the 'syllogism'.

Example:

All humans are mortal	(major or first premise)
Socrates is a human	(minor or second premise)
Therefore Socrates is mortal	(conclusion)

Observe that this syllogism is a pure formal scheme, in which the conclusion by no means depends on the content of the terms involved. This makes it universal. Of course, the use of the common middle term (in this example 'human') is essential to connect the other terms in the propositions. Thus the general scheme has the following form:

all a are b

c is an a

thus c is b

Aristotle described the conditions under which such a syllogism is correct and he also discusses several erroneous uses of it, which he called sophisms. These conditions deal with the fact of the propositions being affirmative or negative, their being true or false, general or particular, and those cases in which the middle term functions as the subject or a predicate. This results in 64 forms, among which Aristotle distinguished the valid and invalid ones.

Note that the essence of its deductive nature is that the conclusion always passes from a universal to a particular proposition. The syllogism has always remained the prototype of analytical reasoning. Still today its different forms are taught on some courses on logic. In modern mathematical and symbolic logic, however, the syllogism is transformed into one of the many formal argument forms, which are called tautologies.

Also Aristotle founded his own school, which was called the Lyceum, after the nearby temple of Apollo Lycius in Athens. The building had a garden and a gallery (peripatos). Aristotle used to teach while walking, which is why his method of teaching is called 'peripatetic' teaching. Besides being a philosopher and a logician, Aristotle was also a very systematic physicist and biologist who was obsessed by descriptive classification and coherence. These were rather based on mental constructions and presumptions than on real experiments. Thus he was a child of his time with an aversion to material handling and contact with physical objects. It is as if he kept his hands clean behind his back when observing the world. In this way it was inevitable that purely speculative theories were proposed. Due to the moral authority of his true genius they initially remained uncontested. Later, the true empiricists of the Renaissance would reveal their contradictions.

According to Aristotle the planets moved in circles around the earth, which itself rested in the middle of a spherical universe due to its own heaviness. This had to be so, since he considered the sphere and the circle as perfect shapes. Although Aristarchus (about 310–230 BCE) had already noticed eighteen centuries before Copernicus that the orbits of the planets, as seen from the earth, were very irregular and put forward the idea that probably the planets circled around the sun, Ptolemy succeeded in saving the view of Aristotle by the elaborate invention of 'epicycles'. Epicycles are curves described by a point on a smaller circle of which the centre is on a greater circle (called deferent), while both circles turn evenly. This way the apparently fanciful motion of the planets could still be explained by a theoretical model based on circles.

Ptolemy's geocentric world view, written down in his *Great Treatise*, which became known to us with its Arabic translation called *Almagest*, survived and lasted for almost fourteen centuries. Only in the sixteenth century would Copernicus replace Ptolemy's system by the 'heliocentric' world view, in which the sun is the centre of the planetary system.

Aristotle's influence on the thinking in the western world has probably been even greater than Plato's. Until the late Middle Ages the neo-Aristotelians and scholastics, in Christian Europe as well as in the Islamic Arabic world, took his method of reasoning as a guideline. Although his nature philosophy was mainly speculative and unrealistic, it was considered infallible in those days and was adopted almost without question.

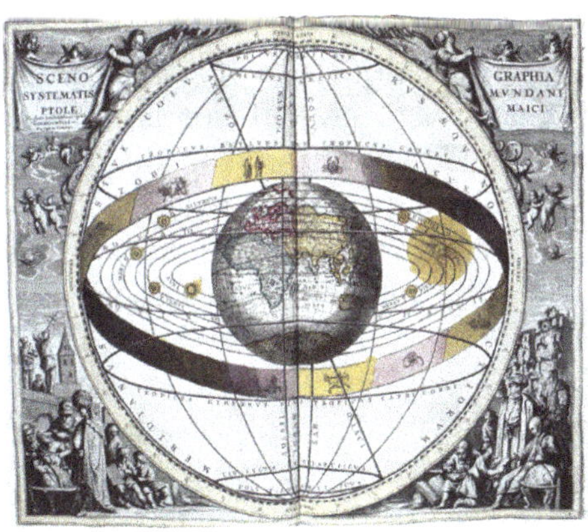

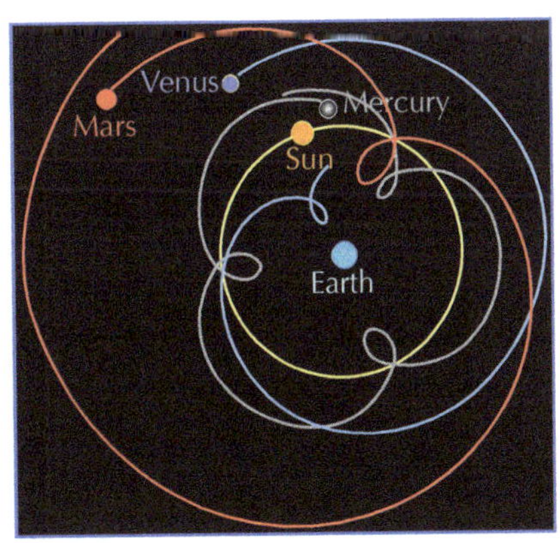

On the left, a picture of Claudius Ptolemy's geocentric world view (about 85–165 CE). On the right, Hipparchus of Nicaea's model of epicycles, in a persistent attempt to explain the apparent irregular orbits of the planets by means of circles.

2.6 The rational model of *The Elements* of Euclid

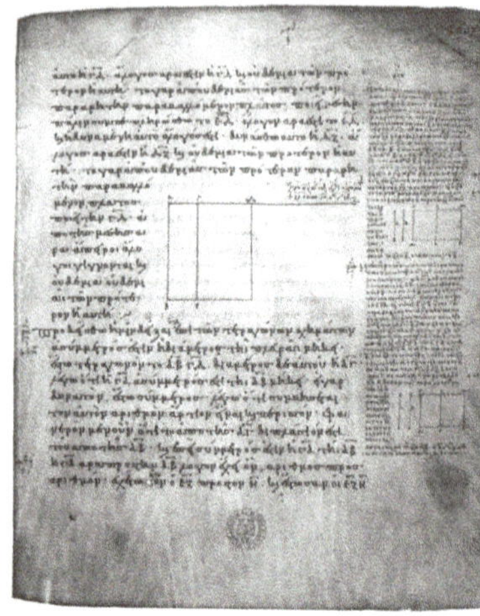

About 300 BCE Euclid summarized the greater part of Greek mathematics, as it was practised in Plato's school in Athens, in one monumental deductively built standard work, called *The Elements* (στοιχεια). He based his synthesis, among others, on the work of Eudoxus (about 408–355 BCE) and of Theaetetus (about 415–368 BCE), both pupils of Plato. With his new theory of equal proportions, Eudoxus had solved the problem of irrational numbers, and with his exhaustive method he had found a means of providing experimental results with a formal proof, as he did for the volume of a pyramid or a sphere. Theaetetus had achieved important results in the domain of irrational numbers and in the domain of three-dimensional geometry.

There is little known about the personal life of Euclid. After the death of Alexander the Great, in 323 BCE, his empire was divided into three parts among his Macedonian generals (the diadochi). Ptolemy I Soter became the king of Egypt and chose Alexandria, founded in 332 BCE, as the capital. There he founded the famous university with its museum and huge library that would remain the intellectual centre of the Hellenistic world for no less than a thousand years. From Proclus' (410–485 CE) comments on Euclid's *Elements*, we know that Euclid was working as a mathematician at the university of Alexandria at the time of Ptolemy I. Proclus mentioned the anecdote about the king who supposedly asked Euclid about an easier way to the knowledge of geometry, to which Euclid is said to have replied 'there exists no royal way to geometry'.

Although Euclid is the author of at least ten more books on mathematics, his worldwide reputation rests on *The Elements*. This manual is without any doubt the best-seller of all times. No other work ever, except for the bible, has been translated and reprinted more often: already more than a thousand editions since the first print in Venice in 1482. The content of *The Elements*, elaborated in thirteen books, not only deals with geometry, but also number theory and algebra, using, however, geometrical interpretations and strategies. It is certainly not a full compendium of the whole of Greek mathematics of that time.

Besides the content, this work evoked even more respect by its form, i.e. its strict logical construction. The axiomatic-deductive method, in which new propositions (theorems) are derived starting from definitions and postulates (elementary propositions acknowledged to be true), remained the standard method for building and organizing mathematical knowledge. Also in other domains of science and thinking, people have wanted to take this model as a guideline. In the seventeenth century, Spinoza even attempted to build a rational ethics 'De more geometrico' ('In the way of geometry').

The Elements starts with 23 definitions, such as 'parallel lines are lines that have no point in common', 5 axioms such as 'entities that are equal to a common entity are mutually equal', and 5 postulates such as the famous parallel postulate: 'if a line intersects two other lines and if the inner angles on its same side sum up to less than two right angles, then those two other lines intersect at a point on the same side of the first line'.

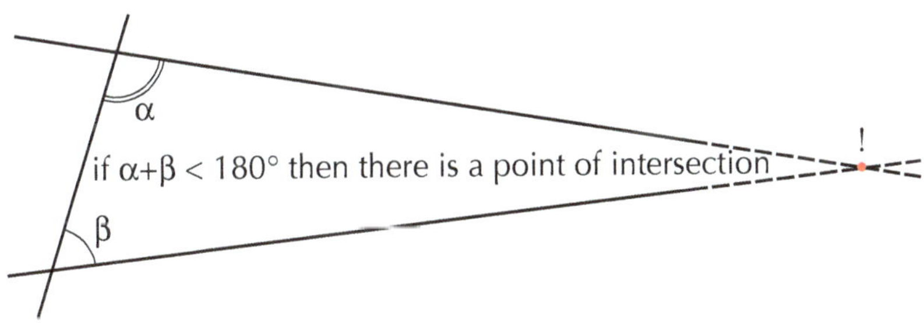

This postulate is called the 'parallel postulate' because it is equivalent to the proposition: 'through a point not on a given line, there exists exactly one line parallel to that line'. In the nineteenth century this postulate was shown to be fundamental to what has since then been called 'Euclidean geometry'.

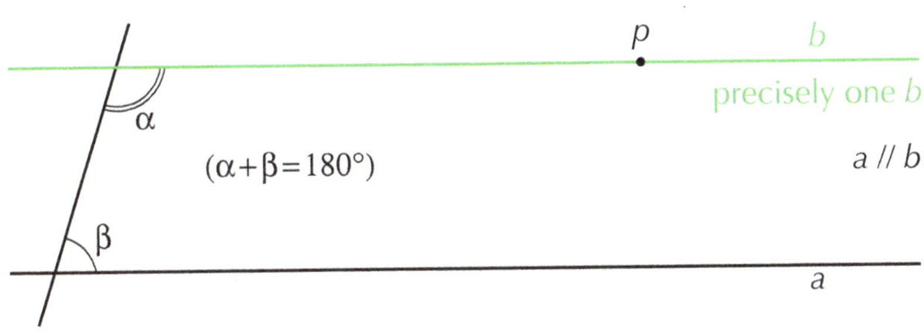

After mathematicians had tried in vain over centuries to derive this postulate as a theorem from the other postulates and the axioms, in the nineteenth century it was finally shown by the construction of non-Euclidean geometry that this is impossible (see chapter 2, 1.2).

The Greeks made a difference between axioms and postulates. For them axioms were rather generally true propositions, while postulates were specific to the domain considered. Present-day mathematicians no longer make this distinction, and speak only of axioms.

From the definitions, axioms and postulates mentioned above, Euclid derived 495 theorems in *The Elements*. We will mention some of the most important ones, for which remarkably elegant proofs were given.

Book I ends with theorems 47 and 48, namely the Pythagorean theorem and its opposite. The proofs are set up by the technique of constructing figures with equal area.

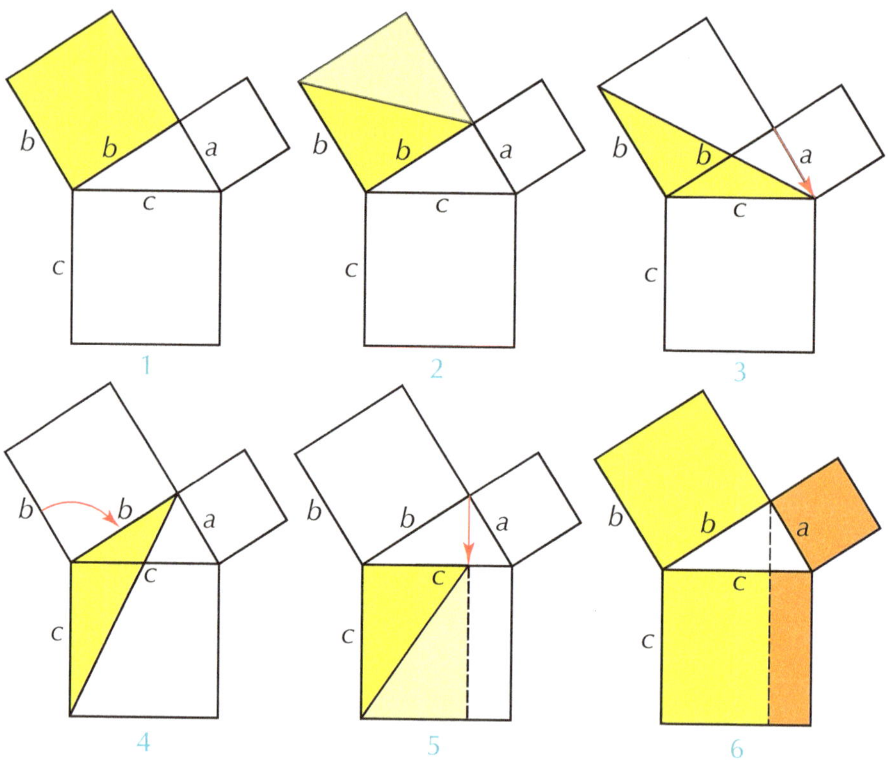

In book VII, theorem 2 gives the so-called 'Euclidean algorithm' for finding the greatest common divisor of two numbers by a succession of divisions that stops as soon as remainder zero shows up. The proof is based on the fact that every common divisor of two numbers also is a common divisor of the smallest of them and the remainder of the division of the greater one by the smaller.

Example

Find the greatest common divisor of 364 and 84, i.e. gcd(364, 84).

364 ÷ 84 gives quotient 4 and remainder 28

From the property mentioned it follows that: gcd(364, 84) = gcd(84,28).

84 ÷ 28 gives quotient 3 and remainder 0

So: gcd(84,28) = 28 and hence also gcd(364, 84) = 28.

	4	3
364	84	28
–336	–84	
28	0	

In book IX we find the two fundamental theorems of the divisibility theory of natural numbers: theorem 14 is equivalent to the property that every natural number can be factorized in a unique way in prime factors; theorem 20 states that there is no greatest prime number, in other words: the sequence of prime numbers is infinite. The proof of this last theorem is so elegant in its simplicity that it became a prototype of beauty in mathematical reasoning. As often in existential problems, it is an indirect proof.

Proof:

Suppose that there exists a greatest prime number with rank n, which means that $p_1, p_2, p_3, \dots, p_n$ is the finite sequence of all prime numbers.

- Compose the number $a = p_1 \cdot p_2 \cdot p_3 \cdot \dots \cdot p_n + 1$.
- This number a is greater than all prime numbers of the given sequence.
- Either a is a divisible number, so it can be factorized as a product of prime numbers. Let q then be a prime factor of a. q cannot be one of the prime numbers p_i in the given sequence, since the remainder of a divided by p_i is always 1 and hence not zero. So q is a prime number greater than p_n which is contradictory to the assumption that p_n is the greatest prime number.
- Or a is itself a prime number, which implies the same contradiction.
- So the assumption that there would be a greatest prime number is wrong.

The last three books XI, XII and XIII deal with three-dimensional geometry and end with a treatment of the five regular polyhedra, i.e. the Platonic solids. The proof that there can be no more than five is again of brilliant simplicity. It is based on the possible shapes of the faces and the sum of their angles meeting in one vertex. This sum must naturally be less than 360°.

Shape of the face	Number of faces at one vertex	Sum of the angles	Shape of the solid
equilateral triangle	3	$3 \cdot 60° < 360°$	tetrahedron
equilateral triangle	4	$4 \cdot 60° < 360°$	octahedron
equilateral triangle	5	$5 \cdot 60° < 360°$	icosahedron
square	3	$3 \cdot 90° < 360°$	cube
regular pentagon	3	$3 \cdot 108° < 360°$	dodecahedron

So, there cannot be more than these five. There can't be less either: by giving construction methods for each of these five cases, it was proved that they actually exist.

Euclid's *Elements* has dominated the teaching of mathematics for more than 2000 years and it has become the emblem of the entire western rational culture.

Euclid can also be considered as the founder of the Alexandrian school, which took over from Athens as the cultural centre of the Greek-Hellenistic world. Many other famous mathematicians and astronomers studied at this school, for example:

Archimedes (we discuss him below).

Eratosthenes calculated the perimeter of the earth by observing a nail in a plank and provided a method for the selection of prime numbers, namely the sieve of Eratosthenes.

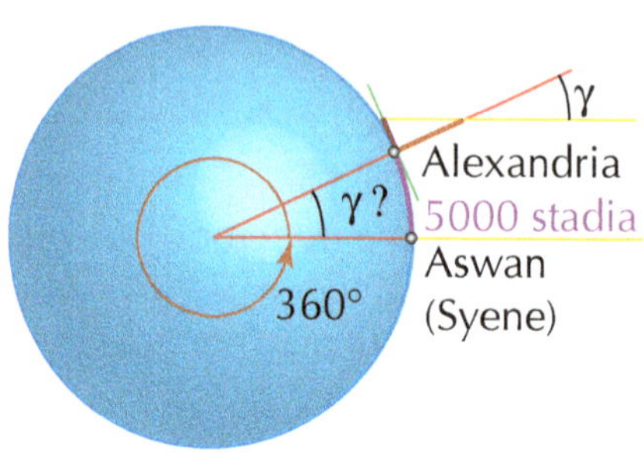

~~1~~ (2) (3) ~~4~~ (5) ~~6~~ (7) ~~8~~ ~~9~~ ~~10~~
(11) ~~12~~ 13 ~~14~~ ~~15~~ ~~16~~ 17 ~~18~~ 19 ~~20~~
~~21~~ ~~22~~ 23 ~~24~~ ~~25~~ ~~26~~ ~~27~~ ~~28~~ 29 ~~30~~
31 ~~32~~ ~~33~~ ~~34~~ ~~35~~ ~~36~~ 37 ~~38~~ ~~39~~ ~~40~~
41 ~~42~~ 43 ~~44~~ ~~45~~ ~~46~~ 47 ~~48~~ ~~49~~ ~~50~~
~~51~~ ~~52~~ 53 ~~54~~ ~~55~~ ~~56~~ ~~57~~ ~~58~~ 59 ~~60~~
...

At the time that the sun is directly above Aswan (Syene), Eratosthenes measured the angle γ by means of the shadow of a nail. With his assumption that Alexandria and Aswan were on the same meridian, he could calculate the perimeter of the earth by means of the proportion of γ to 360° and the distance between the two towns (5000 stadia)

By deleting in the sequence of the natural numbers all multiples of 2 except for 2, and next all still not deleted multiples of 3 except for 3, etc., Eratosthenes could precisely sort out all the indivisible numbers (prime numbers). This principle is called 'the sieve of Eratosthenes'

Apollonius wrote a book on conic sections, like the ellipse, the parabola and the hyperbola.

Heron found the formula that gives the area of an arbitrary triangle as a function of its sides.

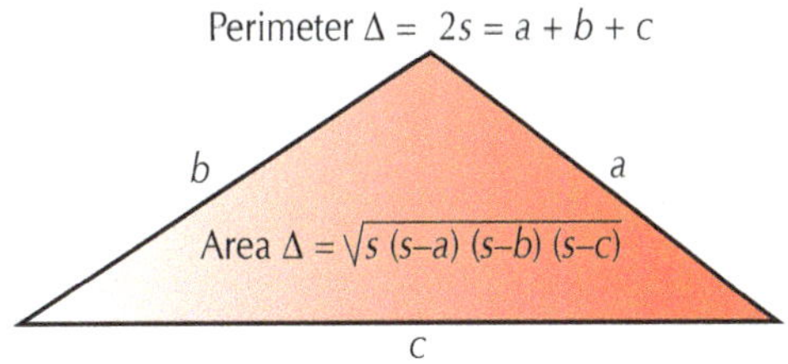

Menelaus wrote about spherical geometry

Ptolemy author of *The Great Treatise* or *Almagest* in Arabic, with the description of the cosmic geocentric world model.

Diophantus author of *Aritmetica*, a little book on general algebraic solutions of indeterminate equations, since then called 'Diophantine equations'.

Pappus author of *Synagoge*, a summarizing of works on geometry including some original work of his own, e.g. his projective 'Pappus's theorem'.

There are still so many others that we cannot discuss here, without running the risk of writing another history of mathematics.

2.7 The Olympian stature of Archimedes

Archimedes (about 287–212 BCE) originally came from Syracuse on Sicily, where he worked the greater part of his life and where he died. He stayed some time in Egypt at the Alexandrian school where he was friends with the librarian Eratosthenes, and with Conon and Disitheus, all of them successors of Euclid. It was to these friends that he wrote about many of the discoveries that he made through his experimental method and for which there was little appreciation in the theoretical Greek world. It was only in 1906 that a letter by Archimedes addressed to Eratosthenos was found in the archives in Constantinople as a palimpsest. In this he reports on his special 'method' for determining areas and volumes. His method is nothing less than the integral calculus of the seventeenth century. Twenty centuries before Newton and Leibniz would receive the highest honours for the design of what since then is called 'calculus', Archimedes had to disguise his method so as to conform to the spirit of the times. To make his impressive results acceptable, he had to show them in an artificial straitjacket of the exhaustive method.

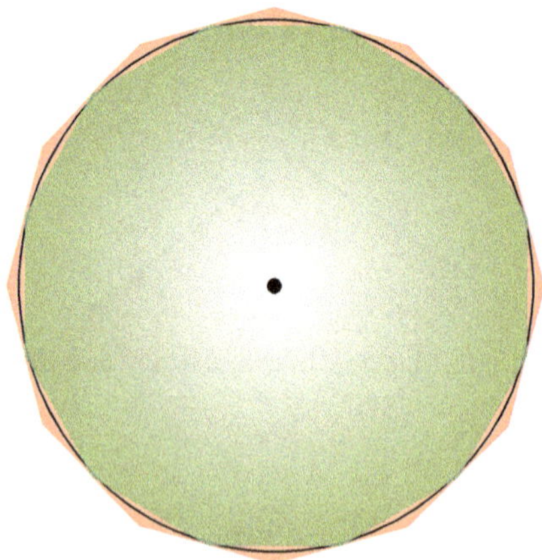

His 'method' consisted in enclosing areas and volumes between ever finer approximations, calculated e.g. from the sum of minuscule planar strips or discs. This enabled him to calculate the area of a disc using inscribed and circumscribed polygons with an increasing number of sides. In doing so, he obtained a value of 3.14... for π !

Area of the inscribed dodecagon
<
Area of the disc
<
Area of the circumscribed dodecagon

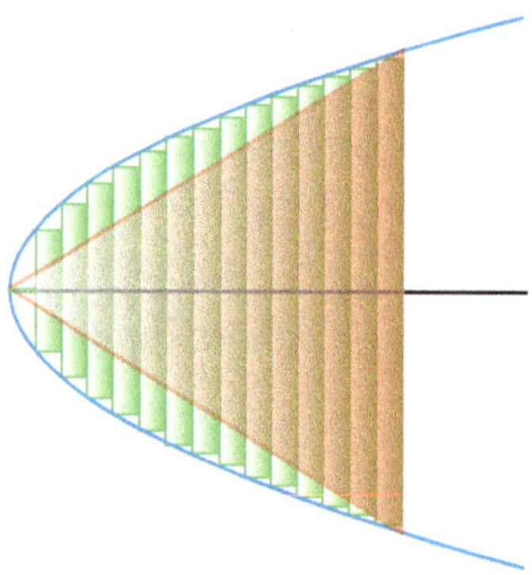

He could also determine the ratio of the area of an arbitrary parabola segment to the area of the inscribed triangle:

$$\frac{\text{Area of the segment}}{\text{Area of the triangle}} = \frac{4}{3}$$

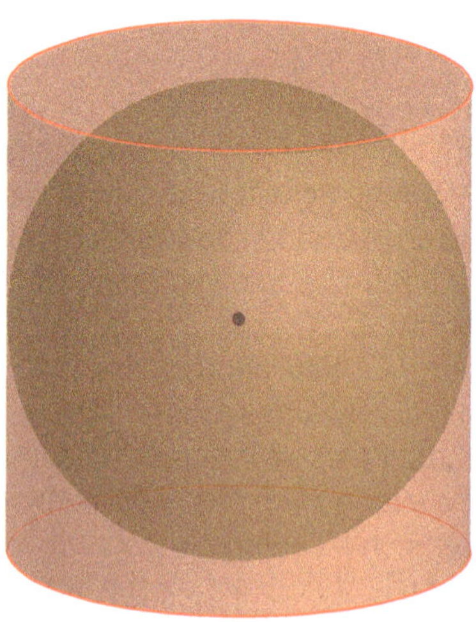

He also found that the volume of a sphere and its circumscribed cylinder are in the ratio 2 to 3. This is called the 'Archimedes theorem' after him. In the same way he succeeded in proving that the area of both solids are in the same ratio as their volumes.

$$\frac{\text{Volume o/t sphere}}{\text{Volume o/t cylinder}} = \frac{2}{3} = \frac{\text{Area o/t sphere}}{\text{Area o/t cylinder}}$$

After finding these results in his 'experimental' way, he proved them to be 'theoretically' correct by proving that both the other assumptions, namely that the correct value would be either greater or smaller than the assumed result, led to contradictions. This way he conformed to the exhaustive method designed by Eudoxus.

Besides being a mathematician, Archimedes was also a great physicist in the most modern sense. He certainly can be considered as the father of the so-called 'laws of nature' that became popular in the seventeenth century. It appears that he discovered the hydrostatic law for floating bodies, when sitting in his bath and screaming in delight: "Eureka, Eureka!" (I have found it!). There is an anecdote that he applied this law when exposing the deceit of a dishonest goldsmith who made a golden crown for King Hiero II, but replaced some gold inside with cheaper metal. The law of the lever was also known to him. This led him to his bold statement: "Give me a fulcrum, and I will lift the earth".

The Archimedean spiral (traced by a point moving away from a fixed point with a constant velocity along a line that spins with constant angular velocity around that fixed point), the Archimedes screw and the weapons he designed for defence against the besieging Roman fleet commanded by Marcellus, are further evidence of his inventiveness and mechanical ingenuity. Often he sat with his head bowed down, thinking over figures he had drawn in the sand. According to a legend, he was killed by a Roman soldier he had ordered not to mess up his figures by walking over them.

Archimedes lifts the earth by means of a lever that has its fulcrum outside the earth, on the moon or on a planet.

Summary

The Greek way of thinking was a complete revolution in the mentality of the ancient cultures until that time. The 'logos' as a rational explanatory principle took the place of traditional divine mythology. Mathematics and its logic formed the guiding principle for this new manner of striving for knowledge.

The only aim of this knowledge was theoretical insight and there was no interest at all in practical applications. The contempt for material experiments, as well as for everything that was related to physical labour, implied that Greek science was very speculative, not so much concerned with reality. It was rather the dialectical way of thinking and the deductive method than the developed theoretical and physical models, that survived the ravages of time. Only from the sixteenth and especially the seventeenth century onwards, when this deductive method was combined with the inductive practice of actual scientific experiments, would it rapidly bear fruit and once more initiate a new wave of civilization.

Especially in the philosophical domain, the Greeks have developed an enormous range of approaches to human existence and to fundamental questions. Nature philosophers, atomists, sophists, stoics, epicureans, ethicists, sceptics, cynics, eclectics, etc. have all been represented in the first place by individuals from this Greek world. One has to move on a lot further in time to encounter new ideas in the field of philosophy. Roman, medieval Christian and Arabic Islamic philosophy have found Greek philosophy to their liking and have remained in its footsteps for a long time.

Of course, beside all this, the splendour of Greek art should not be forgotten: the architecture and sculpture with their famous culminations in the golden century and the immortal epic and dramatic literature linked to the names of Homer, Aeschylus, Sophocles, Euripides and so many others. Rarely has a people, relying on the opinion that truth, beauty and good are just different aspects of the same virtue, given expression to such a cultural coherence, and this in spite of the existence of heathenism and slavery.

3 From an agrarian to an industrial culture

Obviously, history is an ongoing event in which all individuals and all nations have their own place and time. The meaning and the sense of life are at every moment the sum of inherited tradition and the circumstances that to some extent enable us to determine our own destiny. It is not up to later generations to evaluate and compare the life quality of their predecessors in this temporary sublunary world. Every human being must make do with the existence that she or he encounters. Its substantial mysteries have probably always remained the same for everybody throughout the centuries.

However, from a certain point of view and with regard to specific development, some periods in history can prove to be more interesting than others. This is at least true for the remarkable accelerations in the acquisition of knowledge and their causes, which we will take a particular look at in this chapter. To this end, we will take a huge leap in time, right back to the later years of the Renaissance.

After the Romans conquered the Hellenistic territory, only few substantial changes in Greek culture took place. This was partially due to the great respect that the Romans showed for that culture, but also because of the relative peace of the 'Pax Romana'. The Romans had little interest in the theoretical speculative science of the Greeks, but all the more in their ethical philosophy and their visual arts. Thus the practical and organizational ingenuity of the Romans hardly generated new science, but contributed to the wider spread of the Greek mind.

The rising Christian religion, which was initially distrusted and persecuted in the Roman Empire, was legitimatized under Constantine the Great and was elevated to the status of official religion under Theodosius. From then it engaged in a systematic fight against ancient heathenism (paganism). In 529 CE the emperor Justinian closed the academy in Athens, as if it were a heathen bastion. Platonic philosophy is engaged by Augustine (354–430 CE) as a handmaiden of Christian dogmatics. As a result, science was entirely subjected to the authority of the church, as it would still prove to be many centuries later. In the 11th century the Aristotelian way of thinking was rediscovered through Arabic works, which restored attention to empiricism and nature research, yet still within the limits of the Christian lines that Thomas Aquinas laid out in his *Summa Theologica*. The totally changed attitude towards life with the accent on the afterlife and the belief in redemption and grace could only push attention to science and research into the background, if not extinguish it almost completely. Consequently, in the 'dark' Middle Ages in Christianized Europe spectacular scientific progress was no longer achieved. Nevertheless, it was not the existence of god, but the belief in the existence of god which produced the splendid cathedrals as well as the splendour of so much more religious art.

A figurative expression of 13th century Christian scholasticism is the gothic cathedral, as a synthesis of the total organization of the world in which everything had its place and meaning. The **Notre-Dame** cathedral in Chartres is a pearl of this High Gothic architecture. It is especially famous for its outstanding stained-glass windows, such as the **northern rose window.**

Together with the seventh century came the rise of the Arabs and Islam. The Arabic realm rapidly spread both westward as far as the south of Europe and eastward as far as India. In 641 Alexandria also had already been conquered by the Arabs. Familiarization with the science of both the Greeks as well as the Hindus had a stimulating influence on Arab scientists. On the orders of the wise Sultans, important Greek and Indian works, of which the originals later got lost, were translated into Arabic and in this way they were saved for later generations. Between about 766 and 1250 CE prominent Arab scientists were studying and editing these documents. Especially in the field of geometrical algebra they could make interesting progress. One of the most important authors was Mohammed ibn Musa Alchwarizmi, who wrote a book in which he explained the decimal arithmetic system of the Hindus. From its later Latin translation, with the opening clause 'Algorismi de numero Indorum', the term 'algorithm' was derived as a corruption of the name Alchwarizmi. A second remarkable book by the same author with the title *Hisab al - jabr walmuqabala,* deals with solving equations and has led to our term 'algebra'. In this way a great many mathematical terms are loan words from Arabic.

Remarkably, in the Islamic world, where its Prophet and holy book reminds us so strongly of Christianity, virtually the same process took place as in Christian scholasticism. After the Platonic and especially the Aristotelian ideas were implemented in Arabic philosophy by, among others, Avicenna (980–1037 CE) and Averroes (1126–1198 CE), scientific interest gradually decreased and eventually was fully suppressed because of religious concerns.

So, apart from its important conserving function, the initially promising Arabic scientific fervour has not been able to live to the fullest extent. And as much as Christianity, Islamic culture has achieved great heights in its monumental art.

Islamic art is unsurpassed in its decorative aspects. The abstract geometrical structures have an undeniable mathematical background. In the tilings and the paintings in the armoury of the **Baño de Comares** in the Alhambra in Granada (left picture) as well as in the detail of the **muqqarna vault** in the Nasir al-Mulk mosque in Shiraz (right picture), many symmetry groups can be recognized.

In the 15th and 16th century comes the 'Rebirth', the Renaissance. The revival of the study of ancient authors and their culture was stimulated by a number of factors that revived contacts between the west and east: the travelling merchants of the Italian trading cities; the Greek scientists that fled to the west after the fall of Constantinople in 1453; the reconquista (reconquest) and victory over the Arabs in the south of Spain; the crusades; etc. Meanwhile medieval scholasticism came under the attack of critical individuals like Roger Bacon, Duns Scotus and William of Ockham. The separation of religion and science became more and more inevitable, primarily by the changing vision of the acquisition of knowledge by means of experience, and secondly by the nominalist point of view concerning concepts.

Around this time certain inventions caused important social and structural changes. In the first place, the invention of the compass, which enabled navigation on the oceans and paved the way for the fearless explorations by Columbus, Vasco da Gama, Magellan and others like them. Thus it was mainly Western European states bordering the Atlantic that could expand their wealth and power.

Next, the invention of gunpowder that threatened the dominating knights in their fortified strongholds and put an end to feudal social relations. And last but not least, the invention of the art of printing that brought knowledge within the reach of a larger section of the population. In this tumultuous transitional period of a reviving humanism and the bursting abscess of the Reformation, the human individual regained attention and respect. When the Renaissance initially blossomed as an uninhibited renewal of literature and fine arts (*stil nuove*, perspective and portrait painting), a revolution in science and thinking could not lag behind.

In the first place there was the Copernican revolution, i.e. the heliocentric world view. The artificial geocentric world view of Ptolemy contained too many contradictions, revealed by actual astronomic observations. It was the Polish astronomer Nicolaus Copernicus who had the courage to resume the brilliant idea of the Greek Aristarchus, who had already thought of the sun as being the centre of the planetary system. A little later the German astronomer Johannes Kepler would formulate his three famous laws about the orbits of the planets, based on the meticulous observation tables of his teacher Tycho Brahe. These laws are among the most remarkable milestones of Western science. On the one hand, because of the innovative method of inducing laws of nature from an abundance of numerical data instead of deriving them from an abstract theoretical model like the Greeks did. On the other hand, because of the purely mathematical form in which these laws of nature were formulated. Kepler himself admired this coincidence of reality and mathematics and exclaimed enthusiastically: "Ubi materia, ibi geometria" (Where is matter, there is geometry). We will discuss these laws later on.

Just as impressive from those days are the results of the innovator of Western natural science, the Italian Galileo Galilei. He supported Copernicus' point of view, appealing to his observations of the phases of Venus and the discovery of Jupiter's satellites. But the name of Galileo is forever linked to the laws of falling motion. Instead of wondering like Aristotle 'why' bodies fall, Galileo concentrated on the question of 'how' bodies fall. It was for this reason that he performed his specific experiments and, just like Kepler, he derived his well-known laws of falling motion by induction. These laws, which do not have the Greek ambition of explaining the essence of falling motion, but of giving a precise description of its quantitative course, are also embodied in a mathematical formula. It was as clear to Galileo as it was to Kepler: "The great book of nature lies open before us, and to be able to read it we need mathematics". After Galileo the barriers came down and the unequalled triumphal progress of Western natural science began, with a procession of impressive characters in the seventeenth and eighteenth century, among them a host of mathematicians such as Descartes, Pascal, de Fermat, Desargues, Newton, Leibniz, Huyghens, the Bernoullis, Euler, Lagrange, Laplace and so many others. A sequence of names that also contains many prominent philosophers who guided human thinking in new directions.

We finally come to what was the actual intention of this section: illustrating how the development of mathematics in this period played a decisive role in the acceleration of research in the natural sciences and how doing so has paved the way for modern industrial society.

3.1 The brilliant discovery of the decimal fractions

With the growing intensity of economical and financial activities, the variety of different systems of weights and measures, the gigantic calculations in astronomy and the problems of navigation, the beginning of the seventeenth century was marked by an exceptional arithmetical activity for which in the German language they even introduced a particular name: 'Rechenhaftigkeit'. Initially this was mainly practised in a verbal form because of the lack of an appropriate symbolic system.

Under the influence of the works translated from the Arabic about Hindu arithmetic, and especially of the *Liber Abaci* of the travelling merchant Leonardo of Pisa (Fibonacci), the decimal positional numeral system eventually was also accepted in Western Europe. And so the algorists (who made calculations in writing in the positional system) won their battle over the abacists (who calculated on a counting frame).

Simon Stevin (1548–1620) from Bruges considerably extended the usefulness of this system by introducing decimal fractions. In his booklet *De Thiende* (The Tenth or The Disme) he provided the means for arranging the whole metric system on a decimal basis and he showed how one can calculate with fractions in the same easy way as with integers.

With the introduction of the decimal point (later replaced by a comma in continental Europe) by the Scottish mathematician John Napier, the mode of writing of decimal fractions was strongly simplified in the form of decimal numbers. Since then the practical arithmetical rules for calculating with those 'point numbers' have become common property.

Example

Suppose that we have to multiply $5+\frac{3}{4}+\frac{7}{20}+\frac{11}{25}+\frac{31}{50}$ by $2+\frac{7}{8}+\frac{6}{125}$, then it is practical to first convert these sums into decimal numbers:

$$\begin{aligned}5+\frac{3}{4}+\frac{7}{20}+\frac{11}{25}+\frac{31}{50} &= 5+\frac{75}{100}+\frac{35}{100}+\frac{44}{100}+\frac{62}{100}\\ &= 5+\frac{216}{100}\\ &= 5+\frac{200}{100}+\frac{16}{100}\\ &= 5+2+\frac{16}{100}\\ &= 7{,}16\end{aligned}$$

$$\begin{aligned}2+\frac{7}{8}+\frac{6}{125} &= 2+\frac{875}{1000}+\frac{48}{1000}\\ &= 2+\frac{923}{1000}\\ &= 2{,}923\end{aligned}$$

$7{,}16 \times 2{,}923 = 2{,}923 \times 7{,}16$ so

$$\begin{array}{r} 2{,}923\\ \times\, 7{,}16\\ \hline 17538\\ 2923\\ 20461\\ \hline 20{,}92868 \end{array}$$

Simon Stevin (1548–1620)
was born in Bruges in Flanders, nowadays Belgium. As a mathematician, physicist, civil and military engineer he was a 'homo universalis'. In 1581 he escaped from the Spanish rule by fleeing north to Leiden. He entered the service of Maurice, Prince of Orange. Stevin gained an international reputation by his use of decimal fractions in his booklet entitled: *De Thiende* (The Disme). Stevin also proved the equilibrium law on an inclined plane, using an ingenious diagram showing a loop, containing evenly spaced beads, hanging over an incline. He understood that this diagram, which was later engraved on his tombstone, could not produce perpetual motion and therefore had to exist in equilibrium.

All this seems simple from our point of view, but bear in mind that in those days many towns had their own system of weights, measures, and currencies, none of them being decimal. (For instance: a pound could be divided into 20 shillings, a shilling into 12 pence.) It could take some arithmetical acrobatics to convert the weight and price of a batch of goods from one region to another. This was considerably simplified by decimal arithmetic. However, when huge numbers were involved, higher operations like multiplication, division, and especially raising to a power and root extraction still took a lot of time and energy. But an ingenious solution was also found for that problem: logarithmic arithmetic.

3.2 The arithmetical convenience of 'admirable' logarithms

Already in 1554, the German mathematician Michael Stifel (1486–1567) had shown in his book *Arithmetica integra* how a geometric sequence (of successive powers of the same base) could be mapped onto an arithmetic sequence (of the respective exponents). Thus, higher operations on the numbers in the geometric sequence could be performed by the lower operations on the respective terms in the arithmetic sequence.

Example (with powers of base 3)

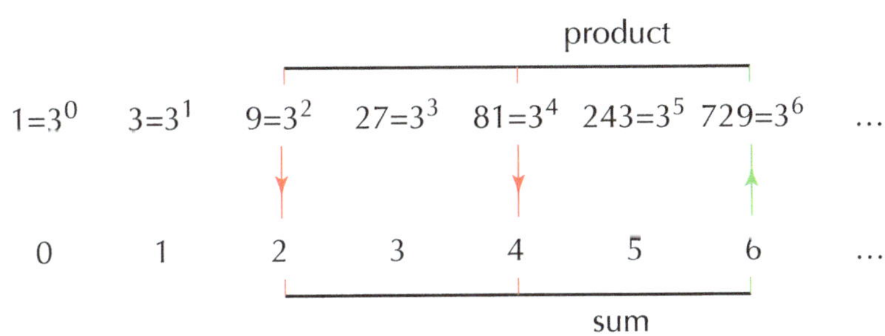

Because $3^2 \times 3^4 = 3^{2+4} = 3^6$, exponential arithmetic shows us that the product of the numbers 9 and 81 in the upper sequence can be found by adding the respective exponents 2 and 4 in the lower sequence and then from 6 passing to the upper sequence again to end up with 729.

In this way multiplication can be done by addition, division by subtraction, raising to a power by multiplication and root extraction by division.

This method of working can be applied to any base. Indispensable is the property that every strict positive number can be written as a power of an arbitrary strict positive base, even if with the help of (approximate) decimal exponents.

Example

$$7 = 10^{0.845098...}\text{ , which also means that }\log 7 = 0.845098...$$

It is a remarkable fact in history that this logarithmic technique was invented and used in a period in which general exponential notation did not yet exist.

However, Stifel has not elaborated on his ingenious idea. The proper invention of logarithms as a system of tables for practical use is attributed to two other mathematicians, namely the Scot John Napier (1550–1617) and the Swiss Jobst Bürgi (1552–1632), who worked independently and elaborated on this arithmetic in a different way. In his book *Mirifici logarithmorum canonis descriptio* (Description of the admirable law of logarithms), published in 1614, Napier starts from a geometrical approach of two points that move dependent on each other (*C* and *F* in the figure below). A mapping is constructed that uses the factor 10^7 and, as we would express it now, that has base 1/*e* (where e = 2.71828182...). This resulted in logarithm tables to 7 decimal places (because of the factor 10^7).

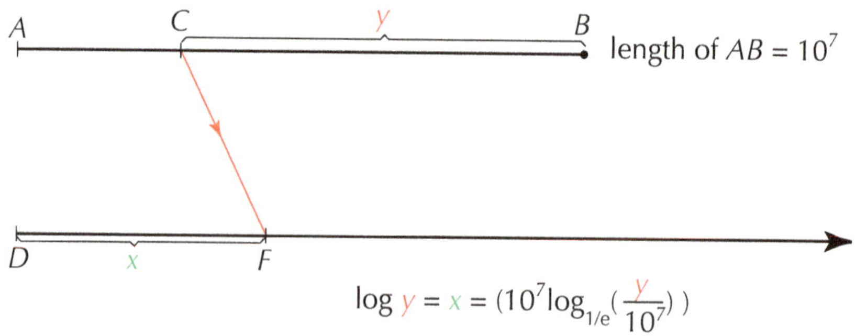

Bürgi started from an algebraic relationship and he published his results only in 1620, so 6 years later than Napier did. He used the factor 10^4 and the base that we represent nowadays symbolically as e (the first letter of the name 'Euler'). His tables gave logarithms up to four decimal places. Because Napier published first, he is generally seen as the father of the 'admirable' logarithms. It is surprising that both Napier and Bürgi designed logarithms that depended on the number e, which at the time didn't even have a definition. But this is just the result of the 'natural' way in which their mappings were constructed, and nowadays we call the base e logarithm the 'natural logarithm' or also, after Napier, the 'Napierian logarithm'. It is an ironical whim of history that Napier's name was attributed to the logarithm that actually was developed by his rival. This incident is, however, not unique in the history of mathematics.

John Napier spent the greater part of his life on the family estate of Merchiston Tower, near Edinburgh in Scotland. He spent a lot of energy participating in the political and religious discussions of that time. He was a fervent opponent of Catholicism and in a book, which he hoped would bestow him with eternal honour, he tried to prove that the Pope was indeed the Antichrist of the Apocalypse and that the end of the world would occur between 1688 and 1700. It wasn't his oddities, but his logarithms that eventually brought him respect and recognition.

In the year following the publication of the *Mirifici* , the English mathematician Henri Briggs (1561–1631) travelled from London to Edinburgh to express in person his admiration for the author. On that occasion he made some proposals to improve the system of Napier. They agreed that base 10 was more suitable for making calculations in the decimal system. Since then base 10 logarithms are called common or Briggesian logarithms.

After many extended publications of logarithm tables, in 1850 the French army officer Amédée Mannheim (1831–1906) would develop a standardised version of a ruler with sliding scales on which logarithmic calculations could be performed. This famous slide rule remained the instrument of engineers and other arithmeticians until the last decades of the twentieth century, when it was made obsolete by the pocket calculator. The seniors among us still remember the delight of exploring logarithm tables and likewise the slide rule.

In the seventeenth century the invention of logarithms, to the enormous joy of the 'Rechenhaftigen', came at exactly the right time to lighten the enormous arithmetical workload. Kepler made no secret of his approval, and later also Laplace would agree: "The invention of logarithms doubles the lifetime of astronomers".

The successive aids in logarithmic arithmetic:

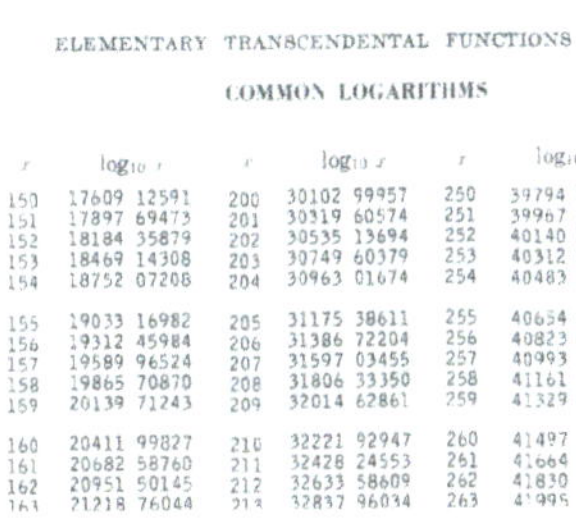

ELEMENTARY TRANSCENDENTAL FUNCTIONS

COMMON LOGARITHMS

x	$\log_{10} x$	x	$\log_{10} x$	x	$\log_{10} x$
150	17609 12591	200	30102 99957	250	39794 00087
151	17897 69473	201	30319 60574	251	39967 37215
152	18184 35879	202	30535 13694	252	40140 05408
153	18469 14308	203	30749 60379	253	40312 05212
154	18752 07208	204	30963 01674	254	40483 37166
155	19033 16982	205	31175 38611	255	40654 01804
156	19312 45984	206	31386 72204	256	40823 99653
157	19589 96524	207	31597 03455	257	40993 31233
158	19865 70870	208	31806 33350	258	41161 97060
159	20139 71243	209	32014 62861	259	41329 97641
160	20411 99827	210	32221 92947	260	41497 33480
161	20682 58760	211	32428 24553	261	41664 05073
162	20951 50145	212	32633 58609	262	41830 12913
163	21218 76044	213	32837 96034	263	41995 57485

logarithm tables

slide rule

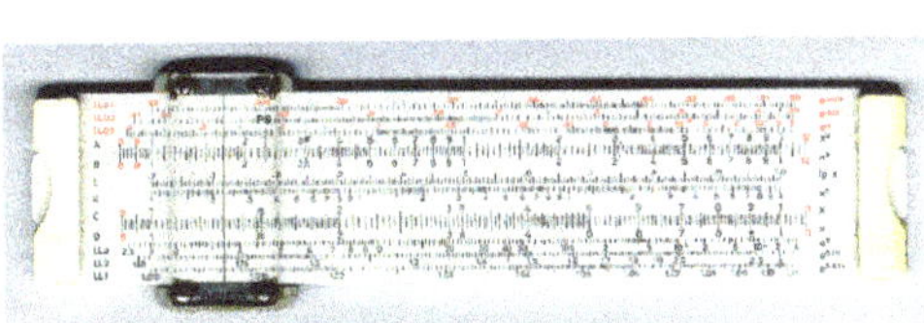

electronic calculator

3.3 A new world view

Already in the sixteenth century, because of observations continuously becoming more accurate, the need to revise Ptolemy's geocentric model of planetary motion was recognized. It was the Polish astronomer Nicolaus Copernicus (1473–1543) who brought about the drastic revolution of the heliocentric world view with his publication: *De revolutionibus orbium celestium* (On the motion of celestial bodies) in 1543, the last year of his life. In this book the three main aspects of the motion of the earth and the other planets respective to the sun were explained: the daily rotation along the earth's axis that explains the phenomenon of day and night, the yearly orbit of the earth around the sun and the inclined position of the earth's axis respective to this orbit that explains the changing of seasons.

However, Copernicus' vision still contained many imperfections, which also caused him doubts. Without the enthusiastic cooperation of his meritorious pupil G.J. Rhaeticus the book would probably never have been published. One of the problems came from the obstinate belief in circular orbits that continued to cause contradictions. Johannes Kepler would apply the appropriate corrections.

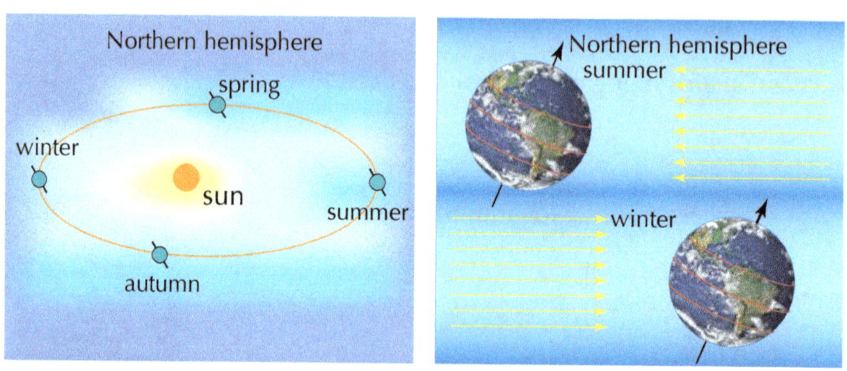

The phenomenon of the four seasons on earth is a benefit of the constant inclination of the earth's axis (from pole to pole) respective to the ecliptic, i.e. the plane of the earth's orbit around the sun (1st figure). This inclination measures 23°. In the summer (June 21st) the North Pole inclines 23° towards the sun and in the winter (December 21st) 23° away from the sun (second figure). In the spring (March 21st) and the autumn (September 21st) the North and South Pole are at an equal distance from the sun.

Kepler (1571–1630) was born near Stuttgart, Germany. After his studies at the university of Tübingen, and a lectureship in Graz in Austria, in 1600 he became assistant to the Danish-Swedish royal astronomer Tycho Brahe (1546–1601) in Prague. This last-mentioned was determined to solve the mystery of the bizarre planetary motions with their apparently retrograde orbits. He had literally and figuratively a golden nose (a prosthesis for a piece that he lost in a duel) for accurate observations and tables. When Brahe died unexpectedly in 1601, Kepler as his successor had all of this data at his disposal. After years of calculations and to his own surprise, these tables revealed the undeniable fact of the elliptic orbits.

In 1609 Kepler published his *Astronomica Nova* containing the first two of the laws named after him, and according to which the orbits of present lunar missiles and of interplanetary explorers are calculated:

- the planets move in elliptic orbits around the Sun, which is in one of the focuses;
- a line joining a planet and the Sun sweeps out equal areas during equal intervals of time (law of conservation of angular momentum).

In 1618 his *Harmonice Mundi* was published with his third law:

- the squares of the orbital periods of planets are directly proportional to the cubes of the semi-major axes of the orbits. (harmonic law).

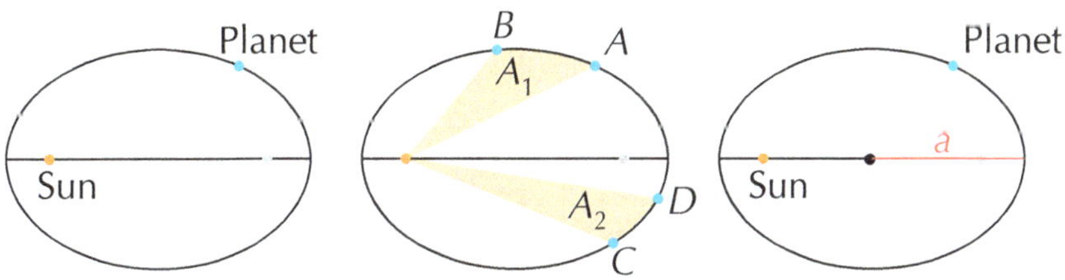

Sun in a focus

If a planet moves from *A* to *B* and from *C* to *D* in equal amounts of time, then area $A_1 = A_2$

t = orbital period
a = semi-major axis
$\frac{t^2}{a^3}$ is equal for all planets

It is surprising that after Apollonius of Perga had theoretically described the conic sections (ellipse, hyperbola and parabola) as intersections of planes with a right circular cone, it took more than 1800 years for a practical application in nature to show up.

Kepler had, just like the Pythagoreans, a deep mystic adoration for cosmic harmony. Already in 1595 he had found that the ratio of the orbits of Saturn and Jupiter, as they were known at that time, was equal to the ratio of the inscribed and the circumscribed circles of an equilateral triangle. This inspired him to the cosmic model of nested Platonic solids and their respective inscribed and circumscribed spheres. That model was presented in his *Mysterium Cosmographicum* of 1596.

Faith in the profound unity of mathematics and the universe kept on inspiring him and was a joyful consolation in his otherwise very tormented life: he lost his favourite child, his first wife became insane and died, his mother was accused of sorcery and he himself of heresy.

3.4 The birth of mechanics

Galileo Galilei (1564–1642) was born in Pisa. He was a professor in mathematics, first in Pisa and later in Padua. According to a legend he is said to have performed his first experiments with falling bodies from the leaning tower of Pisa. So, already in his younger years, he could contradict Aristotle and prove that heavy bodies do not fall faster than light ones, if we disregard the friction caused by air resistance.

In Padua in 1607, Galileo heard about the invention of the telescope by the Dutch lens polisher Johann Lippersheim. He himself succeeded in making powerful telescopes through which he could observe mountains on the moon, the phases of Venus, Saturn's rings and Jupiter's four largest moons. These observations convinced him to support the Copernican theory with the publication of his *Dialogue Concerning the Two Chief World Systems*, to the great dissatisfaction of the Catholic authorities in Rome. Three years later, in 1633, the Inquisition required him to publicly renounce his principle that the earth 'moves' (orbits the sun) on pain of excommunication by the Church. He would, however, not give up his conviction: "Eppur si muove" (And yet it moves) became a metaphor for stubborn holding on to a conviction, however politically incorrect.

According to his own writings Galileo discovered the basic laws of falling motion by experimenting with balls rolling on an inclined plane. He managed to summarize them in a mathematical way with the ingenious formula $s = (1/2)gt^2$ in which s stands for the distance travelled or displacement, t for the elapsed time and g for the acceleration caused by gravity, being 9.81 metres per second, per second. This showed that the distance travelled by a falling body is proportional to the square of the elapsed time, its velocity is proportional to the elapsed time and that the earth's gravitation causes a constant acceleration on all falling bodies. Thus the foundations were laid for general dynamics and mechanics, which Newton would later formulate by means of infinitesimal calculus.

At the end of his life Galileo went blind and in 1642 he died an embittered man. But the modern spirit of harmonic coherence between experiment and theory, induction and deduction that he had shown in his work, would never ever be lost and would generate thousands of scientific successes.

3.5 The intimate embrace of algebra and geometry

Since the ancient Greeks kept pure geometry at a safe distance from number theory after their unpleasant experiences with irrational numbers, for a long time these two domains of mathematics developed separately. In geometry measures were always linked to magnitudes and their dimensions: a number represented a length, a product of two numbers an area and a product of three numbers a volume. Geometrical terms did not allow further thinking, which was a considerable restriction.

Diophantus already gave number theory an algebraic feature by discussing the solutions of indeterminate equations, and later Arabic mathematicians like al-Khowarizmi and Omar Kahyyam made new contributions to this domain. The most spectacular progress in algebra, however, was realized in the sixteenth century, when Italian mathematicians, with the names of Scipio del Ferro, Tartaglia, Cardano, Ferrari and Bombelli, competed with each other to find a general solution to equations of the third and the fourth degree.

Tartaglia (about 1499–1557)

An Italian mathematician who became famous for his algebraic solution of an equation of the third degree. In the first half of the sixteenth century there was a strong rivalry among the Italian mathematicians Mario Fior, Gerolamo Cardano, Niccola Fontana, nicknamed the stutterer (Tartaglia), and others. In a public contest Tartaglia came out as the great winner. Cardano was able to convince him to reveal his method, on a vow of secrecy. Cardano, however, broke his promise instantly and published the formulas in his *Ars Magna*, which made the stutterer furious. 'Cardano's formulas', as they have been called since then, are thus actually Tartaglia's.

The idea of positioning a point on the globe by means of coordinates, namely longitude and latitude, was already expressed by Hipparchys of Alexandria about 140 BCE. Nicole Oresme (1323–1382) already used block diagrams on the plane as to express uniform motion mathematically by means of two variable coordinates which he called 'longitudo' and 'latitudo'.

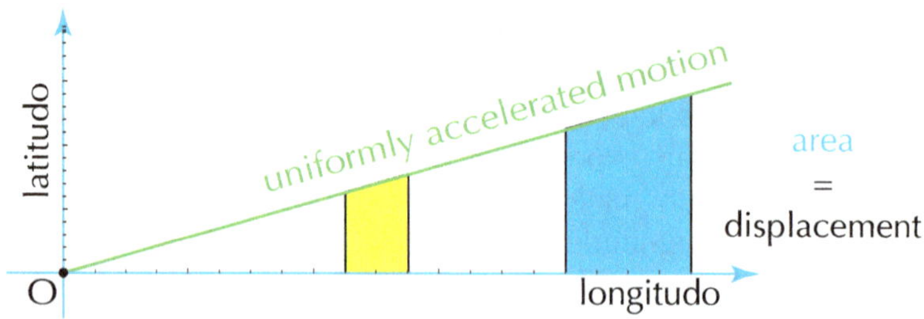

Also Pierre de Fermat had set up equations of straight lines and conic sections respective to a system of axes. Nevertheless, so-called 'analytic geometry' is generally attributed to the French mathematician and philosopher René Descartes. This geometry can be described as coordinate geometry. In this the position of a point is determined by a couple of real numbers (coordinates), which express the respective distances of the point to two perpendicular axes. Specific sets of points like lines, conic sections or other curves can then be determined by means of equations that express the relation between the variable coordinates x and y of a variable point on the curve.

Example

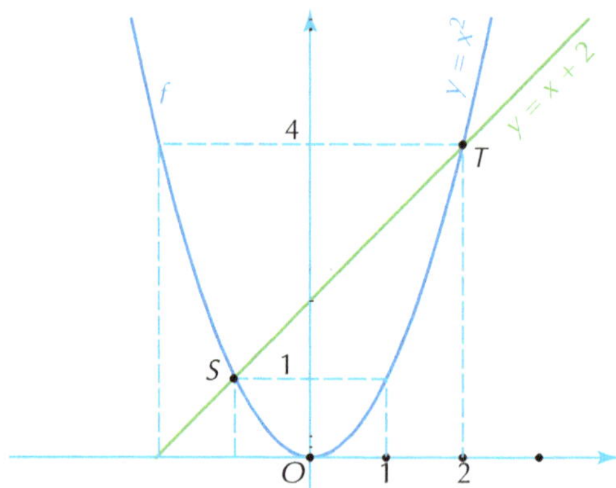

Geometric properties of these curves can then be expressed algebraically and vice versa, algebraic results can be interpreted geometrically. For instance, the points of intersection of the line and the parabola in the figure above can be found by solving the set of simultaneous equations.

From $y = x + 2$

and $y = x^2$

it follows that $x^2 = x + 2$ and thus $x^2 - x - 2 = 0$

The solutions of the last quadratic equation are –1 and 2.

If $x = -1$ then $y = (-1)^2 = 1$ and if $x = 2$ then $y = 2^2 = 4$.

The coordinates of the points of intersection S and T are thus (–1, 1) and (2, 4).

There is a nice story that Descartes, who liked staying in bed and thinking till late in the morning, got this ingenious idea when he watched a fly walking between the beams on the ceiling and was thinking of a way to describe its path by means of (variable) distances of its position respective to the beams.

The method of analytic geometry, as it is now called, makes it possible to translate geometrical problems that generally take a lot of creative thinking into plain algebraic techniques, which in their turn demand more arithmetic skills. Even in those days, when algebra hadn't got its standard symbolic form and the study of equations of a higher degree was scarcely hatched, Descartes could use his 'method' to solve quite a lot of geometrical problems to which the Greeks could offer no reply. No wonder that he enthusiastically reported on this in his *Géometrie*, the last appendix of his famous *Discours de la méthode.*

The main purpose of *Discours de la méthode pour bien conduire sa raison et chercher la vérité dans les sciences* (Discourse on the Method for well conducting one's Mind and searching for Truth in the Sciences) was however mainly philosophical, namely explaining what could be a universal and secure way towards true knowledge. Descartes states that starting from an indisputable base ("Cogito, ergo sum" : I think, therefore I am), it is possible to build a reliable science by means of mathematical deduction.

The notorious 'Cartesian spirit' has unmistakably pushed western thinking further in the rational direction, just like Platonic thinking did, but now with a more pragmatic emphasis. His *Géometrie* has paved the way for the study of functions by their graph, which would later become very efficient in association with the method of infinitesimal calculus.

Descartes (1596–1650) is without a doubt one of the most influential philosophers of western history. The course of his life was, however, erratic and unpredictable. Born in a wealthy family in the neighbourhood of Tours, he lost his mother only a few days later. After his studies at the Jesuit college at La Flèche and later in Paris, he entered service as a soldier in the army of Prince Maurice of Orange at the age of 21. After adventurous rambles in Germany, Denmark, the Netherlands, Switzerland and Italy he returned for some time to Paris and then settled in the Netherlands. There, for twenty years he wrote his most important works. After repeated urging by the nineteen-year-old queen Christina of Sweden he was finally persuaded to move to Stockholm. The queen insisted that, in contrast to Descartes' way of living, he would teach her personally very early in the morning. His feeble health rapidly broke down on a regime of cold and damp mornings in the big library. One year after his arrival in Sweden he died of complications following a pulmonary haemorrhage. Unfortunately, thinking on its own was not sufficient to prolong existence.

Christina of Sweden (1626–1684),
was the second daughter of Gustav Adolf, king of Sweden and of Maria-Eleonora of Brandenburg.
At six she succeeded her father under the regency of chancellor Oxenstiern, who also took care of her further education. At the age of seventeen she became queen of Sweden. Her court became a cultural centre and she was able to persuade Descartes to take part.

3.6 The ways of chance

In times of intense spiritual creativity, mathematics develops not only through the challenges of concrete problems, but also through pure curiosity and playful recreation. In this way it is possible that the sources of important applications of mathematics unconsciously arise already centuries earlier. This was the case in the seventeenth century with probability calculus, pure number theory and projective geometry, which are linked respectively to the names of Pascal, Fermat and Desargues.

Already at an early age Blaise Pascal (1623–1662) showed his ingenious talent for mathematics and was very successful at it. At the age of sixteen he found and proved the theorem that carries his name, of the hexagon inscribed in a conic section: the three pairs of sides intersect in collinear points.

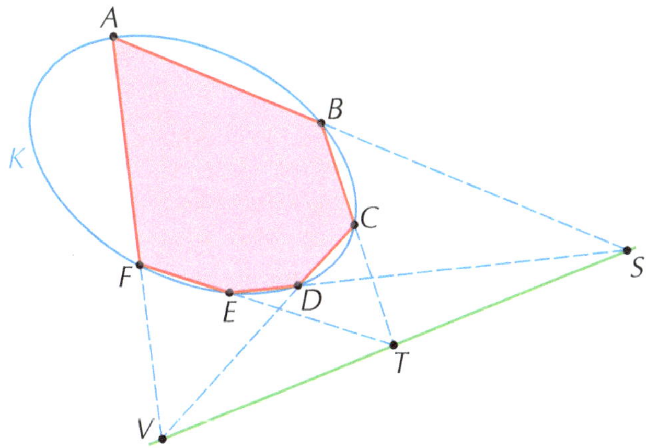

A few years later he constructed the first (mechanical) calculator in the history of mathematics, the so-called Pascaline. And this he did for the convenience of his father, who was an official with … the tax authorities!

In 1654, Prince de Méré, a fervent and skilful dice player, presented the famous problem about sharing the kitty among two players that stop the game prematurely, taking into account the course of the game till that moment. Pascal started corresponding with Fermat to compare and mutually comment on their solutions. In this correspondence the fundamentals for the discerning construction of probability theory were designed.

The mathematical means that Pascal used were based on the so-called 'binomial coefficients'. Those are the sequences of the coefficients that occur in the expansion of the successive powers of a binomial and that can be arranged in a triangular scheme (Pascal's triangle). His 'Triangle Arithmétique' was published posthumously only in 1665.

Examples

1					
1	1				
1	2	1			
1	3	3	1		
1	4	6	4	1	
1	5	10	10	5	1
etc.					

$(a+b)^0 = 1$

$(a+b)^1 = 1a + 1b$

$(a+b)^2 = 1a^2 + 2ab + 1b^2$

$(a+b)^3 = 1a^3 + 3a^2b + 3ab^2 + 1b^3$

$(a+b)^4 = 1a^4 + 4a^3b + 6a^2b^2 + 4ab^3 + 1b^4$

$(a+b)^5 = \dots$

etc.

In this triangle the property applies that the sum of two adjacent coefficients, e.g. 6 and 4, equals the coefficient under the last term of this addition: $6 + 4 = 10$.

It took a long time and a connection with statistics, which grew in interest in the nineteenth century, before the theoretical fundamentals of probability were eventually finalized by the Russian mathematician Kolmogorov.

Pascal, being aware of his feeble health and the shortness of earthly life, returned to a religious and contemplative life with the Jansenists at Port-Royal. There he wrote his famous *Pensées* (thoughts) about the 'Thinking Reed' and 'The heart has its reasons that reason doesn't know'. Only sporadically, since he had noticed that thinking about geometrical problems relieved his chronic toothache, he kept himself busy with mathematics but soon returned to his tormented religious diligence. His painful and perpetual indecision between science and religious devotion exhausted him mentally and physically so that Racine later proclaimed (and with reason): "Pascal died at the age of 39 of ... old age!". But in his short life he made himself immortal as a mathematician, a scientist (Pascal's law of hydrostatics), a writer and a philosopher.

Pierre de Fermat (1601–1665) came from Toulouse and was a lawyer by profession. Purely for the intellectual pleasure he engaged in mathematics in his free time. It is, however, unimaginable what this genius performed as a hobby and this in several fields of mathematics. Fermat can confidently be granted priorities: in analytic geometry, in which he introduced equations of a straight line and of conic sections with regard to a coordinate system already seven years before Descartes; in differential calculus, in which he mentioned a method for calculating tangents to curves already before Newton and Leibniz. We already mentioned that, together with Pascal, he stood at the cradle of probability calculus. But the name of Fermat has probably been linked most of all to what he performed in the field of number theory and the solution of indefinite (Diophantic) equations.

In 1621 Fermat got hold of a Latin translation of Diophantus' *Arithmetica*. Fascinated by this discourse on indefinite equations, he enriched this copy with marginal notes. For more than three centuries, some of them have kept mathematicians busy for their whole life. The famous problem 'Fermat's Last Theorem' was eventually solved in 1995 by the English mathematician Andrew Wiles. Even the worst lay mathematician has once come up against the formula of the Pythagorean theorem, namely $x^2 + y^2 = z^2$. This formula can also be considered as an equation with three unknowns x, y, and z. It is peculiar that this equation has an infinite number of solutions in the set of natural numbers. Such a solution is called a Pythagorean triple. For instance, (3, 4, 5) and (5, 12, 13) are Pythagorean triples. In one of his marginal notes Fermat states that the equation $x^n + y^n = z^n$ has no solution in the set of natural numbers for any n greater than

2, except for *x* or *y* being zero. He wrote that he had found a truly marvellous proof of this "which this margin is too narrow to contain". A record of a possible proof was never found, which saddled the mathematical world with an intriguing problem. A whole army of mathematicians desperately sought a proof. Some of them, like the German mathematician E. Kummer (1810–1893) spent their whole life working on it with partial success, but could not find a complete solution in spite of many innovative developments in mathematics.
Finally, in 1995, A. Wiles found a proof, by unexpected means and in relation to theorems of a totally different domain than number theory.

Whether Fermat himself actually got a general proof is strongly doubted and will probably never be resolved. But the fact of finding a conjecture on an equation with four unknowns in an infinite set is already surprising enough in a time when computers were not yet available.

Gérard Desargues (1593–about 1662) was an engineer-architect and for some time an officer in the French army. By virtue of his profession he was very interested in perspective, on which he wrote a booklet. In 1639, nine years after Kepler's death, he also wrote a theoretical discourse on conic sections with the long title: *Brouillon projet d'une atteinte aux événements des rencontres d'un cone avec un plan* (Draft of a trial on the events of a cone meeting a plane). It contained a draft of the development of what is now called 'projective geometry', including the study of the properties of figures that remain invariant under central projections. It contained a lot of new concepts that were named using an unusual 'botanic' terminology. At the time when it was published it hardly drew attention and soon the few existing copies had disappeared and the content forgotten.

Two centuries later, in 1845, the French land surveyor Michel Chasles found a transcription by Philippe de la Hire, a skilful pupil of Desargues. Soon thereafter it was acknowledged as an important and innovative work in the development of projective geometry, of which the torch was taken over by Gergonne, Poncelet, Brianchon and others.

The name of Desargues is forever linked to a remarkable theorem from another of his booklets. It is the theorem about two perspective triangles: 'if the pairs of respective vertices of two triangles are on three concurrent lines, then the three intersection points of the pairs of respective sides are collinear and vice versa.'

Briefly: 'two triangles are centrally perspective if and only if they are axially perspective'.

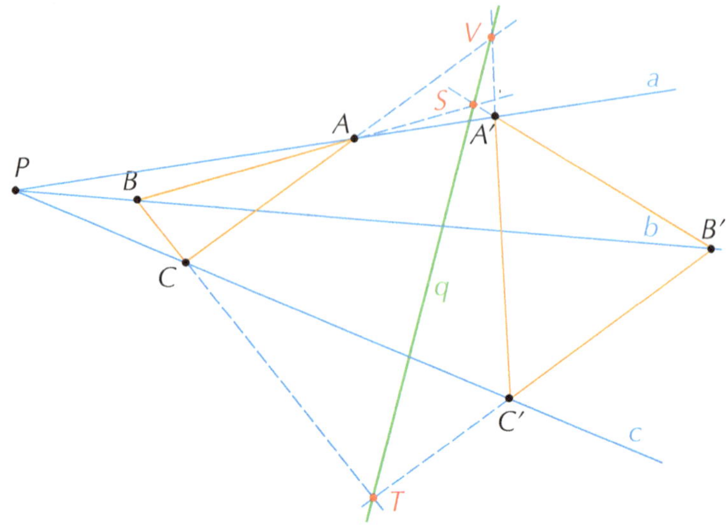

a, *b* and *c* are concurrent in *P*

AB and *A'B'* intersect in *S*

BC and *B'C'* intersect in *T*

CA and *C'A'* intersect in *V*

S, *T* and *V* are collinear on *q*

This theorem has a playful application in the following problem:

'How to plant ten trees in ten rows so that on each row are precisely three trees and every tree is on precisely three rows?'

(One possible solution: look at the points *A*, *B*, *C*, *A'*, *B'*, *C'*, *S*, *T*, *V* and *P*.)

The examples in this item show that fundamental and useful mathematics can bloom on pure curiosity and mental entertainment and this long before any applications are found for it.

3.7 The revealing of the 'Heavenly Mechanic'

We now come to the design of what became without any doubt the most pioneering mathematical instrument in the success story of physics, namely differential- and integral calculus. A meticulous analysis of motion and the underlying forces revealed both the secrets of orbiting planets and falling bodies as phenomena of the same laws. In the following centuries numerous applications of this so-called 'calculus' will stimulate technological progress and contribute to the launching of unstoppable industrialization as the next phase of civilization.

Although some mathematicians had successfully tackled problems related to integral- and differential calculus, only the circumstances of the seventeenth century provided the context in which this new arithmetic could develop so as to become a powerful aid.

Already Archimedes' mechanical technique of splitting areas into a sum of narrow strips can be seen as a primitive form of integral calculus. Also Kepler had applied a similar method in 1615 in his *Nova stereometria doliorum vinarium…* (New measurement of the volume of wine casks…) by splitting a body into a sum of small composing parts. In his *Geometria indivisibilibus continuorum nova* of 1635, Bonaventura Cavalieri, professor at the University of Bologna, developed an analogical technique of adding up 'indivisibles' (smallest particles) for calculating areas and volumes, just like a piece of cloth can be split up into threads and a book into pages.

Bonaventura Cavalieri (1598–1647)
Italian clergyman and mathematician. He developed his own method of calculating areas and volumes: 'Cavalieri's Principle'. This principle relies on the property that bodies with the same height and having the same planar section at each level, have an equal volume. There is an analogical principle for areas.

On the other hand, in 1638 Fermat had given a method for finding local minima and maxima of a curve by means of horizontal tangents. This procedure corresponds to the technique of differential calculus.

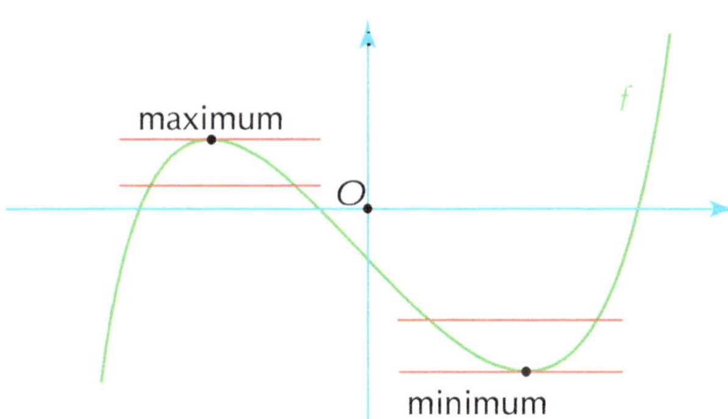

It is remarkable that the first ideas on differential calculus were developed only two thousand years after those on integral calculus. It shows that these techniques were not seen as being reciprocal.

Copernicus' new world view, successfully formulated in Kepler's and Galileo's laws, asked for precise definitions of the basic principles of motion, namely velocity, acceleration and force. Once Descartes had given an algebraic basis to geometry so that curves were interpreted in equations and vice versa, the possibilities were created for an algebraic interpretation of these elements of motion. In his *Arithmetica infinitorum* of 1655, John Wallis, an Oxford professor, had already paved the way by his treatment of infinite sums and products. Starting from this point of view, Newton and Leibniz have designed the proper differential- and integral calculus, assumedly independently from each other and in any case in a different way. This is why they are both acknowledged as the inventors of calculus.

The genius of Isaac Newton is linked in the first place to the discovery of the law of universal gravitation: the attraction that two bodies mutually exert. This law states that this force is proportional to the product of the masses of these bodies and inversely proportional to the square of the distance between them. This law constitutes the basis of the motion of orbiting planets as well as of falling objects. Newton is said to have been inspired when he saw an apple falling from a tree. It probably wasn't this simple. However, it all originated from his profound analysis of motion, which resulted in the so-called 'fluxion' calculus, Newton's version of differential- and integral calculus.

Newton studied the continuous motion of a point by means of the graph of a curve in a Cartesian coordinate system. From the equation of the variable coordinates of a moving point he could determine at each moment both the speed and the acceleration at this point, by a technique of calculation that he called 'first and last proportion'.

Let f be a function expressing the displacement in function of the elapsed time. Let P be a fixed point with coordinates t_0 and s_0, and let Q be a moving point with coordinates $t_0+\Delta t$ and $s_0+\Delta s$, then the slope of the line PQ is given by the proportion $\Delta s/\Delta t$ (the 'rise' divided by the 'run'). This equals the average velocity of an object moving over a distance Δs in the time span Δt.

If the point Q approaches the point P, then the line PQ approaches T, the tangent at the point P. In other words, if Δt approaches zero, then also Δs, but the proportion $\dfrac{\Delta s}{\Delta t}$ approaches the slope of the tangent T.

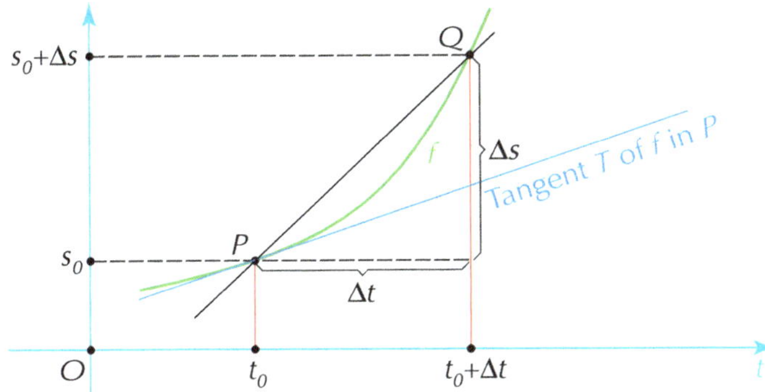

At the moment that $\Delta t = 0$ then also $\Delta s = 0$ and thus $\frac{\Delta s}{\Delta t} = \frac{0}{0}$, this means that 'the last proportion' of $\frac{\Delta s}{\Delta t}$, disappears. Newton quoted this as $\dot{s}$ and called it the fluxion of s (at the point P). Leibniz quoted it as $\frac{ds}{dt}$, a symbol that all mathematicians still use at present. The meaning of the 'differential quotient', $\frac{ds}{dt}$ (the derivative of s respective to t), is the velocity of the motion at the moment t_0.

In the same way Newton derived 'the fluxion of the fluxion' (second derivative of s respective to t), which he quoted as $\ddot{s}$, and which is a measure for the increase of velocity at the point P, i.e. the acceleration at the moment t_0.

It was also discovered that reciprocally one could calculate the displacement from the velocity by measuring the area under the graph of the velocity function, like Nicole Oresme had already shown for uniform motion. If P is a fixed point of the graph of the function f and Q a moving point, then the function F that measures the area under the graph of f over the interval between a and t is the integral of f. In other words, f is the derivative of F. This is the fundamental theorem of calculus.

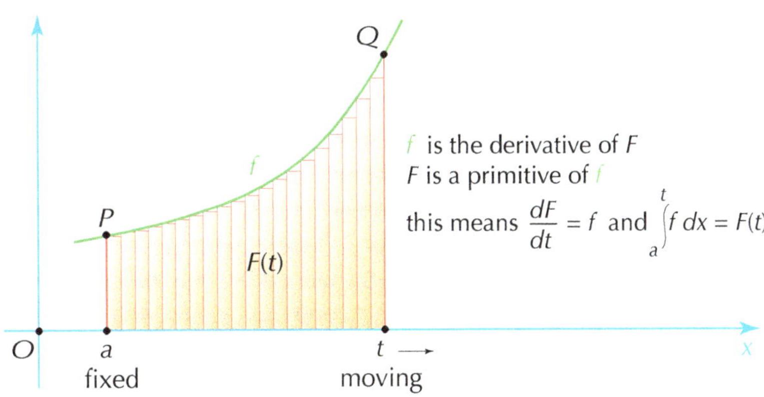

The area under the graph of f over the interval between P and Q was approximated by the sum of the areas of the rectangles with width Δt shown in the figure above. By narrowing the rectangles better approximations are produced. This procedure could be performed repeatedly.

Just as the derivative was found by a limit procedure on differences (hence the name differential calculus), the integral was obtained by a limit procedure on sums and both techniques proved to be reciprocal. By this, a fundamental understanding of the differential- and integral calculus was obtained. This doesn't mean that its theoretical fundamentals were already finalized. Although the practice of calculating 'last proportions' yielded undeniable successes, these premature limits were still vague concepts. The fact that the proportion 0/0 could evoke concrete numbers was fervently contested. Bishop and philosopher George Berkeley, a contemporary of Newton, was annoyed by the fact that the new science seemed to make religious doctrine more and more superfluous. He reacted to the 'calculus of fluxions' with the sneering remark that these 'last proportions' rather resembled 'ghosts of deceased magnitudes', the validity of which could be as easily doubted as the dogmas of Christianity.

Newton (1642–1727) was born on Christmas Day of the same year Galileo died. His father, a prosperous farmer, died before he was born. Although he showed interest in physical problems already at a young age, it was not until his admission as a student to the famous Trinity college in Cambridge at the age of eighteen, that he applied himself to mathematics. When the university of Cambridge closed temporarily in 1665 and 1666 because of a plague epidemic, Newton spent exceptionally creative days at home. In this period he laid the foundation of his fluxion calculus that he would complete in 1671. Because of bad experiences earlier with publications, the first edition appeared only in 1704. This was long after Leibniz had published his version of differential- and integral calculus in 1684 and 1686. On the other hand, Leibniz developed his version between 1673 and 1676, i.e. after Newton did. These circumstances created a painful and fruitless discussion concerning the priority of the invention of 'calculus' between English mathematicians, who sided with Newton, and continental mathematicians, who supported Leibniz. This created a long-lasting separation that was not advantageous for the development of mathematics in the United Kingdom. Leibniz's vision and symbolic were more practical and more open to generalization. Moreover, the complete version of the *Method of fluxions* appeared no earlier than 1736, after Newton's death. Newton was also only willing to publish his main work *Philosophiae naturalis princpia mathematica* with his findings concerning gravitation in 1687, after repeatedly being urged by his friend, the astronomer Edmund Halley. In 1669, after he had succeeded Isaac Barrow, who ceded his professorship to his brilliant student, Newton would still teach for 18 years at Cambridge. After that he was still showered with honours. He became Master of the Mint and President of the 'Royal Society', the English academy for science. In 1705 he was knighted. He died at the age of 84 after a painful and drawn-out illness, embittered by the dispute with Leibniz. He was buried in Westminster Abbey.

Newton has given his name to what is now called 'classical physics'. The mathematical foundation of the laws of dynamics, still formulated within the framework of absolute space and time, has established the mechanical world system. In the metaphor of 'The Celestial Clockwork' it has for always been given a paradigmatic appearance, just as the poet Alexander Pope expressed it beautifully:

> *„Nature and Nature's laws lay hid in night;*
> *God said 'Let Newton be' and all was light."*

Gottfried Wilhelm Leibniz (1646–1716) was born in Leipzig. His father died when he was only six years old. By self-tuition he became one of the greatest universal geniuses of his time. Still only a child he could already read Latin and Greek, and before he was twenty he had a command of the main ideas of mathematics, philosophy, theology and law. Nevertheless, he also felt drawn to diplomacy. So, in 1672, he found himself on a diplomatic mission in Paris where he met Christiaan Huyghens, who gave him further lessons in mathematics. The next year he spent on a political mission in London where he had the opportunity of demonstrating to famous mathematicians the calculator he had invented, which in contrast to Pascal's could also carry out multiplication and division. Later Newton's supporters would claim that during that stay Leibniz found out about the fluxion calculus and was prepared to commit plagiarism. He was very embittered about this because his version of differential- and integral calculus was indeed original.

In 1675, for the first time, Leibniz used the modern symbol for integration in his texts, namely $\int$, being a stretched character *S* as the first letter of the word 'Summa' (Sum). About the same time he introduced the symbols *dy* and *dx* for the differentials of the dependent variable *y* and the independent variable *x*. These symbols proved to be more efficient than Newton's fluxion notations and are still in general use today.

Leibniz, who never taught at any university, nevertheless founded an academy of sciences in Berlin and also tried to do so in Dresden, Vienna and St Petersburg. Together with Otto Mencke in 1682 he established the famous and widely circulated periodical *Acta Eruditorum*. As the chief editor he published his main mathematical discoveries in it. He also harboured great dreams and pursued projects. Thus he hoped to devise a universal logical language, *Characteristica Generalis,* in which each form of reasoning would be performed by some kind of automatic calculation. An idea that some centuries later would take shape in the symbolic logic of George Boole (1815–1864) and its application in the modern computer. Despite his international fame and esteem, only his faithful secretary was present at his funeral.

The Leibniz calculator (1671)
which he took with him on his visit to the Royal Society in London. This device could perform the four main operations (addition, subtraction, multiplication and division) mechanically by means of a system of stepped gear, i.e. cylinders with 0 to 9 cogs at appropriate levels.

Summary

With Newton and Leibniz the mathematics of the seventeenth century reached a peak of efficiency, which resulted in the continuing success story in the development of sciences. In the following century, the torch of these new mathematics would be taken over and carried by equally legendary persons: the brothers Jakob and Johann Bernoulli, Leonhard Euler, Joseph Louis Lagrange, Pierre Simon Laplace, d'Alembert, Monge and many more.

Under the protection of enlightened emperors and kings, such as Frederik II of Prussia, Catherina of Russia, Louis XV and Louis XVI of France, scientific activity flourished mainly at the important academies of St Petersburg, Berlin and Paris. In the first place differential- and integral calculus was further developed and was successfully applied to the problems of terrestrial and celestial mechanics. Also in the improvement of the production process, military strategy and navigation, mathematics would play a more decisive role. All this led to euphoria among philosophers of the Enlightenment who promised humanity a rosy future.

In his monumental work, *Mécanique Céleste,* Laplace presented a synthesis of the work of his predecessors. Among other things, he gave an explanation for the shape of the globe, analysed the complex problems of the interactive motion of the planets and their satellites, and carried out an investigation on the stability of the entire solar system. He overconfidently formulated the then deterministic vision of the world in the well-known witticism: "An intelligence knowing all the forces acting in nature at a given instant, as well as the momentary positions of all things in the universe, would be able to comprehend in one single formula the motions of the largest bodies as well as the lightest atoms in the world, provided that its intellect were sufficiently powerful to subject all data to analysis; to it nothing would be uncertain, the future as well as the past would be present to its eyes." Not surprisingly, then, that on Napoleon Bonaparte's question "why God didn't appear anywhere in his books", he answered "Sire, I had no need of that hypothesis".

And while the real wave of industrialization in the nineteenth century still had to get into its stride, a new acceleration in the development of mathematics was already taking place in that same century, which once more would give a formidable thrust to the history of civilization in the following century and would lead it in totally unexpected directions.

4 Entering modern times

After the explosive development of mathematics in the seventeenth and eighteenth century, under the influence of the flourishing natural sciences, there comes a time in the nineteenth century when mathematics rather turns back on itself. Gradually it frees itself from what till then had been a truly fertile framework of mechanics and astronomy to dedicate itself to theoretical deepening and pure abstraction. This process will end in a critical study of the fundamentals, the philosophical aspects of which have led to different points of view on the mathematical method. Together with this, the relation between mathematics and logic was given particular attention. The argument forms of Aristotle that had undergone hardly any modifications in twenty-three centuries of use and research, would be given a simple arithmetical structure with the work of George Boole and Gottlob Frege, exactly as Leibniz had hoped for in his *Characteristica Generalis*. Although this provided the theoretical basis for algorithmic processing of information, it would nevertheless still take till the second half of the twentieth century before the technological conditions for its realization in the modern computer were satisfied. Thus, almost one and a half centuries after the esoteric thinking of brilliant minds, we now experience day after day the astonishing advance of automated information and communication culture.

4.1 Plunging into the abstraction of non-Euclidian geometry

Throughout the centuries, the logical construction of Euclid's *The Elements* was considered to be the archetype for the deduction of mathematical 'truth'. In spite of the hypothetical 'if... then...' nature of the geometrical theorems, there was no doubt about the correctness of the postulates on which the proofs were founded. The link with reality was considered to be obvious and unquestionable.

However, critical spirits had already several times expressed objections to the unclear formulations of some basic concepts and postulates. More specifically concerning the fifth postulate, which states: 'if a line intersects two other lines and if the inner angles on its same side sum up to less than two right angles, then those other two lines intersect at a point on the same side of these angles'. This postulate is equivalent to the proposition: 'through a point not on a given line, there exists exactly one line parallel to that line'. Therefore this axiom was given the name 'parallel postulate'.

A lot of mathematicians supposed that possibly this axiom could be deduced as a theorem from the other (simpler) postulates.

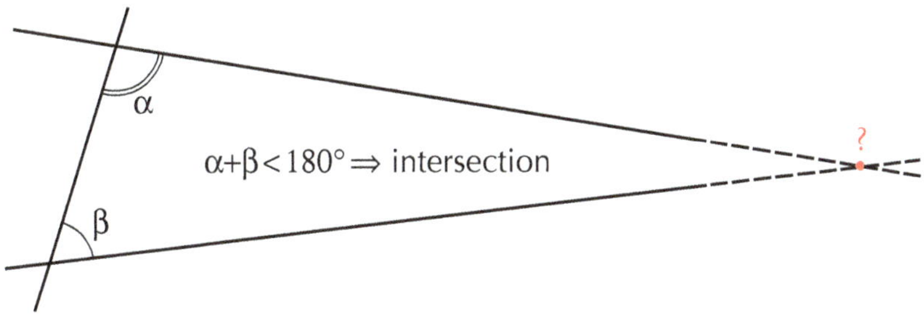

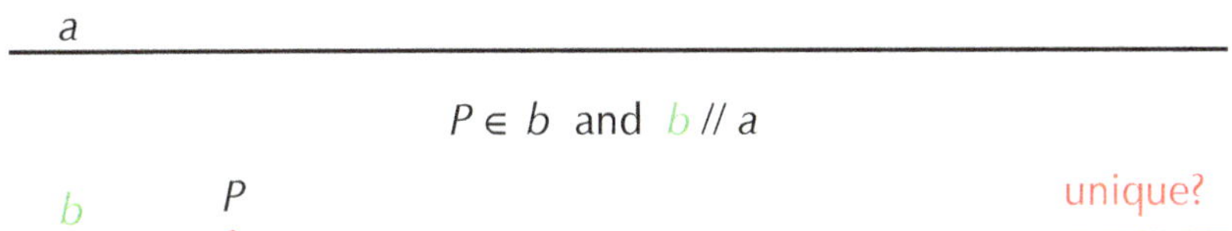

After fruitless attempts and despairing comments by, for example, Arab mathematicians like Omar Kahyyam, the Italian mathematician Girolamo Saccheri (1667–1733) made a remarkable attempt to find a proof. He tried to develop an indirect proof by starting from a denial of the postulate. But however far he got in his deductions, he could not obtain a contradiction. Which means that in fact he initiated non-Euclidean geometry. Nevertheless he remained convinced of the correctness of the parallel postulate. He could, however, not succeed in achieving his intended objective.

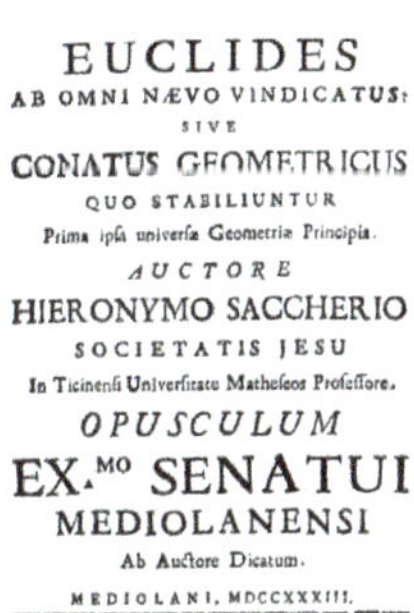

EUCLIDES
AB OMNI NÆVO VINDICATUS:
SIVE
CONATUS GEOMETRICUS
QUO STABILIUNTUR
Prima ipsa universæ Geometriæ Principia.
AUCTORE
HIERONYMO SACCHERIO
SOCIETATIS JESU
In Ticinensi Universitate Matheseos Professore.
OPUSCULUM
EX.MO SENATUI
MEDIOLANENSI
Ab Auctore Dicatum.
MEDIOLANI, MDCCXXXIII.
Ex Typographia Pauli Antonii Montani. Superiorum permissu.

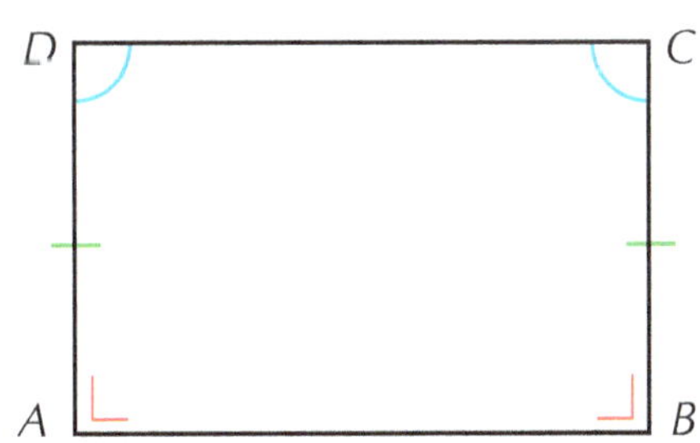

Title page of Girolamo Saccheri's main work from 1733: *Euclides ab omni naevo vindicatus* (Euclid vindicated of all errors). In this book, he tried to deduce the correctness of the parallel postulate from the previous postulates. On the right, the famous Saccheri quadrilateral, which shows that from the facts that *A* and *B* are right angles and the equality of *AD* and *BC*, one can prove the equality of the angles *D* and *C*, but not that they are right angles as well. In other words, it can't be proved that *AD* and *BC* are equidistant (i.e. parallel).

At the beginning of the nineteenth century, mathematicians suspected more and more that the fifth postulate was likely independent of the other axioms and that therefore it might be possible to create a consistent geometry in which this postulate was replaced by another. For instance, by a denial of it, such as: 'through a point not on a given line there exist several lines parallel to that given line' or even 'there exist no parallel lines at all'. These alternative constructions were worked out around 1830, more or less simultaneously and independently from each other, by Nikolai Ivanovitsj Lobachevsky, János Bolyai and Karl Friedrich Gauss.

Nikolai Lobatschewski (1793–1856)

János Bolyai (1802–1860)

Karl Friedrich Gauss (1777–1855)

Initially these ideas were rejected as being absurd and, in any case, useless. There is a touching historical anecdote about Farkas Bolyai, father of János and also a mathematician who had dabbled in this ungrateful subject earlier, and advised his son against publishing on this matter. Nevertheless János, full of hope, turned to Gauss to hear his opinion of these discoveries. To his deep disappointment, Gauss let him know that he found it difficult to praise his work because he would be praising himself then, since he had already made similar discoveries long ago, but because of the expected lack of comprehension he had preferred not to publish them.

Bernhard Riemann (1826–1866) has incorporated both Euclidean and non-Euclidean geometry in the general theory of spaces. Also for the new geometries models were constructed that linked the consistency of non-Euclidean geometries to the consistency of Euclidean geometry (see Chapter 1, 5). From that time onwards, however reluctantly, they were acknowledged as pure theoretical alternatives. After Einstein's publications on relativity theory in the beginning of the twentieth century, it was shown that precisely the model of a non-Euclidean geometry was suitable to adequately describe this theory. This is one more example of a creation of pure mathematics that only later finds an application. In this discussion on the right of alternative geometries to exist, mathematicians were confronted with questions that belong to the meta-theoretical domain, such as:

> *'How can one prove that something can't be proved?'*
> *'How can one show that certain propositions don't depend on each other?'*
> *'How can one know that a set of axioms is consistent?'*
> ...

Questions that would soon also appear in other domains of mathematics and that would require profound digging into the fundamentals.

4.2 The birth of modern algebra

Since the creation of numbers in ancient cultures, numeracy has been refined by the knowledge of the properties of the operations and their application to solving equations. So as to provide more complicated types of equations with solutions, the set of numbers had to be extended several times. Thus were constructed successively negative, rational, irrational and imaginary numbers.

Examples

-2 is a solution of $x + 3 = 1$	(-2 is a negative number)
$\frac{3}{7}$ is a solution of $7x = 3$	($\frac{3}{7}$ is a rational number)
$\sqrt{2}$ is a solution of $x^2 = 2$	($\sqrt{2}$ is a real number)
i is a solution of $x^2 = -1$	(i is a complex number)

Although the concept of 'real number' and certainly of 'complex number' (as an ordered pair of real numbers) was still very vague in the nineteenth century, nevertheless people smoothly calculated with them, and the fact that the usual properties of calculation with integers remained valid, gave a familiar appearance to real and complex arithmetic. Both addition and multiplication were commutative and associative, and multiplication distributed over addition (see Chapter 1, 6). Although in algebraic symbolism letters were used for the unknowns as well as for the coefficients, one always had numbers in mind. Just like Euclidean geometry had always had an expected and familiar nature, also algebraic arithmetic seemed to work out in a familiar structural framework. Until, to general consternation, suddenly between 1843 and 1857 objects were created such as 'quaternions' and 'matrices', for which multiplication proved not to be commutative.

Quaternions are an extension of the complex numbers and can be considered as quadruples of real numbers (a, b, c, d), for convenience written as $a + bi + cj + dk$, in which the symbols i, j and k represent imaginary units so that $i^2 = j^2 = k^2 = -1$ and $i \cdot j \cdot k = -1$. All properties of addition and multiplication that apply to complex numbers also apply to quaternions except for the commutativity of multiplication. Therefore, the set of quaternions is a non-commutative algebra.

Example

From $i \cdot j \cdot k = -1$ it follows after left-side multiplication of both terms of the equation by i, and because of the associativity of $\cdot$ that: $i^2 \cdot j \cdot k = -i$ and hence $-j \cdot k = -i$ and also $j \cdot k = i$. (1)
From this we can deduce in an analogous way the following results:

$j^2 \cdot k = j \cdot i$	hence	$-k = j \cdot i$,		(2)
$-k \cdot i = j \cdot i^2$	hence	$-k \cdot i = -j$	and $k \cdot i = j$,	(3)
$k^2 \cdot i = k \cdot j$	hence	$-i = k \cdot j$.		(4)

From (1) and (4) it shows that $j \cdot k \neq k \cdot j$.

likewise: $(2+3j) \cdot (1-3k) \neq (1-3k) \cdot (2+3j)$,
because the left term $(2+3j) \cdot (1-3k) = 2+3j-6k-9j \cdot k = 2-9i+3j-6k$,
and the right term $(1-3k) \cdot (2+3j) = 2-6k+3j-9k \cdot j = 2+9i+3j-6k$.

Quaternions were introduced by **William Rowan Hamilton** (1805–1865). One October day in 1843, when he was walking over Broom bridge near Dublin, he suddenly got the bright idea of the ingenious multiplication rule as described. This bridge now carries a stone plaque with the inscription: "Here as he walked by on the 16th of October 1843 Sir William Rowan Hamilton in a flash of genius discovered the fundamental formula for quaternion multiplication $i^2 = j^2 = k^2 = i \cdot j \cdot k = -1$, and cut it on a stone of this bridge".

Matrices are rectangular tables of numbers and may represent, among other things, linear mappings. The composition of linear transformations is related to the product of their matrices, which is carried out by the row-column rule (see Chapter 1, end of 6). In particular, square matrices form a non-commutative algebra since their multiplication is not commutative.

Example

$$\begin{bmatrix} 1 & 3 \\ 4 & 0 \end{bmatrix} \cdot \begin{bmatrix} 2 & 0 \\ 1 & 5 \end{bmatrix} \neq \begin{bmatrix} 2 & 0 \\ 1 & 5 \end{bmatrix} \cdot \begin{bmatrix} 1 & 3 \\ 4 & 0 \end{bmatrix}$$

since

$$\begin{bmatrix} 1 & 3 \\ 4 & 0 \end{bmatrix} \cdot \begin{bmatrix} 2 & 0 \\ 1 & 5 \end{bmatrix} = \begin{bmatrix} 1\cdot 2 + 3\cdot 1 & 1\cdot 0 + 3\cdot 5 \\ 4\cdot 2 + 0\cdot 1 & 4\cdot 0 + 0\cdot 5 \end{bmatrix} = \begin{bmatrix} 5 & 15 \\ 8 & 0 \end{bmatrix}$$

and

$$\begin{bmatrix} 2 & 0 \\ 1 & 5 \end{bmatrix} \cdot \begin{bmatrix} 1 & 3 \\ 4 & 0 \end{bmatrix} = \begin{bmatrix} 2\cdot 1 + 0\cdot 4 & 2\cdot 3 + 0\cdot 0 \\ 1\cdot 1 + 5\cdot 4 & 1\cdot 3 + 5\cdot 0 \end{bmatrix} = \begin{bmatrix} 2 & 6 \\ 21 & 3 \end{bmatrix}$$

Once Arthur Cayley (1821–1895) had developed his matrix algebra in 1857, the gate to abstract algebraic structures was opened once and for all. Mathematicians had understood that properties of operations could be taken as independent axioms and that the objects to which these operations applied didn't necessarily need to be numbers. So, abstract structures were designed of which the elements, depending on the appropriate context, could be interpreted as polynomials, functions, vectors, propositions, etc... . Since then, mathematics has operated mostly within the framework of such an algebraic structure, constructed as a deductive system of which the basic properties are the axioms from which theorems are deduced. One of the simplest but richest structures of algebra is the 'group' (see Chapter 1, 6). The father of group theory is the French mathematician Evariste Galois (1811–1832). He died at the age of twenty-one in a pistol duel. On the eve of the duel he had an anxious foreboding and hastily wrote the report of his discoveries to a friend, "lest it be forgotten for posterity". This document, full of scrawls, finishes with the dramatic words "I have no more time, I have no more time". It was only later assessed to its full value and its content has become one of the most fertile theories in mathematics, namely the Galois theory. This theory examines, among other things, the general problem of the possibility of solving equations of a higher degree in an algebraic way. He found the surprising result that as from the fifth degree a solution on the basis of root extraction is no longer possible. So the Italian mathematicians of the fifteenth century had travelled to the furthest frontier in this field, possibly without realizing it themselves.

4.3 Laying the solid foundations of analysis

Since the development of differential and integral calculus by Newton and Leibniz, there was criticism from several sides of the vague nature of the infinitesimals. Both bishop Berkeley and burgomaster Nieuwentyt had already made things hot for Newton and Leibniz respectively on this issue and had stirred up scepticism with regard to the fundamentals of the new calculus. Nevertheless, the continued undeniable successful results constantly encouraged and established the practice of the new arithmetic. In the course of the eighteenth century however, mathematicians themselves would start to realize the paradoxical results that were produced by simplistic and improper manipulations of infinite sums and products. Even a mathematical genius like Euler obtained results that were totally unacceptable.

Example

By continued division one gets the following expressions:

$$\frac{1}{1+a} = 1 - a + a^2 - a^3 + a^4 - \dots \qquad (1)$$

$$\frac{1}{1-a} = 1 + a + a^2 + a^3 + a^4 + \dots \qquad (2)$$

By substituting a by 1 in (1): $$\frac{1}{2} = 1 - 1 + 1 - 1 + 1 - \dots \qquad (3)$$

By substituting a by 2 in (2): $$-1 = 1 + 2 + 4 + 8 + 16 + \dots \qquad (4)$$

Some people got even more reckless and rearranged (3) in:

$$\frac{1}{2} = (1-1) + (1-1) + (1-1) + \dots = 0 \qquad (???)$$

or in: $$\frac{1}{2} = 1 - (1-1) - (1-1) - \dots = 1 \qquad (???)$$

Encountering so many contradictions, consolation was sought in the fact that $\frac{1}{2}$ is indeed the average of 0 and 1.

People were aware that these anomalies originated in the infinite number of terms in these expressions. Because, when ending the division in (2) at the fifth term in the quotient and taking into account the remainder a^5, one gets:

$$\frac{1}{1-a} = 1 + a + a^2 + a^3 + a^4 + \frac{a^5}{1-a}$$

and this is correct for every value of a, e.g. for $a = 2$:

$$-1 = 1 + 2 + 4 + 8 + 16 - 32$$

In the course of the nineteenth century mathematicians became increasingly aware that if they wanted to secure the neat results of calculus, they had to define its fundamentals very sharply. The merit for this belongs to:

Augustin-Louis Cauchy, (1789–1857) Karl Weierstrass (1815–1897) and Richard Dedekind (1831–1916).

This process is called 'the arithmetizing of analysis'. It is reflected in the modern definition of the concept of 'limit' by Cauchy, the specification of the system of the real numbers by Dedekind and the theory of convergent sequences by Weierstraß. This way, infinitesimal calculus was freed from its initial geometrical intuitive framework and could fall back on the solid axiomatic system of number theory. The manipulation of infinite series was subjected to the conditions of relative and absolute convergence. Placing brackets or moving terms cannot be done unconditionally in an infinite sum. Departing from these insights the paradoxical results were put aside and the applications cleared of every blemish.

4.4 The fascination of infinity

Already in ancient Greek times there arose problems and discussions about the nature of the infinite. Both in mathematics and in philosophy, Zeno's paradoxes had highlighted the pathological aspects of the continuum and of infinite constructions. Although Aristotle declared himself against the 'actual infinite', many mathematicians and philosophers didn't want to share his opinion for various reasons. In the seventeenth century the discussions flared up violently again as a result of the use of infinitesimals in differential and integral calculus and the anomalies in infinite series.

In his study of falling motion, Galileo had run into another singularity of the infinite. From the consideration that each natural number has exactly one square, he decided that the sequence of the squares must be as numerous as the sequence of the natural numbers itself, although it is but a part of it.

Example

1	2	3	4	5	6	7	8	9	...
1	4	9	16	25	36	49	64	81	...

Keeping in mind his difficulties with other obvious observations, he found it safer not to make this public. After him Bernard Bolzano (1781–1848) made similar discoveries showing that the number of points on a line segment with length 1 obviously contained as many points as whatever other line segment with another length, e.g. the double of it. In an analogous way one can even show that an (open) line segment contains as many points as the whole line of which it is a part. This means also that there are, for example, as many real numbers between 0 and 1 as there are real numbers altogether.

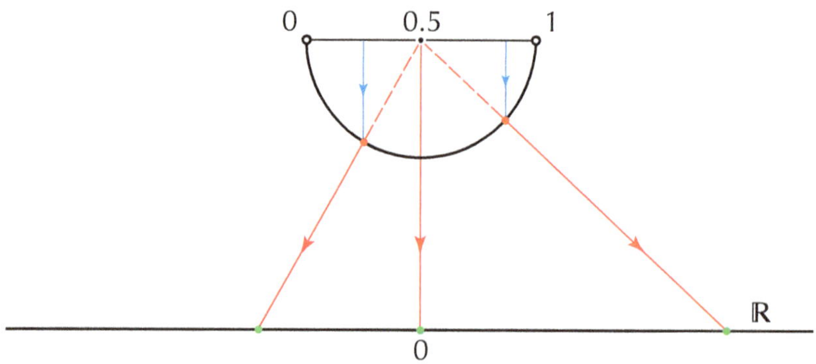

These remarkable results inspired Dedekind to take this property as a defining characteristic of infinite sets. If it is plausible that the same numerousness is assigned to two sets, of which the respective elements stand in a one-to-one relationship, then it implies that it is precisely an infinite set that has the characteristic of being equinumerous to a proper subset of it. For instance, the set of the natural numbers is equinumerous to the set of their squares.

The German mathematician Georg Cantor (1845–1918), founder of modern set theory, fully agreed with Dedekind's definition. By argumentations that would both astonish and bewilder the mathematical world, he obtained strange results. In the proof of the famous proposition that carries his name, he could show that the set of all subsets of a set always has a strictly higher numerousness than the original set, and this not only for finite sets but also for infinite sets (see Chapter 2, 2.2.3).

Example

Let $V = \{a,b,c\}$, then the set of all subsets of V is:

$$\mathcal{P}V = \{\{\ \},\{a\},\{b\},\{c\},\{a,b\},\{a,c\},\{b,c\},\{a,b,c\}\}$$

Observe that V contains three elements and that $\mathcal{P}V$ contains eight and $3 < 8$.

If we denote the number of elements of a set V as $\#V$, then it generally applies that: $\#V < \#\mathcal{P}V$. This is Cantor's famous theorem. This theorem has the incredible result that there exist different types of infinite sets with a different numerousness (cardinality) and that there is no such thing as the most numerous set. For instance, if we represent the set of natural numbers by $\mathbb{N}$, then:

$$\#\mathbb{N} < \#\mathcal{P}\mathbb{N} < \#\mathcal{P}\mathcal{P}\mathbb{N} < \#\mathcal{P}\mathcal{P}\mathcal{P}\mathbb{N} < \ldots$$

Like many other famous mathematicians, Leopold Kronecker (1823–1891), who had even been Cantor's professor, couldn't reconcile himself with such aberrations of mathematical deduction. He considered them as a personal insult by his alumnus, whose further career he would obstinately thwart. Other prominent mathematicians from their side saw in set theory a favorable means of giving a common basis to the different fields of mathematics and of unifying them. So Hilbert exclaimed enthusiastically: 'From the paradise that Cantor has created for us, nobody will ever chase us'. However, from the reasoning in Cantor's style and the foolhardy diligence for creating ever-bigger sets, soon there emerged annoying paradoxes that raised many doubts and even despair in those who had cherished the holy house of mathematics and its applications.

As a striking example we can once again quote Russell's paradox (see Chapter 1, 4.7). Most sets that we can imagine have the property of not being an element of themselves. For instance: the set of the vowels of the alphabet is in itself not a vowel and thus not an element of itself. Russell wanted to define a set V of which the elements are precisely those sets that do have the property of not being an element of themselves, so:

$$V = \{\, X \mid X \notin X \,\}$$

If V were really a set, then it is appropriate to put the question: 'Is V an element of V?'

This would be true if V satisfied the selection criterion for the elements of V, namely $V \notin V$. This would mean then:

$$V \notin V \Leftrightarrow V \in V \qquad \text{(Oops!)}$$

If it were not true, in other words if V were not an element of V, then exactly V would satisfy the selection criterion and it has to be accepted as an element of V, which means as well:

$$V \notin V \Leftrightarrow V \in V$$

The paradox was thus inevitable and, to make things worse, this appeared not to be an isolated case. The whole bunch of paradoxes, however, all proved to have the same characteristic of self-reference, as in the statement of the Cretan who says 'I lie', or in the proposition 'this sentence is not true'. Whatever truth-value one adopts, inevitably one always ends up denying what one had supposed to be correct.

Because in this way the whole construction of mathematics, as well as the arsenal of its successful applications, seemed to be situated on a logical minefield, mathematicians could only dig down to the roots of their reasoning so as to exterminate these intolerable antinomies. In particular, the relation between logic and mathematics was thoroughly unravelled. As a result several competitive schools arose that can be distinguished, according to their presented solutions, as logicians, intuitionists and formalists.

The logicians, with such distinguished representatives as Bertrand Russell and Alfred North Whitehead, authors of the monumental *Principia Mathematica*, tried to show that the whole of mathematics can be reduced to pure logic. An attempt which Kurt Gödel in his earlier mentioned article of 1931 would show to be not feasible (see Chapter 1, 7).

The intuitionists, with Luitzen.E.J. Brouwer as their flag-bearer, defended a point of view in which infinite constructions are only considered as 'potential', i.e. as a possible continuation of successive finite constructions, and not as 'actual', i.e. as a completed process. By also rejecting the law of the excluded third possibility they clipped the mathematical wings so much that a lot of the precious fruits of mathematics threatened to get lost. Many fellow-mathematicians were not willing to go that far.

Bertrand Arthur William Russell (1872–1970)

British mathematician, logician and philosopher. At the age of three, he lost his mother and his sister, one year later his father and two years later his grandfather. After his studies in logic, he got in contact with Peano and Frege. Together with Alfred North Whitehead he wrote the 'Principia Mathematica', a project in which they tried to prove that mathematics and logic have the same basis. In 1901, in this work he formulated his famous paradox that is mentioned above.

The programme of the formalists, who had David Hilbert as their leader, existed in designing an axiomatic system that enables the manipulation of in themselves meaningless symbols. The hope of integrating a guarantee of absolute consistency within such a system itself was, however, swept aside also by the results of Kurt Gödel. Since then mathematicians have learned to live with undecidability, incompleteness and temporariness. This has not prevented the restoration of the possibility of reliable and steadily progressing research.

4.5 The sources of the flowing stream of information

In contrast to what the logicians had tried, namely to show that mathematics was a part of logic, it turned out to be the opposite: mathematics can structure logic. In 1854 George Boole showed in his book *The Laws of Thought* how Aristotle's logical argument forms could be converted into a simple arithmetic calculus. Also the work of Gottlob Frege, professor at Jena, continued the formalization of logic. Although at that time the mathematical basis for an automation of the logical processing of information was available, it would nevertheless take until the second half of the twentieth century before the technological conditions for the realization of a full computer were available.

When humanity progressed from the stage of counting to the complicated stage of calculating, auxiliary tools were welcome. Finger work was replaced by counting frames and numeral systems. But their use still required skill, knowledge and much effort. In the course of the seventeenth century, several attempts to construct mechanical computers were undertaken. In 1624 Wilhelm Schickard, a friend of Kepler's, succeeded in building a machine that could multiply one number by another. In 1642 Pascal introduced the Pascaline mentioned earlier, which could make additions and subtractions by means of cogwheels that operate in the way a milometer works. Then in 1674, Leibniz introduced a calculator that could infallibly carry out the four basic operations, including multiplication and division. At the same time he introduced the binary numeral system. But automation was still out of the question.

After the French weaver Joseph-Marie Jacquard had developed a loom that was controlled by means of punched cards (the punches determined the variation in the weaving patterns), the English mathematician Charles Babbage (1792–1871) adopted the idea in an attempt to design an automatic calculator. This design of the 'Analytical Machine', with its functions of input, memory, control and output, would form the basis of the modern computer. In spite of the financial and moral support of his assistant Lady Ada Lovelace, daughter of the well-known Lord Byron, this project was not completed because of the inadequacy of technology at that time. Between 1884 and 1890 the American statistician Herman

Hollerith also employed a system of punch cards to automate the input and the processing of census data. He used the facilities of electricity by transforming the mechanical movement of the tracker pegs over the punches into electric pulses. With this technology the gigantic processing of the census was considerably shortened from ten to three years. After a number of improvements the machine could be used for other tasks, too.

The great breakthrough of electronic computers, however, only started in, and mainly after, the Second World War. This was partly due to military needs like encoding and decoding secret messages. Once the young English mathematician Alan M. Turing had described in 1936 the elementary operations of an abstract model of a general logical machine, the so-called Turing machine, the realization of these principles and the ideas of Babbage and Hollerith became more and more successful.

An important milestone was the creation in 1944 of the MARK I at Harvard university by the American physicist Howard Aiken in association with IBM (International Business Machines). This 'device' weighed five tons and contained 3304 mechanical relays that were still controlled by a programmed punched tape. Calculations were indeed already performed in fractions of seconds, but this was still considered as being too slow.

The first full electronic computer was the ENIAC (Electronic Numerical Integrator And Computer), developed between 1943 and 1946 at the university of Pennsylvania by the two American scientists Prosper Eckert and John Mauchly. It was an enormous construction too, weighing 30 tons, containing 18,000 vacuum tubes and filling a room of 10 by 16 meters. The idea of representing numbers in the binary numeral system, in which the digits 1 and 0 are coded as electric power being switched on or of, speeded up calculations to 300 multiplications per second. It was an electronic calculation brain of the first generation that was still extremely vulnerable. Some tube would always burn out somewhere or other. Moreover, the tubes consumed a lot of power and therefore developed tremendous heat.

Driven by technological progress computers of the second, third and fourth generation soon followed, based respectively on transistors and integrated circuits (microchips). This was accompanied by a steady increase in calculation speed, extension of memory capacity and especially also incredible miniaturization. Current microprocessors may be no bigger than a mere matchbox.

In front of the ENIAC, Aberdeen Proving Ground, April 4th 1950, on the occasion of the first digital weather forecast executed by means of a computer. From left to right: H.Wexler, J. von Neumann, M. H. Frankel, J. Namias, J. C. Freeman, R. Fjortoft, F. W. Reichelderfer, and J. G. Charney.

In the meantime the 'personal computer', the 'pocket calculator' and the 'cellular phone' have become common. At a pace almost impossible to keep up with, increasing complexity and user-friendliness offer the whole arsenal of data processing and communication technology for use like a toy in the hands of the average man in the street. Nobody can predict the further spectacular development of cybernetics, automation and the fauna of artificial intelligence in the near future. The impact of this process can, however, already be noticed everyday and everywhere in our super-fast changing culture.

Conclusion

In this chapter we have tried to identify the innovative aspects in the history of mathematics that contributed to some remarkable social and cultural changes. These aspects certainly do not stand apart from the spirit and the social context of that time. However, mathematical activity relies on few material conditions. Reflecting and, if necessary, a stick for writing in the sand does not demand big investments or lobbying for subsidies. Also, the mathematical universe is mainly esoteric, platonic or even poetic in the sense that it steadily expands because of creativeness and imagination. For this reason its continuous applicability and efficiency are all the more remarkable.

For the development of agrarian cultures, numerous practical problems had to be solved with the help of calculation and measurement. But besides that, the fascination of pure numeric relations and the aesthetic of abstract forms were attractive enough for also occupying oneself in a purely recreational and creative way without having any expectation of usefulness or profit.

The Greek mind was not at all oriented to practical experimental science. Only striving for understanding and insight was held in high esteem. The Greeks were content with speculative, purely idealistic models offering a framework for possible explanations of phenomena. Reality was not the main concern; a coherent interpretation was sufficient to come to terms with one's existence. And it was precisely in this culture that the foundation was laid for our entire Western rational attitude, free from political or economic considerations. It would be difficult to maintain that the mystic obsession of the Pythagoreans, with their amicable numbers and pentalpha symbolism, originated from the concrete needs of the then social context, and no more than Euclid's titanic work and the intellectual sport of the Diophantine equations did.

On the other hand, it is unmistakable that in the latter days of the Renaissance, with the emergence of experimental science under the influence of philosophers such as Francis Bacon, John Locke and David Hume, the mathematical developments found their inspiration in practical problems of physics and astronomy. Nevertheless, Galileo, Kepler and Newton immediately looked for some 'mathematical accommodation' for their findings, so that the inductive nature of their experiments was again given the prestige of deductive causality. Although the efficiency of the differential- and integral calculus made its indispensable contribution to the study of almost all dynamic processes during the following centuries, the progress of the industrial revolution took place at the margins of pure science for a long time still. In the meantime, in the burning brilliance of the Enlightenment, several domains of intrinsic mathematics, such as analytic geometry, calculus, probability, projective and descriptive geometry, were further elaborated without continuously taking their direct usefulness into account. Both the three body problem of the mutual attraction of the sun, the earth and the moon as well as the stability of the solar system, are beautiful inspiring questions that certainly have contributed to a lot of research on solving differential equations, but their contribution to everyday life is rather negligible. And over and over again, it is the pure mathematical interest that holds the mind in its spell, before some concrete practical application is possibly to be found.

The extent to which an apparently esoteric crazy thought can change the appearance of our world sometimes only centuries later is illustrated in our present culture. Who would have predicted that by thinking about the provability of the parallel postulate, the paradoxes of infinite sets and the consistency of an axiomatic system, which forced mathematicians to do research on the fundamentals and to poach on logic, the foundation has been laid for the miraculous phenomenon of the computer and its relatives? Our so-called 'modern' mathematics, on which all this is based, undeniably goes back to the nineteenth century; an era in which the industrial revolution could still abundantly benefit from the science and the mathematics of the two preceding centuries. And using the inconceivable calculation speed and the simulation possibilities of the computer, present-day mathematics can push its own frontiers forward with the exploration of the wondrous world of fractals and deterministic chaos, which in the fields of non-linear dynamic systems throw light on a new science in the making.

And so the tale of the torch of understanding, which according to the analysis of its own nature is inadequate to illuminate the entire universe with its glow, continues to light up the frescoes of yet unexplored corridors.

"Theory attracts practice like a magnet attracts iron"

Karl Friedrich Gauss

Chapter 4

Mathematics in nature and art

Life is complex: it has a real and an imaginary part.

Tom Potter

In this chapter we give some examples of how several aspects and structures of mathematics appear in natural phenomena and in various art forms.

On the one hand, the phenomena that the cosmos has produced since the `Big Bang', including the different forms of life, are marked by processes determined by what we call the laws of nature, but on the other hand they are the result of accidental choices from a number of combinatorial possibilities available at certain moments of evolution. Hence it should be no surprise that mathematics, which has proved to be particularly suitable for describing both the laws of nature and probability theory, can contribute to a better understanding of what takes place in nature.

Why do things have the shape they have?

What is the importance of symmetry?

Where does the predilection for efficiency and optimization come from?

What is the deeper meaning of all those constants that press down upon us, such as:

- the speed of light (the maximum speed in the universe),

- the Planck constant (the minimum energy of an elementary quantum particle),

- the Hubble constant (the proportional factor of galactic nebulae expansion)?

So many questions with answers that could throw some light on the mysteries of the cosmos and its inescapable destiny. Mathematics evidently reveals itself in every realm of reality:

- the cone-shaped umbrae of an eclipse;
- the changing seasons;
- the growth pattern of a plant;
- the shape of shells on the beach, etc.

R. Verhulst, *Under the Spell of Mathematics*, https://doi.org/10.1007/978-3-031-63871-8_4

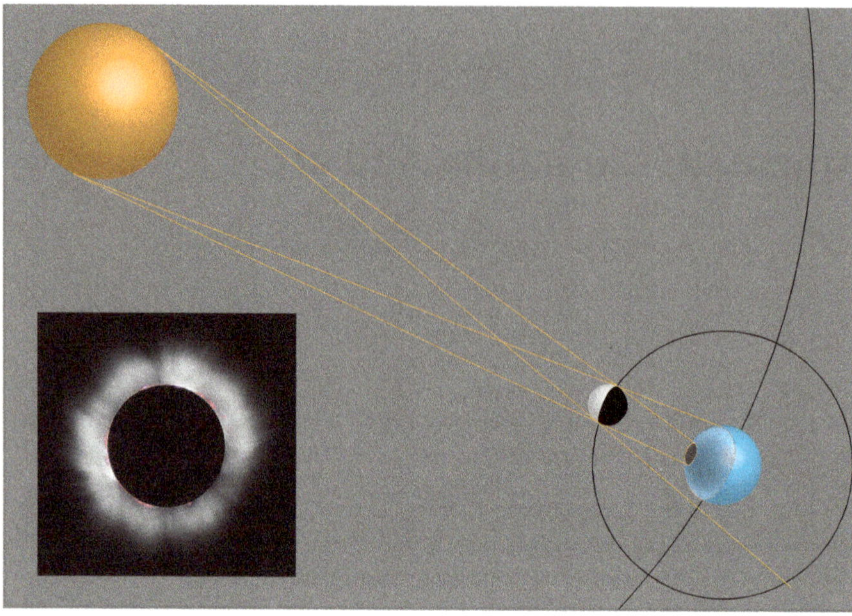

A solar eclipse occurs when the Moon comes into alignment with the Earth and the Sun. The common tangents of the Sun and the Moon determine the umbra (light radiation completely blocked) and penumbra (light radiation partially blocked).

Shells acquire the form of a logarithmic spiral because their growth is proportional to their actual size. This form possesses the golden ratio.

While the functioning of nature is not supposed to take autonomous and conscious control of itself, things are of course different in the world of arts. Art is the product of the artist, and what is expressed is the fruit of his or her entire personality, sometimes kept within deliberate boundaries, but often purely intuitive. There is a very real danger that an art lover with a subjective view imagines structures and meanings in an artistic work that are totally unintended by the artist. In Piet Mondriaan's work, one would easily be inclined to assume mathematical principles. When he was asked if he was in the habit of starting his compositions from basic squares, he answered that he himself did not notice any squares at all in his works and that his only intention was to break with the notion of volume. Nevertheless, one should not conclude from this that there are never deliberate rules in an artwork that may originate, for example, in mathematics. For instance, a painter who wants to realistically represent a three-dimensional scene on a planar surface, will have to take into account the rules of perspective from projective geometry. Even so, architects cannot restrict themselves to the mere aesthetic aspect of a design but they are obliged to take into account all kinds of physical conditions, such as gravity, which was mathematically streamlined by Galileo and Newton. This is why the shape of a building also reflects the hidden beauty of all the calculated forces that contribute to its stability.

So far there is no unambiguous criterion for aesthetics, even though some philosophers and art critics have made attempts to formulate one. Still, even a person without a musical education can hear the difference between the random tinkling of piano keys and the performance of a score that takes into account the rules of harmony and composition, bound by proportions and periodicity. Therefore, in this domain we will restrict ourselves to examples that will not arouse controversy.

The Pantheon in Rome, originally built by Agrippa about 28 BCE. After being destroyed by a fire in 80 CE, it was rebuilt between 118 and 125 by Hadrian.

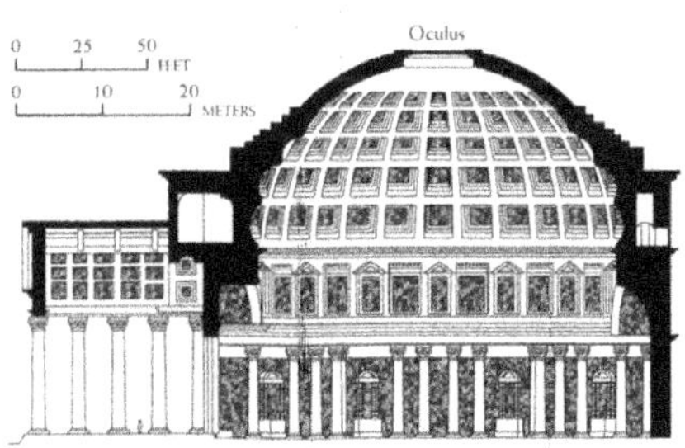

Section of the Pantheon. The cupola has a 9-metre wide opening *(oculus)* at the top. The building's diameter equals its height of 43 metres.

1 Mathematics in natural shapes

1.1 The equations of the cosmos

When primitive human beings were contemplating the dark nocturnal sky, not at all hindered by an abundance of street lamps and neon signs, watching the constellations of an uncountable number of pin-pricks of light that each night slowly turned around the Pole Star in invariable coherent figures that moreover were traversed from time to time by the waxing or waning moon, they certainly must have been impressed by the grandeur of this spectacle. The fact that this sight did not inspire them to imagine the earth as being a sphere probably had little influence on their day-to-day worries. This, however, does not affect the true fact of the earth being a sphere that revolves around its own axis as well as around the sun, motions that already by themselves explain a lot of natural phenomena. Their feelings could not have had the added value that can only result from understanding all those phenomena with the models that are available now. Eclipses, shooting stars, thunder and lightning no longer scare the living daylights out of us, but nevertheless we are silenced by the marvellously beautiful connection between them: the mass and the proximity of the planet Jupiter protect us from a rain of harmfully large meteorites; the presence of the moon contributes to the stability of the obliquity of the earth's axis with regard to the ecliptic and through that to the regularity of the seasons, etc. Moreover, the laws of gravitation make their own precise calculations and we can just go on living and being grateful for so much refinement and discernment. From our present point of view the universe is a gigantic energy system in evolution.

1.1.1 The Big Bang model

On a macroscopic scale we enjoy the spectacle of galactic nebulae, dust clouds, gasses and of stars taking shape, exploding and collapsing while a carousel of planets, satellites, comets and meteorites accompany them.

The horse head nebula

in the constellation of Orion. The dark cloud consists of molecular gas, the red glow is hydrogen gas, ionised by a nearby star. Galactic nebulae of this kind are like a delivery room for new stars.

All this is grandiosely described by the theory of general relativity. In the microscopic world we observe in astonishment the atom components' bizarre flashing and the hotchpotch of electromagnetic phenomena, successfully described in their turn in the models of quantum theory. Both these levels are brought together in the catching story of a big bang at the origin of matter and radiation as well as space and time. This story is nicely summarized in some mathematical equations, which do not only provide a model for all kinds of things that are observed but also enable extrapolations over time. The sceptic will raise the fact that the Big Bang model is but a hypothesis and that anyhow the equations do not precede the phenomena but offer only the means to deliver some post hoc overview and synthesis. However, since the day Galilei had succeeded in catching the falling motion in a formula and exclaimed: "The great book of nature lies open before our eyes, and to be able to read it we need mathematics", so many convincing examples have already been revealed. So it seems that the book metaphor becomes more and more appropriate.

The theory of the Big Bang is of course not just a made up variation on some story of the Creation, but a consequence of calculations and observations. The strongest arguments for the usefulness of the Big Bang model lie in the field of three crucial measurements.

1 Hubble's constant

Edwin Hubble had discovered a linear correspondence between the distance of a galaxy and the velocity with which it moves away from us. This he did by measuring the red shift in the colour spectrum of the galaxy, caused by the Doppler effect.

The Hubble telescope

orbiting around the earth. Shortly after its launch in 1990 it became clear that the mirror was wrongly adjusted regarding the incidence of light. The error was 2/5 of the thickness of a human hair! In 1993 a team of astronauts was sent on a special mission to adjust this error, and the telescope could start its amazing performances.

There seemed to be a constant factor involved that since then has been called Hubble's constant. From the precise value of this constant the age of the universe can also be derived. Because of the importance of a correct measuring of this constant, in 1990 the Hubble telescope was sent into orbit around the earth so as to refine the observations. From the calculated value it became clear that the universe is approximately fourteen billion years old.

2 Background radiation and the temperature of the universe

Cosmic background radiation is a remnant of the electromagnetic radiation that was released because of the extremely high temperature caused by the Big Bang. The theory of quantum physics can describe the interaction between matter and radiation. It seems that the temperature and the wavelength of the maximum amount of radiation are inversely proportional, and because of this the current temperature of the universe can be derived from the actually measured background radiation. Moreover, an inversely proportional link was also observed between the dimension of the universe and its temperature: the bigger the universe gets, the lower its temperature. The universe, which during its expansion kept on cooling down, has now a (mean) temperature of –270.42° Celsius, only some degrees above the absolute minimum of –273.15°. From the spectrum of the background radiation, which was observed by the satellite COBE, a number of characteristics appeared that support the Big Bang theory: isotropy (the same radiation in all directions with negligible irregularities); the perfect correlation of more than 100 measuring points on the radiation graph of a presumed 'black body' (a source with only thermal radiation) and that with a particular constant temperature.

3 The distribution of matter in the universe

The Big Bang model also offers a satisfying explanation for the peculiar distribution of matter in the universe (72% hydrogen, 26% helium and only 2% for all other elements), as well as for the fact that there is almost twice as much radiation (photons) as matter (atoms). The bond of the protons, neutrons and electrons and the amounts of them that were available in the first moments after the Big Bang explains why the lightest two elements, hydrogen and helium, take such a much bigger portion than the other elements, among which only iron stands out. The quantitative proportion of radiation to matter would follow from the reciprocal destruction of matter and antimatter particles. Also, the density of matter in the universe (the proportion of the total quantity of matter to the volume of the universe) has a critical value related to it, on which the eventual destiny of the universe depends. Below that critical density the universe keeps on expanding at an ever increasing speed (expansion model). Above that value, however, the expansion would be slowed down by the overall gravitation of matter and would subsequently change to a process of shrinkage, finishing in a 'Big Crunch'. For a value exactly equal to the critical density, the expansion would gradually slow down and proceed in a stable situation (steady state model). A definite answer has not yet been given because spectral analysis only reveals matter from which the light reaches us. How much more matter there is, for instance, in the 'black holes' or in galaxies of which the light has not reached us yet, can only be guessed and so far isn't clear. Considering our lifespan on the cosmic time scale, we need not worry about a possible disaster scenario.

Although alternative theories on the evolving universe circulate, the Big Bang model has for just and fair reasons many supporters among prominent physicists, who attribute equal authority to it as, for instance, to Darwin's theory of evolution. It was presented for the first time in 1927 by the Belgian physicist Georges Lemaître

Georges Lemaître

1.1.2 Theoretical deductions and prophecies

Since the development of the natural sciences from the seventeenth century, mathematics has always stood by their side and has provided the concepts and techniques for a clear description of the different fields of its research. In natural phenomena the laws of nature do not catch the eye, otherwise they would have been noticed across the ages. Similarly, neither the formulas nor the equations that gave them a simple appearance. It is, however, remarkable that at all levels where physics wrapped itself in mathematics, innovations appeared that initially weren't included nor expected. More than once it has turned out that if the equations and calculations required something to exist, it truly did exist or had existed. At present, theoretical physicists have even come so far that they have more trust in their beautiful equations than in the experiments that sometimes seem to contradict their prophecies. They rather assume that the experiment was carelessly executed than give up their successful equations at the very first setback. We'll now present some historical examples of this.

1 The parabolic trajectory of a projectile

By means of his formula for the falling motion, Galilei could also predict the parabolic trajectory of a fired projectile that is rather hard to trace during its flight. Ballistics gained quite some 'efficiency' this way.

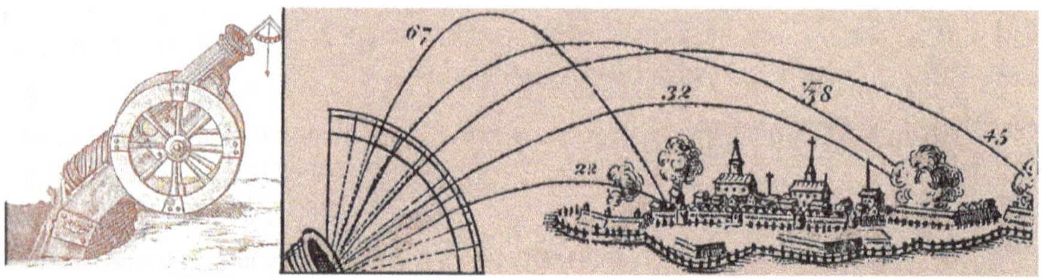

Galilei was the first to deduce the parabolic trajectory of a projectile from the law of falling motion. One gets a maximum range with an angle of 45° and equal ranges with angles that differ equally from 45°, whether greater or smaller..

2 The discovery of the planet Neptune

Thanks to calculations based on Newton's gravitation law and Kepler's theory of planet orbits, planets were discovered that hadn't been noticed before. After the discovery of the planet Uranus, there seemed to be a difference between its calculated orbit and its actual position in the sky. Because this difference appeared time and time again, astronomers presumed there had to be a yet unknown planet that disturbed the orbit of Uranus. Two students, the Englishman Adams and the Frenchman Le Verrier, independently of each other, started the complicated calculations in order to determine the position of this unknown planet. Adams finished first and sent his findings to the Greenwich observatory in 1845 with the request to make the appropriate observations. The work of the student was, however, not taken seriously and got hidden away in a drawer. Le Verrier had more luck; his letter to the German astronomer J.G. Galle got an immediate response. On September 23rd 1846, in the calculated position, the telescope found a planet unknown until then, namely Neptune, the eighth planet in our solar system. Also the minuscule and remotest dwarf planet Pluto has a similar history.

3 The companion of Sirius

F.W. Bessel

The parallax-method for determining the distance between a star and the earth consists in the measurement of (half) the angle under which the star is observed from two opposite positions of the earth on its orbit around the sun. This method was finalized in 1838, by the German mathematician and astronomer F. W. Bessel. In this way, for a whole year, he followed the positions of the star Sirius in its apparent motion respective to the background of the firmament. This motion should occur similarly to the motion of the earth (see the figure). But he noticed irregularities. Sometimes the motion was slower than expected, sometimes faster. He refused to even assume that his calculations might be erroneous and concluded that Sirius had to have a companion that caused these aberrations. With the power of the existing telescopes the search was unsuccessful. Bessel died without having seen the star concerned, but fully convinced that it did exist. Twenty years after his death it was detected by the more powerful telescope of the Yerkes observatory, just like Bessel had predicted.

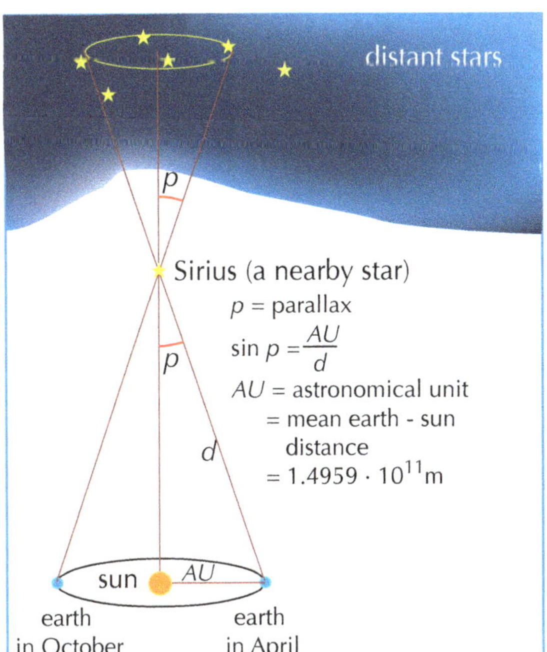

Determination of the parallax of the star Sirius from the positions of the earth in April and October. The smaller ellipse represents the apparent motion of Sirius respective to the background of the night sky. This is the result of earth's orbit around the sun and is therefore observed in the opposite direction.

4 The curvatures of space

A. Einstein

Einstein's theory of general relativity, with its amazing aspects like the curvature of space in the neighbourhood of gravitation fields, the shrinking of time when velocity is high, a universe with four dimensions and not least the 'explosive' relation between energy and matter ($E = mc^2$), was mainly the fruit of mere experiments in thinking. Telescopes and particle accelerators were wasted on him. Thinking and scrawling some equations on a piece of paper was enough to set up the theory that became the framework for the macroscopic spectacle of the cosmos. In order to test his prophesy about a light beam deflecting in the neighbourhood of the sun's field of gravitation according to the curvature of space in a field of gravitation, the Royal Society of London sent an expedition to Principe off the coast of Africa and Sobral in Brazil to carry out observations on the occasion of the eclipse of May 29th 1919. When the photographs confirmed a displacement of the lightspots (see figure), it meant an enormous triumph for him, but nevertheless he stayed calm, cool and collected. The observed displacement (1.64 seconds of a degree) differed very little from the value he had calculated before (1.75 seconds of a degree). But this caused him no doubt at all and he did not think of redoing his calculations. "Let the observers take a little more care next time" was his laconic comment. Also his prophecies about the shrinking of time were confirmed by experiments and, alas, the very regrettable concrete aspects of the formula $E = mc^2$, being the basis of that fatal mushroom shape.

Photo of the memorable sun eclipse of May 29th 1919 that confirmed Einstein's theory of gravitational fields. The apparent displacement of stars caused by the deflection of light beams in the gravitational field of the sun is indicated by the arrows.

5 The discovery of the anti-electron and the anti-proton

P. A. M. Dirac

Quantum mechanics is the counterpart on the microscopic scale to the theory of relativity and it describes the world of the atom and its components. In this miniature cosmos, it is not gravitation but the electromagnetic forces, the weak and the strong interaction, that keep things stable or let all hell loose in them. An important equation describing the quantum mechanical world is the Schrödinger equation. In this field Schrödinger is Einstein's equal. Also the Scottish physicist Paul.A.M. Dirac has set up important equations in this field. One of these equations described the behaviour of an electron and postulated in this respect the existence of an opponent, namely an electron with a positive charge (instead of negative), hence also called 'anti-electron' or 'positron'. One and a half years after the incredible notion of an 'anti-electron' was abstracted from an equation, it was experimentally discovered by the physicist C.D. Andersen. The 'anti-proton' that was predicted by the same equations, was discovered only twenty years later. Werner Heisenberg called the prophecy and the confirmation of the existence of antimatter the biggest leap forward of physics in present times.

6 The ultimate theory of all?

We will not dwell too long on the successes of so many other number theoretical models and equations like, for instance, the design of the periodic system of the elements by Mendeleev which proved so fruitful for chemistry, or Maxwell's equations that gave the whole phenomenon of electromagnetism such an elegant form. We can, however, not refrain from mentioning something about the *nec plus ultra* of contemporary theoretical physics. Just like the theory of relativity is successful in explaining the large structures and processes in the universe, quantum mechanics is extremely successful in the description of the world of the utterly small. While the theory of relativity, in spite of the term `relative', is really deterministic, quantum mechanics has, on the other hand, probabilistic characteristics that, for instance, take shape in the Heisenberg uncertainty principle, which states that a particle's speed and position cannot be measured simultaneously. This gave rise to the epitaph: „Er liegt irgendwo hier." (He lies somewhere here).

That the observer himself, moreover, could influence perception was to Einstein one of the unacceptable aspects of quantum theory. "God does not play dice" is his well-known negative comment. At the end of his life he actually undertook attempts to get the truth on his side, albeit in vain.

Since then attempts have been made to unite both theories in a sort of 'theory of all' that in a way of speaking would bring science 'face to face with God'. This theory, which is in the pipeline as a purely mathematical construction, is called 'string theory' after the new most elementary particles out of which all matter is supposedly built. One shouldn't even hope to be able to undertake significant experiments on these things considering their extremely small dimensions and the unachievable temperatures required in the particle accelerators to discover a trace of them. Therefore the theory can only acquire legitimacy and esteem by its pure elegance and its mathematical consistency, and very ironically for the evolution of science, it will have to be practised as a new kind of belief. But just like sometimes a deeper truth comes to the fore in the expression of an inspired poet, also the beauty of the universe undeniably resonates in the elegance of the equations of the inspired scientist.

In a particle accelerator, like this one at CERN in Geneva, charged atom components are brought to high energy levels by accelerating them to velocities close to the velocity of light. Some circular accelerators, for instance cyclotrons and synchrotrons, can have a diameter of several kilometers. Collisions drive out all kinds of 'exotic' particles, the traces of which can be photographed.

1.2 The shapes of the things around us

Just as there are many people who never read any poetry, there are many who aren't bothered by equations when they look at the sky, which is why we'll give some concrete examples of the presence of mathematical patterns in the shape of the things around us.

1.2.1 About rabbits and growth spirals

L. Fibonacci

Leonardo Fibonacci (his family name means: son of Bonaccio) lived around the turn of the twelfth and the thirteenth century in Pisa, which explains the other name by which he is known: Leonardo of Pisa. When travelling in Egypt, Sicily, Greece and Syria, he came into contact with the Hindu-Arabic decimal arithmetic method. Soon after coming home in 1202 he wrote his famous book *Liber Abaci*, which we already mentioned in chapter 3. This book made a contribution to the final introduction of the decimal system in the western world. This book also mentioned a problem that became forever famous. It is about the (theoretical) growth of a rabbit population starting from one pair, and observing reproductive rules that are rather strict from the point of view of rabbits.

'How many pairs of rabbits does one get after n months if:

- *one starts with one newly born pair*
- *this pair becomes fertile after one month, and one month later produces a new pair*
- *the same applies to every new pair*
- *all rabbits are supposed to stay alive and active during at least these n months?'*

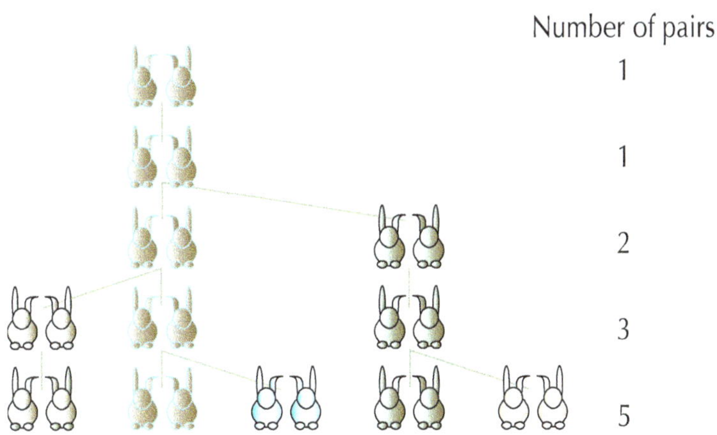

Because the solution of this problem resulted in a remarkable sequence that isn't felt as being unnatural by nature, we will deal with it somewhat more extensively. The analysis of the procedure leads to a recursive formula:

after 0 months:	1 pair (the ancestor pair)
after 1 month:	1 pair (the same ancestor pair that is now fertile)
after 2 months:	2 pairs (the new pair will have to wait a month to get 'active' themselves)
after 3 months:	3 pairs (the two pairs remaining from the previous month plus the new pair begot by the fertile pair of two months ago)
after 4 months:	5 pairs (the three pairs remaining from the previous month plus the two new pairs begot by the pairs that already were alive two months ago and hence were fertile one month ago)
after 5 months:	8 pairs (the five pairs remaining from the previous month plus the three new pairs begot by the pairs that were already alive two months ago and hence were fertile one month ago)

...

It is clear how this goes on: to find the number in the present month one has to add up the numbers of the two previous months. This results in the Fibonacci sequence: 1, 1, 2, 3, 5, 8, 13, 21, 34, 55, 89, 144, ...
Within a year we get a total of 144 rabbit pairs.

If P_n is the number of rabbit pairs after n months, then the following recursive formula applies: $P_n = P_{n-1} + P_{n-2}$.

Now there exists a close correlation between the Fibonacci sequence and the 'golden ratio' that we will discuss in the next section. When we divide a term of the Fibonacci sequence by its previous term, then in fact we divide the sum of two consecutive terms by the bigger of them:

$$\frac{P_n}{P_{n-1}} \text{ is indeed equal to: } \frac{P_{n-1}+P_{n-2}}{P_{n-1}}.$$

The value of this quotient varies when progressing in the sequence, but it converges closer and closer to a fixed value that is referred to as φ [phi].

$$\frac{1}{1}=1;\ \frac{2}{1}=2;\ \frac{3}{2}=1.5;\ \frac{5}{3}=1.66...;\ \frac{8}{5}=1.6;\ \frac{13}{8}=1.625;\ \frac{21}{13}=1.615...;$$
$$\frac{34}{21}=1.619...;\ \frac{55}{34}=1.617...;\ \frac{89}{55}=1.618...;\ ...$$

This mysterious number φ, on which these quotients converge, is called the 'golden ratio' or 'golden section'.

Although you can't expect to produce the Fibonacci numbers' sequence with a real pair of rabbits, it is remarkable that the terms of the sequence are encountered in various natural shapes that deal with a process of growth.

In the pattern of the seeds of a sunflower one can notice two sets of spirals that turn in opposite directions. For some varieties of sunflowers, the number of spirals in each set is 21 and 34, for other varieties it is 34 and 55, but in all cases two consecutive terms of the Fibonacci sequence! The same occurs in the scales of a pine cone, with 5 spirals in one direction and 8 in the other. Also on the rind of a pineapple the number of spirals in both directions seems to show Fibonacci numbers, namely 8 and 13.

This seems to have an explanation in the way small parts of organic tissue grow. In order to maximally occupy the available space, so that the structure becomes dense and solid, the particles develop mostly under a constant angle of 137.5°. The remaining angle respective to a full turn is then 360° – 137.5° = 222.5° and so it applies that $\frac{222.5}{137.5} = 1.618... = \varphi$. Clever nature!

A growth spiral has a very specific characteristic: the tangent at an arbitrary point of the spiral and the line connecting that point to the centre make a constant angle. Consequently all tangents to the spiral in all the intersections of the spiral with a fixed line through its centre are parallel. This is also the case in the spiral shape of the nautilus shell. In its X-ray picture below, you can admire the beautiful proportional succession of its chambers.

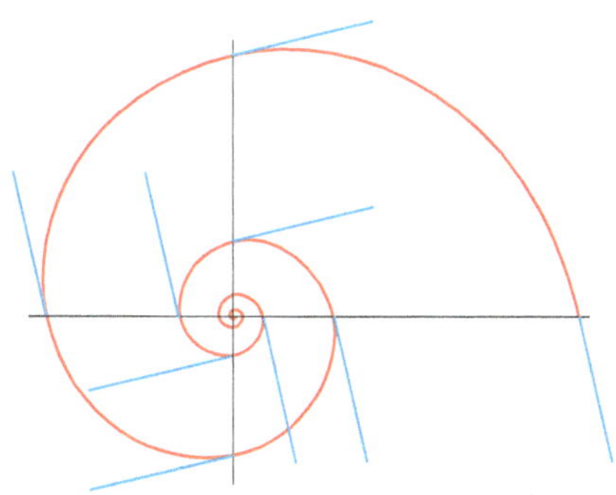

The shape of the growth spiral results from the fact that the growth of an organism is proportional to its actual size and that is why it is a logarithmic spiral. This was Jakob Bernoulli's favourite curve because it keeps on reproducing itself under various transformations. He had it depicted as a symbol on his gravestone together with the inscription 'Eadem mutata resurgo' (Changed and yet the same, I rise again). Also in the branching or in the arrangement of the leaves of certain plants, numbers of the Fibonnaci sequence are revealed.

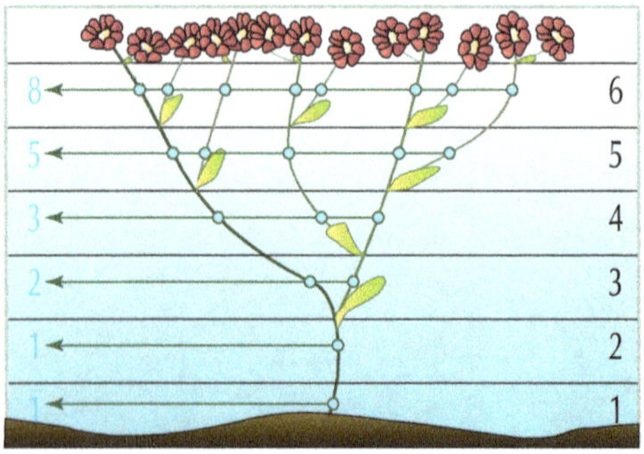

1.2.2 The golden section and Da Vinci's navel

When a line segment (or any other magnitude) is split into two unequal parts x_1 and x_2 so that the ratio between the total length a and the longer part x_1 is the same as the ratio between the longer part x_1 and the shorter x_2, then this ratio is the specific number we met before as the 'golden ratio' and that we referred to as φ. It is also said that the segment is split into 'the mean and extreme ratio'.

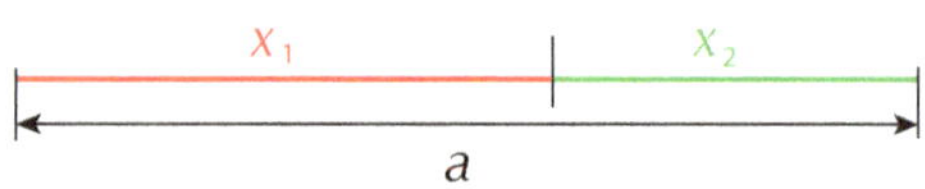

If $x_1 > x_2$, then $\frac{x_1 + x_2}{x_1} = \frac{x_1}{x_2} = \varphi$. If $x_1 + x_2 = a$, then $x_2 = a - x_1$.

If we substitute the last two in the first, we can calculate x_1 as a function of a:

$$\frac{a}{x_1} = \frac{x_1}{a - x_1} \Leftrightarrow a^2 - ax_1 = x_1^2 \Leftrightarrow x_1^2 + ax_1 - a^2 = 0 \Leftrightarrow x_1 = \frac{-a \pm \sqrt{a^2 + 4a^2}}{2}$$

$$\Leftrightarrow x_1 = \frac{-a \pm a\sqrt{5}}{2}$$

$$\Leftrightarrow x_1 = \frac{-1 \pm \sqrt{5}}{2} a$$

Since x_1 should be positive we conclude:

$$x_1 = \frac{-1 + \sqrt{5}}{2} a \qquad \text{hence} \qquad \varphi = \frac{a}{\frac{\sqrt{5} - 1}{2} a} = \frac{\sqrt{5} + 1}{2} = 1.618...$$

$$\text{moreover} \quad \frac{1}{\varphi} = \frac{\sqrt{5} - 1}{2} = \varphi - 1$$

The number $\frac{\sqrt{5}-1}{2} = 0.618...$ is the 'silver ratio', although sometimes it is taken as the golden ratio. Then $\frac{\sqrt{5}+1}{2} = 1.618...$ would be its reciprocal. It just depends on what you prefer as numerator and denominator. Anyhow, both values are related by the splitting of the segment into the mean and extreme ratio. According to Thomas Aquinas, the harmony of certain proportions is particularly pleasing to the senses. He was convinced that there had to be a calculable connection between our impressions of beauty and mathematics. It seems that precisely the golden ratio gives a balanced and pleasing aesthetic impression. That is why it is consciously or unconsciously applied by many artists.

Also nature seems to dearly love this ratio. Several examples of the golden section have been found in several physionomic proportions of living things, in the first place in human beings. In an ideally built person, the navel would divide their length as a golden section (you can sneakily measure it for yourself). Also in other parts of the body this ratio should (approximately) apply: the length of a hand to its width; the length of the forearm (hand included) to the upper arm, etc. Don't jump too easily to the conclusion that you belong to the army of 'deviants'. Somewhere or other you will find 'gold' in your body.

It was Leonardo da Vinci who suggested this ratio in some of his drawings, for instance in the various dimensions of the profile of a head (most likely his own).

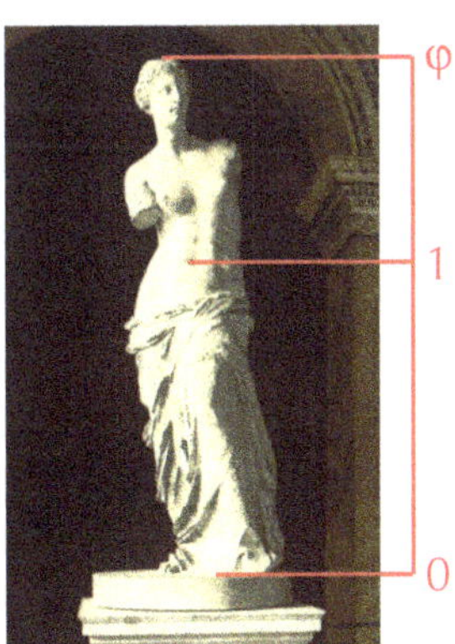

Venus of Milo
(Louvre, Paris)

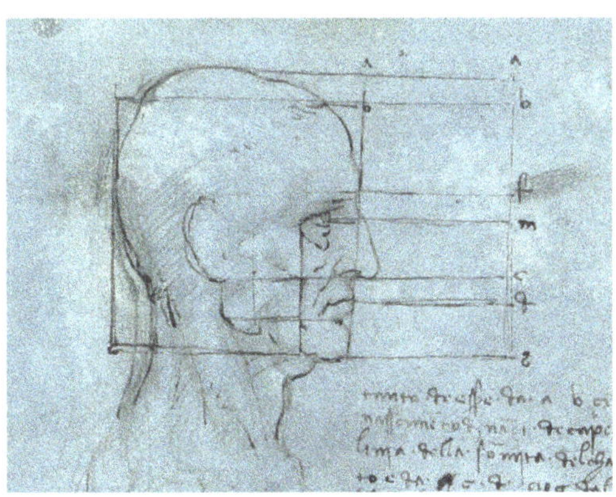

Sketch of a head profile
by Leonardo da Vinci

1.2.3 Golden figures

The golden ratio occurs in many particular geometrical figures that can easily be constructed by ruler and compass, like golden rectangles, golden triangles, regular penta- and decagons, etc.

1 Golden rectangles

A 'golden rectangle' is a rectangle of which the length is to the width as φ is to 1. The construction of such a rectangle can be made starting from a square, as follows:

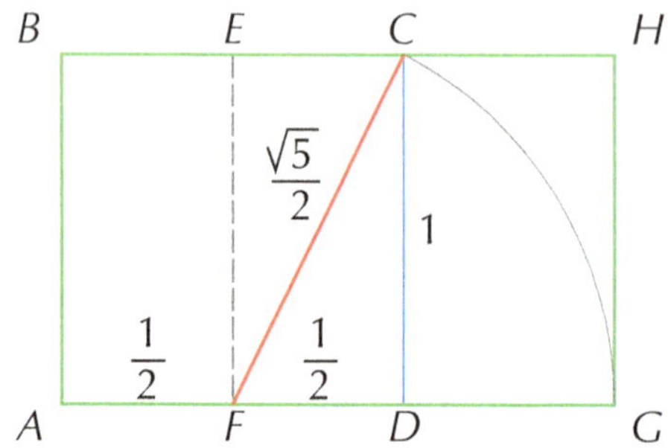

If, for convenience, we take the side of the square *ABCD* to be 1, then according to the Pythagorean theorem, the diagonal *FC* of the rectangle *ECDF*, in which *E* is the middle of *BC* and *F* the middle of *AD*, equals $\sqrt{1^2+\left(\frac{1}{2}\right)^2}=\frac{\sqrt{5}}{2}$ and the length of *AG* equals $\frac{1}{2}+\frac{\sqrt{5}}{2}=\frac{\sqrt{5}+1}{2}=\varphi$.

The rectangle *ABHG* is then a golden rectangle.

Now a golden rectangle has the property of rapidly generating other golden rectangles. If one cuts off a square from a golden rectangle, of which the side equals the width of the rectangle, then a rectangle is left over that is again a golden rectangle. For instance, in the figure above, the rectangle *DCHG* is also a golden rectangle. The ratio of the length *GH*, namely 1, to the width *DG*, namely $\frac{\sqrt{5}-1}{2}=\frac{1}{\varphi}$, indeed equals $\frac{1}{\frac{1}{\varphi}}=\varphi$. We can keep on repeating this procedure.

In an analogous way one gets a bigger golden rectangle by adding a square to the longest side of a given golden rectangle.

By drawing quarters of circles in the successive squares, one can get an almost spiral shaped curve that approximates to a logarithmic spiral. A well-built hand should fit in a golden rectangle when laid flat. When developing in the uterus, a human embryo should curve in the form of a golden spiral.

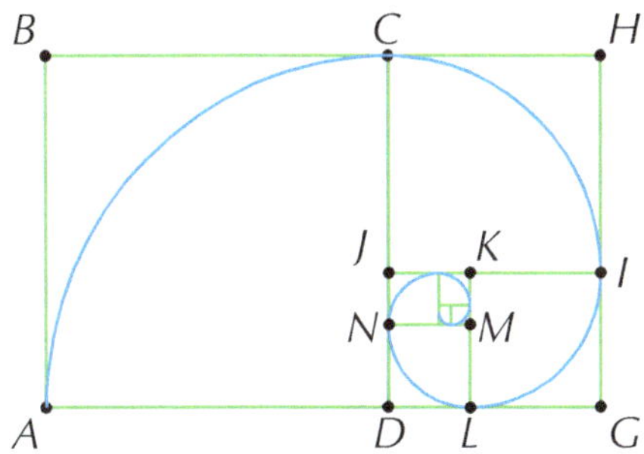

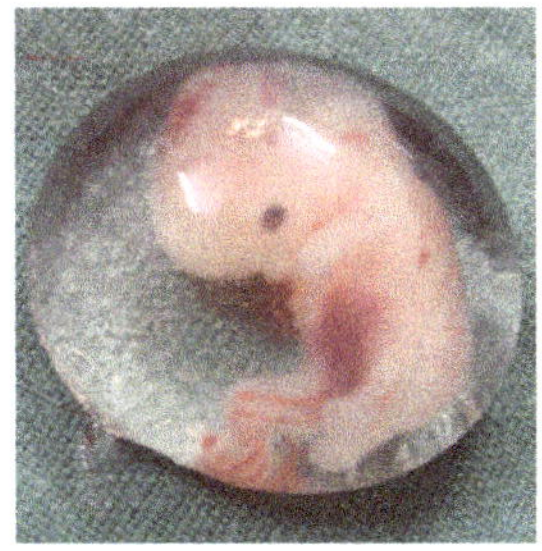

2 Golden triangles

There are two kinds of golden triangle: acute and obtuse.

- in an acute isosceles triangle with an apical angle of 36°, the upright side is to the base as φ is to one.
- in an obtuse isosceles triangle with an apical angle of 108°, the base is to the upright side as φ is to one.

The proof of these propositions is based on similarity of triangles. We restrict ourselves here to briefly calculating these proportions:

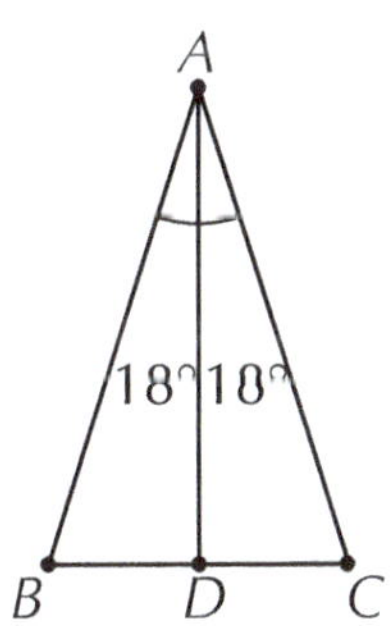

Proposition 1: let $AB = 1$, then

$$\frac{AB}{BC} = \varphi \Leftrightarrow \frac{BC}{AB} = \frac{1}{\varphi} \Leftrightarrow BC = \frac{1}{\varphi} = 0.618...$$

Proof: $\frac{BD}{AB} = \sin 18°$

so $BD = \sin 18°$

and $BC = 2\,BD = 2 \sin 18° = 0.618...$

Proposition 2: let $AB = 1$, then

$$\frac{BC}{AB} = \varphi \Leftrightarrow BC = \varphi = 1.618...$$

Proof: $\frac{BD}{AB} = \sin 54°$,

so $BD = \sin 54°$

and $BC = 2BD = 2 \sin 54° = 1.618...$

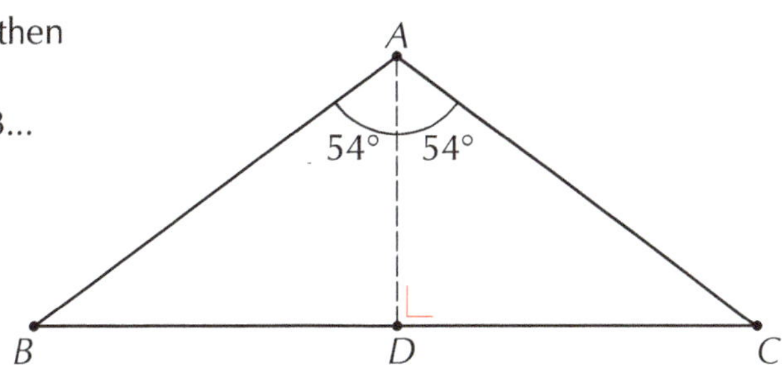

These golden triangles are easily reproduced, as shown in the following figures:

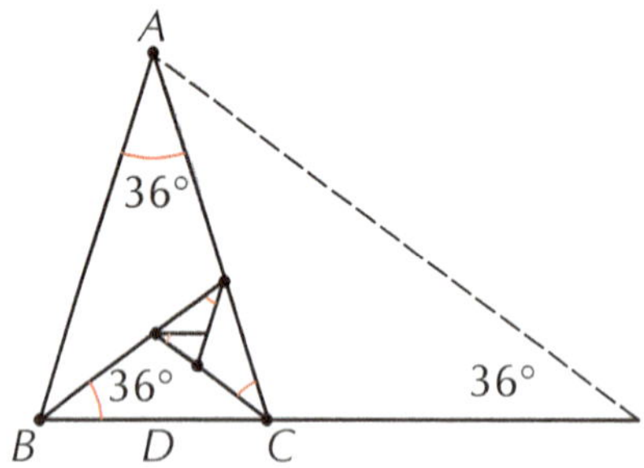

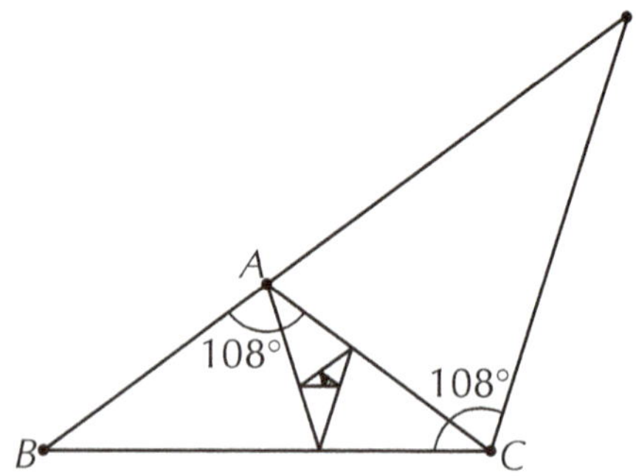

3 Regular penta- and decagons

Because in regular penta- and decagons provided with their diagonals, isosceles triangles with apical angles of 36° and 108° appear, these figures are teeming with golden ratios.

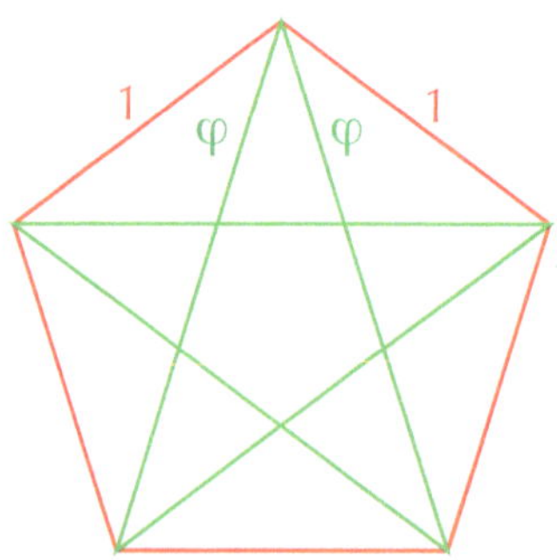

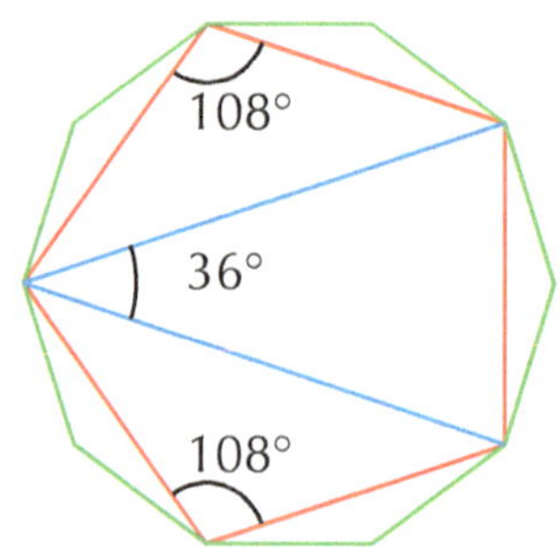

In five-armed starfishes, in which the shape of a regular pentagon appears, these golden ratios can actually be found.

4 The powers of φ and the Fibonacci sequence

We now come back to the close relationship that exists between the Fibonnaci sequence and the golden ratio.

Besides the property that the sequence of the quotients of two consecutive terms of the Fibonacci sequence converges to φ, there is still another remarkable relationship, namely between the successive powers of φ and the terms of the Fibonacci sequence that miraculously make a return appearance there.

Examples

$$\varphi^1 = 1\cdot\varphi$$

$$\varphi^2 = \left(\frac{\sqrt{5}+1}{2}\right)^2 = \frac{5+2\sqrt{5}+1}{4} = \frac{\sqrt{5}+1}{2}+1 = 1\cdot\varphi+1$$

$$\varphi^3 = \varphi^2\cdot\varphi = (1\cdot\varphi+1)\cdot\varphi = \varphi^2+\varphi = (1\cdot\varphi+1)+\varphi = 2\cdot\varphi+1$$

$$\varphi^4 = \varphi^3\cdot\varphi = (2\cdot\varphi+1)\cdot\varphi = 2\cdot\varphi^2+\varphi = 2\cdot(1\cdot\varphi+1)+\varphi = 3\cdot\varphi+2$$

$$\varphi^5 = \varphi^4\cdot\varphi = (3\cdot\varphi+2)\cdot\varphi = 3\cdot\varphi^2+2\cdot\varphi = 3\cdot(1\cdot\varphi+1)+2\cdot\varphi = 5\cdot\varphi+3$$

...

hence

$$\varphi^1 = 1\cdot\varphi$$

$$\varphi^2 = 1\cdot\varphi+1$$

$$\varphi^3 = 2\cdot\varphi+1$$

$$\varphi^4 = 3\cdot\varphi+2$$

$$\varphi^5 = 5\cdot\varphi+3$$

...

$$\varphi^n = P_n\cdot\varphi+P_{n-1}$$

So, all powers of φ can be expressed as functions of the first degree in φ of which the coefficients are consecutive terms of the Fibonacci sequence. This means that the sequence appears in two ways in relationship to the golden ratio. Although the sequence of the powers of φ can also be found in the human body, these values are (intuitively) less recognizable than the golden ratio itself. But it remains a miraculous fact that the Fibonacci sequence and the golden section refer to one another. This must be due to the Playboy side of these naughty rabbits!

1.2.4 Symmetry and a 'super formula'

Symmetry frequently occurs in nature and is also undeniably an aspect of beauty in forms. From experiments, in which a very heterogeneous sample of persons were asked to select the most attractive faces from a variety of photographs, it showed that all prefer those that most closely approach perfect symmetry. But apart from the aesthetic, symmetry also seems to have a functional aspect: thanks to a couple of eyes we can assess distances by telemetry in an analogous way as the parallax method does in astronomy; two ears give sound a stereo effect; because of two legs our walking is more balanced than if we hoppod along on one leg; with one wing flying probably would only succeed in small circles, just like rowing with one oar; etc.

Symmetry does not only occur as reflections in respect to a line, as in a butterfly, or in respect to a plane, as in the build of the human and animal body, but also in the form of rotation symmetry, like in a flower, and in other kinds of mathematical transformations.

In this context we cannot keep from referring to the 'super formula' of botanist Johan Gielis. In his splendidly illustrated book *Inventing the Circle* he shows how, by varying the values of the six parameters in that formula, a whole range of symmetrical and 'super symmetrical' forms arise that actually also appear in nature. It is fascinating how one formula, which is to be understood as a generalization of the equation of a circle, forms the basis of an infinite variation of natural patterns.

1.2.5 Tessellations

Tessellations are tilings of the plane by specific polygons and fillings of the three-dimensional space by polyhedra of the same type. Their theory makes a separate branch in geometry. It not only examines what is possible or impossible, but also what the optimizations under several criteria such as firmness, maximum occupation of space, etc. may be. While mathematicians have to ponder to get some insight, nature seems to be ingenious enough to find the correct solutions spontaneously. In a beehive the buzz-buzzes diligently build their hexagonal honeycombs without anyone having to explain to them in advance that this form not only offers the maximum covering of space but also gives maximum firmness. Also particular forms arise during the formation process of crystals that are interesting from the mathematical point of view, be it minimum refraction of light in the prism or optimal hardness of the configuration. It is not only diamonds that sparkle and are 'for ever'.

1.2.5 Fractal forms

The diversity and the fancifulness of most forms in nature do not immediately bring to mind the perfect ideal figures of Euclidean geometry such as the circle, sphere, triangle or prism. Therefore these forms remained beyond the reach of the mathematical approach for a long time. But, as was shown in chapter 2 section 2.7, things have changed with the theory of fractals. Coastlines, clouds, trees, snowflakes or whatever comes to your mind can be imitated by a computer by means of the iteration of simple algorithms or by the recurrence of elementary basic patterns. This exposes the fractal structure of all these phenomena and also makes it possible to understand the complexity of these forms.

dendrite

fern

clouds

1.2.6 The double helix of life and the genetic code

Since its origin, life on earth has developed an inconceivable diversity of species that have appeared and sometimes also disappeared. It took until the end of the nineteenth century before the mechanism that lies at the basis of that variety was understood. Evolutionary theory, formulated by Darwin and Wallace, gives a plausible explanation for the origin of the multiformity of species. With the disentangling of the DNA structure in 1953 by the English physicist Francis Crick and the American biologist James Watson, many of the pieces of the puzzle of reproduction and heredity gradually fell to their place. Since then genetic biology has reached a level of insight that enables people now to intervene in these processes. After a history of millions of years, life has taken itself in hand, together with a crushing responsibility as the price for scientific knowledge.

All living organisms are made up of cells. In the nucleus of the cell are threadlike structures, called chromosomes, which occur in pairs, one copy of which comes from the mother and the other from the father. The chromosomes are the bearers of the hereditary information that determines the properties of the cell and the organism to which it belongs. That hereditary code is stored in the double helix of the DNA (Desoxyribo Nucleic Acid), which consists of a combination of four different bases: adenine (A), cytosine (C), guanine (G) and thymine (T). The structure of these bases is formed by the elements oxygen (O), carbon (C), nitrogen (N) and hydrogen (H). In these structures cytosine and thymine have single rings whereas guanine and adenine have fused double rings.

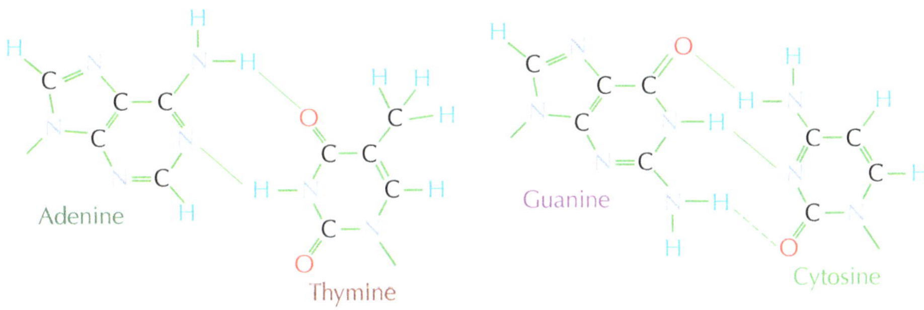

The bases A and T are always opposite to one another on the two different strands that make the double helix of DNA. The same goes for C and G, so that the mutual distance between the two strands consists of three rings everywhere. The bases on the same strand are mutually linked by a sugar and a phosphate group. This last bond is stronger than the weak hydrogen bonds (two between A and T and three between C and G) of the paired bases on the opposite strands. By radiation, heating or chemical reactions started by enzymes, the strands can separate (denaturate) and with reversed impulses they can couple again (renaturate).

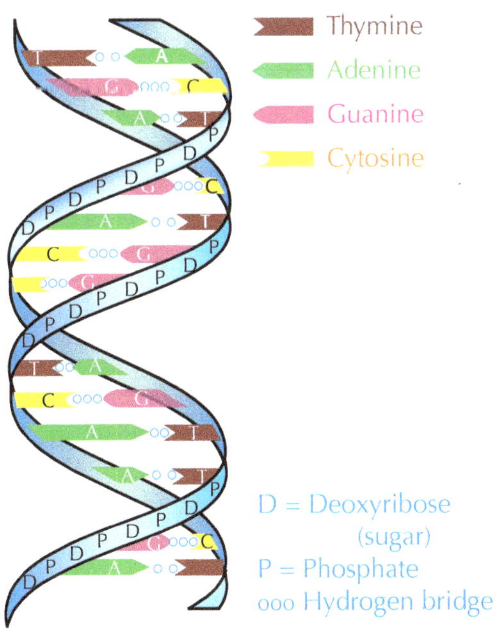

Every living organism has a specific DNA, completely defined by the sequence of the successive bases A, C, G and T, in an endless variety of combinatorial possibilities. According to the order of the pairs of bases in groups of three (codons) certain bits of the DNA form amino acids, which are the building components of the proteins and enzymes. Such a fragment of DNA that forms the code for a protein is called a gene. Each gene contains the instruction for the production of a protein that in its turn is responsible for the coding of a specific characteristic, such as the colour of the eyes or the shape of the nose.

The entirety of the genes in the DNA is called the genome, which is therefore a kind of blueprint for the organism. The human genome consists of approximately 25000 genes, spread out over a DNA helix with a length of approximately two meters. These are divided into 23 pairs of chromosomes and their duplicates, so all together 46. When producing gametes (reproduction cells), only one copy is selected from each type of chromosome, altogether 23, to be linked to 23 chromosomes of the gamete of the partner. A child therefore inherits, by means of the genes of the mother and father, characteristics of both parents.

In ordinary cell division (mitosis), which is needed for the growth of the organism, the DNA will at first copy itself (DNA replication). This happens by the uncoupling of the strands and addition of new bases according to the principle of fixed pairs of bases A – T and C – G. As a result two identical double helices arise that after the cell division pass on the properties of the mother cell to the daughter cells. This way the cells know whether they have to become an elephant or a mosquito. If nevertheless a modification happens in the order of the bases in the strands, a mutation of the species occurs. Although being relatively stable, the diversity of the species of life sometimes depends on random factors.

The system of genetic coding, being based on the succession of the four bases A, C, G and T in the DNA strands and the 64 ($4 \cdot 4 \cdot 4$) possible combinations of them in codons, is equal for all forms of life. This means that the borders between vegetable, animal or human life are crossable in any direction. Consequently, in the so-called DNA recombination technology one awaits the possibility of exchanging the properties between plants, animals or people. Many consider this a risky venture and perhaps they are right: imagine if you combined the genes of a kangaroo with those of the cactus in your living room! And you certainly wouldn't even dream of incorporating a bit of your beautiful lawn into your family. One doesn't need to go that far in giving one's imagination free rein to having reservations about certain gene manipulations. But probably a lot of positive possibilities will also see the light of day, such as eliminating hereditary defects and improving desirable properties. Why should only chance make the favourable choices in nature?

And so it becomes clear that the miracle of the variety of life and its reproduction is based on simple arithmetical models: nuclear bonds according to the rules for the number of electrons on the outer shell (valence electrons); structural formulas of molecules built of only some elements; paired linking of four bases according to the available hydrogen bridges and the principle of ring formation; combinatory arrangements in codons that define the genes as bearers of properties; the unlimited number of possibilities of the sequential arrangement of the genes in the double DNA helix. Life as a symphony of chords composed on the four tones A, C, G and T.

Summary

In nature there are physical laws active that unmistakably have a mathematical background. They give shape to a variety of forms, of which the complexity as well as the eye-catching order fill us with a feeling of amazement and delight. Since the mind is precisely geared to analysing such order structures, these examples from nature give it the inspiration and the impulse for transcending these forms in the different fields of art.

In its appearance the romanesco is flower as well as cabbage as well as broccoli. In its spirals the numbers of the Fibonacci sequence appear hand in hand with the golden ratio. In the self-similarity of the ever smaller pinnacles that keep on repeating the overall pattern, it reveals its fractal structure. The beauty of nature has unmistakably been drawn with mathematical fingers. We do not always realize how much mathematical refinement we are eating when the cook has once again arranged our dish so nicely.

2 Mathematical structures in art

It is probably superfluous to defend a thesis stating that in the different fields of artistic activity mathematical resources and structures are often applied. Also people who cannot always explicitly point out these mathematical backgrounds generally consider this to be obvious. What follows might therefore seem like labouring an obvious point. Already, from the earliest primitive cultures onwards there was a tendency for people to embellish their daily life with objects and jewels, with often no other objective than pleasing their senses or expressing their own mood. This is why the historical attributes of what we now call art have made up an immense collection. It is impossible to present a concise inventory of all this, even in broad outline. The following examples are only meant to illustrate some aspects of these mathematical backgrounds without the intention of describing each artwork in all its other aspects.

2.1 The shapes and techniques of architecture

Architecture differs from the other fine arts by the inescapable physical requirements of stability and its aim of functionality. The structure of a building therefore depends not only on the aesthetic design, but also on a combination of load and the material available to balance these forces. Moreover, a building also has a specific function of a civil, military or religious nature, for instance as a house, a concert hall, a sports stadium, a reinforced citadel or a cathedral. For this reason the architect, besides being a stylist, is necessarily also an engineer. Both the development of technology and the progress of mathematics have determined the changing appearance of actual building styles in the different phases of history. To show this we will go through the characteristic forms that were time-bound because of the restrictions imposed by the available material and the mastery of the complexity of the calculations.

2.1.1 The successive building styles

From the very first attempts that people in ancient cultures undertook to erect constructions, be it houses or monuments for religious or mythical purposes, initially simple geometrical forms were chosen: the huts in Africa had a cylindrical body with a conical roof; the clay huts in the European plains had a cuboid form roofed by a triangular prism; the graves of the pharaoh in Egypt are quadrilateral pyramids.

As technical knowledge and the diversity of materials increased, imagination and inventiveness were given free rein, as is shown in the successive styles that we distinguish in construction. We restrict ourselves to the most striking characteristics that can be found in the following illustrations.

1 Antique architecture

In ancient architecture there seemed to be a preference for simple forms and monumentality, as can be seen in the Egyptian pyramids and the Babylonian ziggurats. Both peoples had sufficiently developed mathematics to make the required calculations. In the constructions numerous specific proportions can be found in which they represented their mythical interpretations of the cosmos. The materials used were collected in the immediate surroundings. For the Egyptians it was cut stone and for the Babylonians sun-dried clay. The dimensions of the blocks for the pyramids were calculated so that they fitted together without mortar. Babylonian clay blocks simply stuck together while drying.

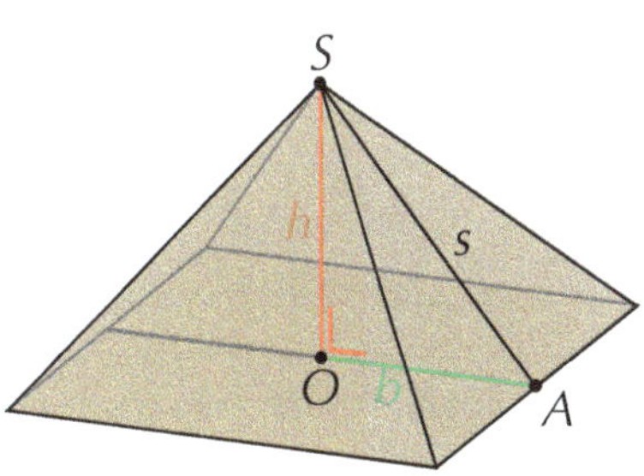

In the dimensions of a section of the pyramid of Cheops ('cheops section'), we recognize the golden ratio, namely: SA/OA = 1.618 (rounded).

The layerwise built ziggurat in Babylon, with its decreasing dimensions of the successive floors, is known as the Tower of Babel from the Bible.

2 The classical architecture of the Greeks

In Greek classical architecture, particularly in the temples, we recognize a preference for straight lines. Much attention is paid to the subtle proportions and all kinds of optical corrections, such as the bulge (entasis) of the columns and their judicious placement for optimizing the ever-changing angle of incidence of sunlight on the marble.

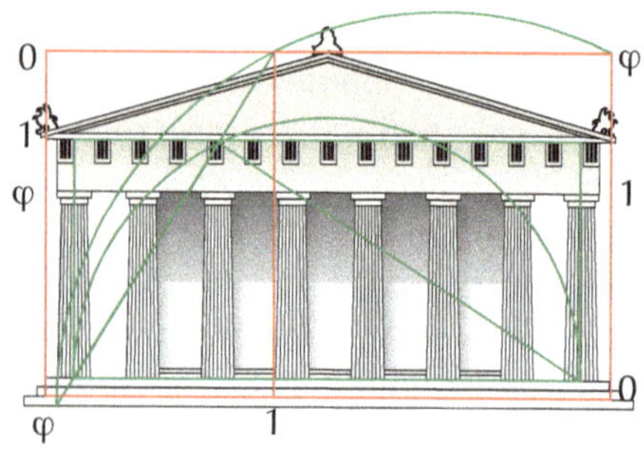

The frontal view of the Parthenon in Athens fits precisely into a golden rectangle. Also in many other proportions, such as the height of the pillars and of the fronton, φ can be recognized.

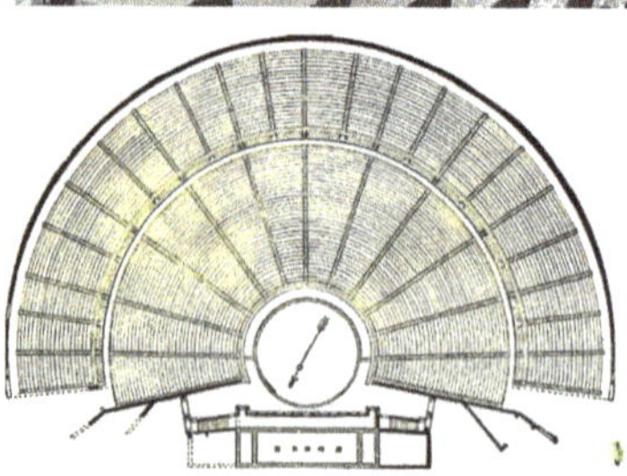

The theatre of Epidaurus has two separate bench areas, 34 and 21 rows respectively, which are consecutive terms of the Fibonacci sequence and therefore related to the golden ratio..

3 Roman architecture

Roman architecture adds a frequent application of round shapes to the Greek style. Arches, vaults and cupolas make their entry. The load of the spanning is lightened, for instance, by the use of hollow bricks. Later, in the East Roman empire, the cupola will become a distinctive characteristic of Byzantine churches and basilicas. In his ten books *On Architecture*, the Roman architect Vitruvius describes the basic rules for all kinds of proportions, inspired by those of the human body, which later also influenced Leonardo da Vinci.

The Arch of Titus in Rome is based on cheops sections and therefore on the golden ratio

The spherical cupola of the Hagia Sophia (Holy Wisdom) in Istanbul is supported by two half spheres, four pendentives and pillars.

4 The Romanesque style

The Romanesque churches of the Middle Ages are characterized by heavy walls and small windows in order to support the dominating vaults. These impressive structures required a lot of construction material and labour. The high costs resulting from this often imposed modest dimensions.

The arches of the triforium of the cathedral of Rochester in England are semi-circles whose areas are in a beautiful proportion. The area of the remaining crescent equals precisely half of the area of the large semi-circle, and thus also of both small semi-circles together.

The rosette of Patrixbourne church in England represents a symmetrical rotation group of eight elements.

5 The Gothic style

The lighter constructions of the gothic style resulted in higher vaults that transfer the load to the pillars by means of a web of ogives and high pointed arches. By using flying buttresses, the load on the bearing walls was lessened so that they could include large glass windows. The details were still subordinate to the harmony of the whole.

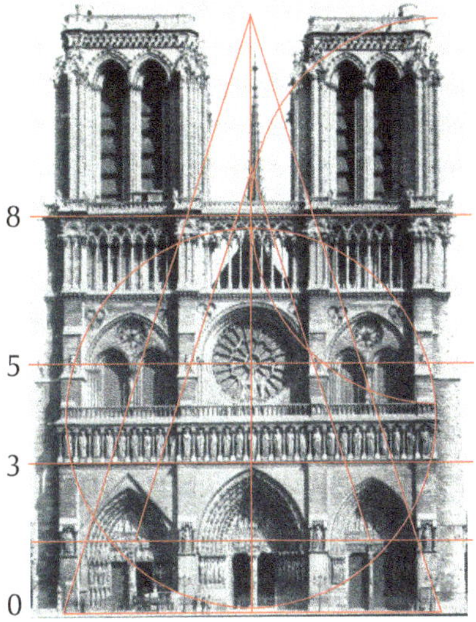

Notre Dame in Paris is constructed in a golden triangle with an apex of 36°.

The pillar of Saint Blasius Cathedral in Braunschweig, Germany, is a strand of equidistant helices.

6 Renaissance and Baroque

With the incentive of the rebirth of classical sculpture and the renewal of the art of painting, the architecture of Renaissance also arrives at a synthesis of new building styles and classical concepts. In the civil projects the decorations receive much attention, culminating in the subsequent baroque with its exuberant palaces.

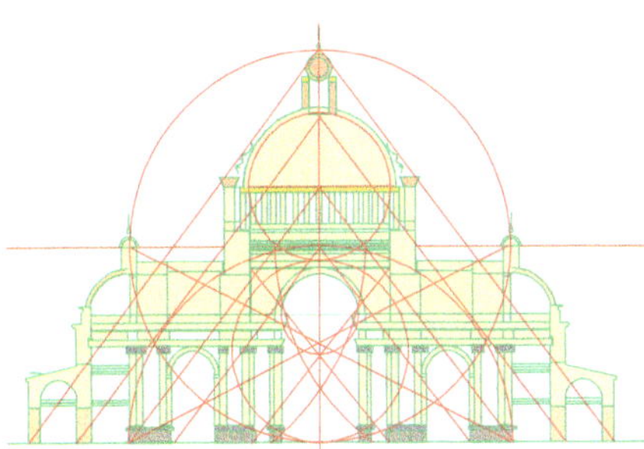

The Basilica of Saint Peter in the Vatican City is built on a pattern of several cheops sections.

The double-helix staircase in the Château de Chambord (France) was designed by Leonardo da Vinci.

7 Modern times

In modern times, new materials appear such as cast iron, steel and concrete and these inspire several neo-styles such as Art Nouveau, Cubism, the art movement *De Stijl* and the Rationalism of American architecture. Also a kind of organic architecture of assembled prefabricated components and simple modules emerges.

The Eiffel tower in Paris is an interweaving of iron and steel that is based on an abundance of mathematical curves and figures.

The stairs and the hallway of the 'Hotel Tassel' in Brussels (Belgium), designed by Victor Horta, is a playful application of his 'whiplash' curve.

The Opera building in Sydney, Australia, has shell-shaped roof segments, the form of which was calculated by integrals.

The Guggenheim museum in New York is shaped like a double three-dimensional spiral from both the inside and outside.

A house near Paris by Le Corbusier that is consciously designed according to the proportional scales of his 'modulor'. The entire building, the windows and the terrace openings are golden rectangles.

The Seagram Building by Mies Van Der Rohe in New York displays a transparent simplicity, a purely modular piling up of steel, concrete and glass elements.

The 'Modulor' is a model by Le Corbusier that contains two scales based on the golden ratio and the successive powers of it. The first (red) scale starts from the height of the navel (113 cm) of an average man (of length 183 = $\varphi \cdot 113$ cm). Divided by successive powers of φ this makes the sequence: 183; 113; 70; 43; 27; 16.5; 10; …

The second (blue) scale starts from the height of the lifted arm of the same man. This height proved to be exactly double that of the navel, namely 226 cm. In the same way, he obtained the second sequence 226; 140; 86; 54; 33; 20; …

By shifting the index one place to the right, the terms in the second sequence are double the terms in the first. Le Corbusier related these scales to all kinds of proportions of body parts. He consciously applied these proportions in the design of his buildings.

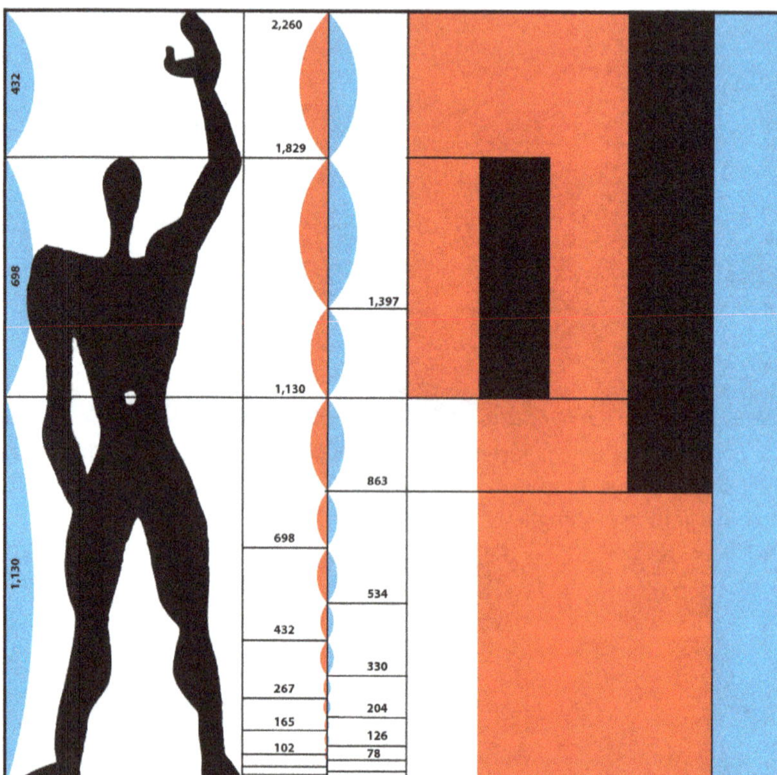

2.1.2 The dynamics of geometrical transformations

Besides the elegance of certain proportions, the method of repetition and variation also makes a strong aesthetic impression. This makes use of several geometrical transformations such as symmetry, translation, rotation, congruence and similarity. Just as the simultaneous swinging of a leg by a group of ballerinas accentuates the movements of each one of them and its repetitive nature enhances our sensual impression, these geometrical transformations also bring rhythm into the physiognomy of buildings.

1 Symmetry

The Taj Mahal in Agra (India) is the world famous mausoleum built between 1630 and 1652 by Shah Jahan in memory of his favourite wife Mumtaz Mahal. It is a jewel of perfect symmetry, congruence and similarity. The dynamics of these transformations are a treat for the eye.

2 Translations

In the Mesquita of Cordoba in Spain the position of the pillars is an intriguing play of combined translations in two directions.

3 Glide reflections

A glide reflection is the composition of a translation and a reflection in respect to an axis that is parallel to the translation. For example, a glide reflection transforms an M into a W, and this again into an M, so that the square of a glide reflection is a translation. This transformation is frequently used in the design of friezes, as in the following example from Greek art.

4 Rotations

In the stained glass rosette on the north side of the cathedral of Chartres in France we see the effect of a rotation group of the order twelve.

5 Congruences

The congruences of windows, bays, arches and decorative surfaces lend stately serenity to the building of the Château de Chenonceau on the river Cher (tributary of the river Loire) in France.

6 Similarities

The similar modules that get smaller as they near the top of the impressive cupola of the Pantheon in Rome, affirm its curved shape and deepen its dimensions.

With this, the arsenal of geometrical transformations is far from being exhausted. There are also projections, dilatations, torsions, etc. that have their applications in ornamental aspects of architecture. Unfortunately, we cannot present all of them here.

2.1.3 Ruled surfaces

A ruled surface is a three-dimensional surface that can be generated by nothing but straight lines. It can be thought of as being obtained by a movement of just one line, e.g. a rotation of it around an axis or a displacement along one or more curves that support the tracing line. These curves are called directing curves. Simple examples of ruled surfaces are the lateral surfaces of solids of revolution like a cylinder or a cone. More complicated forms are, for instance, a corrugated surface (of which the directing curve is a wave) or a conoid (of which the directing curves are a line and a planar curve).

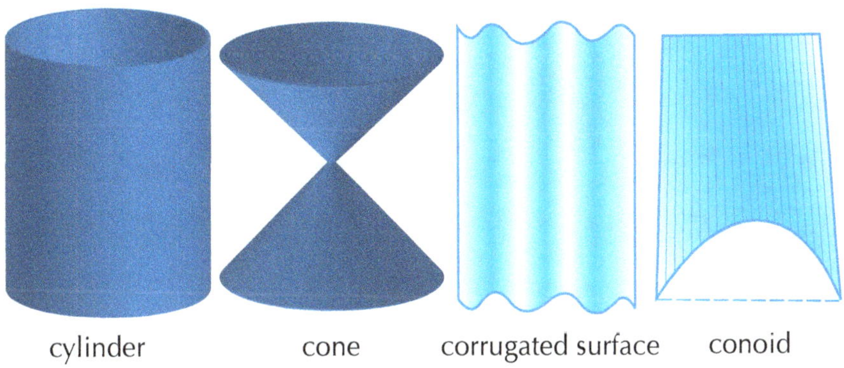

These forms are used in architecture not only for their aesthetic features but also for functional purposes like sturdiness and inflexibility.

1 Cylindric forms

The exhibition hall in the *La Défense* building in Paris has a roof consisting of three assembled cylinder segments with an equilateral triangle as a base.

2 Parabolic cylinders

The railway station 'Gare Satolas' in Lyon (France) has a roof that looks like the spread wings of a bird taking off. Both wings have the shape of a parabolic cylinder, i.e. a ruled surface with a parabola as its directing curve.

3 Conoids

The wall of the *Atlántida* Church in Uruguay is shaped like a conoid. The directing curves are a straight line at zero height (the ground) and a wave curve in a horizontal plane at a height of seven metres.

4 Hyperboloids

A one-sheeted hyperboloid is a ruled surface that can be obtained by rotating a line around an axis that is not parallel and that does not intersect the rotating line that describes the surface.

It is a form that we often recognize in water- and cooling towers.

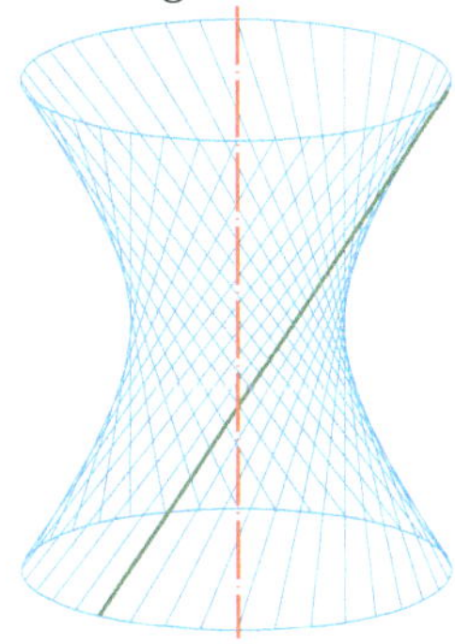

The light openings in the church *Sagrada Familia* in Barcelona, designed by Gaudi, are based on intertwined hyperboloids.

2.1.4 Minimal surfaces

Minimal surfaces are three-dimensional surfaces that result from processes of optimal curvature and the aim for a minimal surface tension. The elementary forces of nature realize them, for instance, in soap film obtained by dipping a wire frame into a soap solution. Just as an ordinary soap bubble adopts a spherical form to achieve a state of minimal potential energy, for the same reason these fanciful soap films in wire frames also aim for a minimal area. In architecture a minimal surface has the evident advantage of a minimum need of building material, resulting in lighter weight and lower cost.

The German architect Frei Otto used these forms for the construction of membrane roofs. The thin wires of soap film models are imitated by a network of steel cables and the soap film by membranes of a synthetic material. They are frequently applied in (self bearing) shell structures, cable-spannings and hanging roofs. These are roof constructions that are not supported by pillars but hang from a network of cables.

Examples

The roof of the Olympic Stadium in Munich (Germany) is a minimal surface.

For the design of the arched roof of the new railway station in Stuttgart, Otto Frei also started from the soap film model. The figure below shows the network of cables that imitates the wire frame of the soap film model.

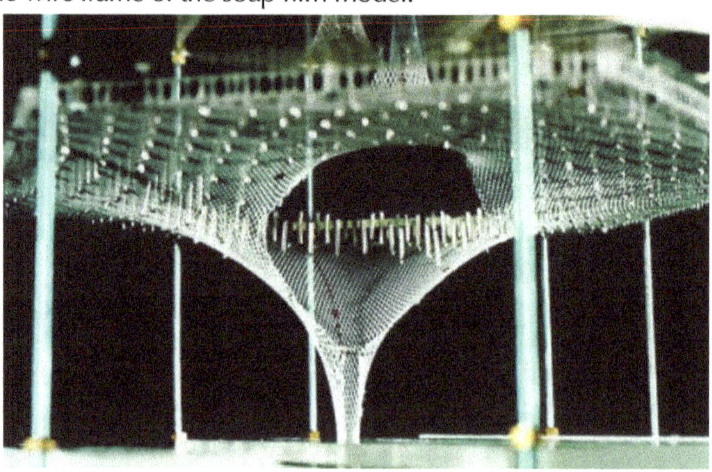

Summary

Because of its multitude of aspects, architecture is without doubt the mother of all visual arts. It uses a whole arsenal of mathematical methods for both the design of the shapes and the necessary calculations. This is why building constructions also reveal in their appearance their internal logic as part of the intended overall harmony.

2.2 Visual arts

It is obvious that in most other visual arts, mathematics does not as explicitly catch the eye as in architecture, which is why we treated it separately and first. If it may be true that in most museum works there is no direct trace of a mathematical background, surely it is not difficult to find examples in which there is, and this applies to all domains of the graphic arts, such as painting, plastic arts, sculpture, assemblages and collage. Especially in some examples of modern art it is sometimes difficult to make the distinction between the artistic and the purely mathematical aspects.

2.2.1 Again those harmonic proportions

Also in sculptures and paintings the composition is often based on the rules of harmonic proportions, such as the golden section, golden triangles and cheops sections.

Examples

1 In Boticelli's painting *The birth of Venus* the navel is at the 'golden height'. Moreover, the group of the Winds (on the left) and the figure of the Grace (on the right) are inscribed in golden rectangles, *AEGD* and *FBCH*.

2 The width of Boticelli's painting *Spring* is based on two golden triangles *AED* and *BFC* with an apex of 36°.

3 Also Leonardo da Vinci was obsessed by the golden section. In *The Vitruvian Man* not only is the navel at the right height, but numerous golden rectangles can also be recognized. The *Mona Lisa* is teeming with golden rectangles in which respectively the face, the forehead, and the other parts of the body are inscribed.

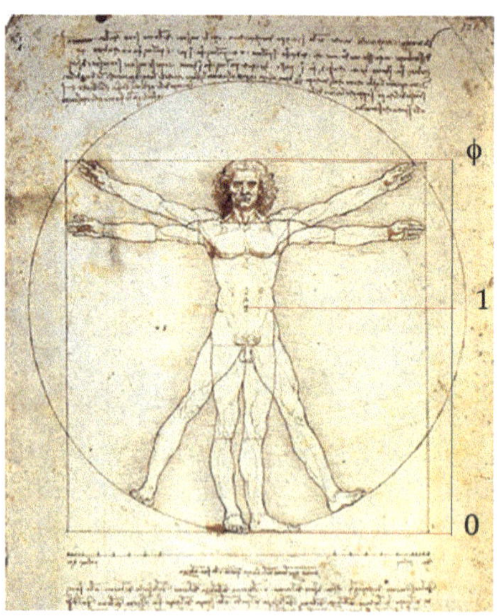

Also in *The Last Supper*, golden proportions can be found in the partition of the height of the table, the window openings and the overall height, as well as in the proportion of the width of the middle part of the table to the overall width and the width of the central window to the width of the other two.

4 In 1509, Fra Luca Paccioli wrote a booklet on the golden section, namely *De Divina Proporzione* (On the Divine Proportion). In the painting below he is depicted as a lecturer on... the golden section! The book itself is in the picture and the index finger of his left hand points to the divine proportion.

5 Starting from the Middle Ages, the use of a 'golden ratio compass' was a frequent practice in graphic work. Because of the similarity of triangles, the hinge of this compass extends its legs in a golden proportion in all its positions.

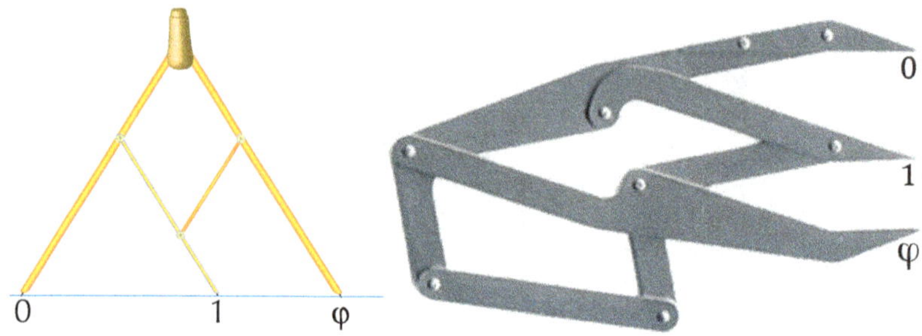

2.2.2 The rules of perspective

Until the introduction of the method of perspective in the Renaissance, graphic artists had always had difficulties in representing three-dimensional scenes on the picture plane in a realistic way. Many medieval pictures are still endearingly naive in this respect. Using the rules of central projection from projective geometry a satisfying solution could be found to this problem. From the (single) viewer's eye concurrent lines start projecting the various points of the scene onto a planar section, in which the respective proportions are represented in the way that the eye perceives them. The most important line is the optical horizon on which the so-called *vanishing points* are situated. Lines that converge at a vanishing point of the picture plane are in reality parallel lines, just like we perceive that the tracks of a railway seem to converge. Lines parallel to the picture plane, however, keep their direction, but their length shrinks as their distance from the observer increases. This we recognize in the appearance of the railway sleepers and the telegraph posts if we stand between the tracks of a railway line.

If a circle is centrally projected onto a plane not parallel to the plane of the circle, then its image is another type of conic section, such as an ellipse, a parabola or a hyperbola, depending on the position of the circle plane respective to the projection plane. In other words, it depends on how the projection plane intersects with the projection cone. The different cases can easily be simulated by means of a lamp with a circular lampshade and by letting it shine on a wall from different angles.

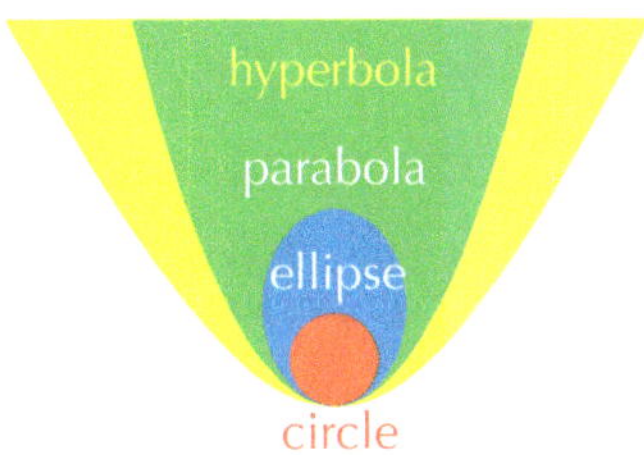

Examples

1 In the picture below we see Albrecht Dürer applying the method of perspective by means of a grid placed between the artist and the scene to be pictured (the reclining woman).

In the following picture, the eye of the observer is replaced by the eye of a screw in the wall from which a thread is pulled out to a detail of the scene (a lute). The intersection of the thread with the picture frame is recorded and then marked on the movable picture canvas.

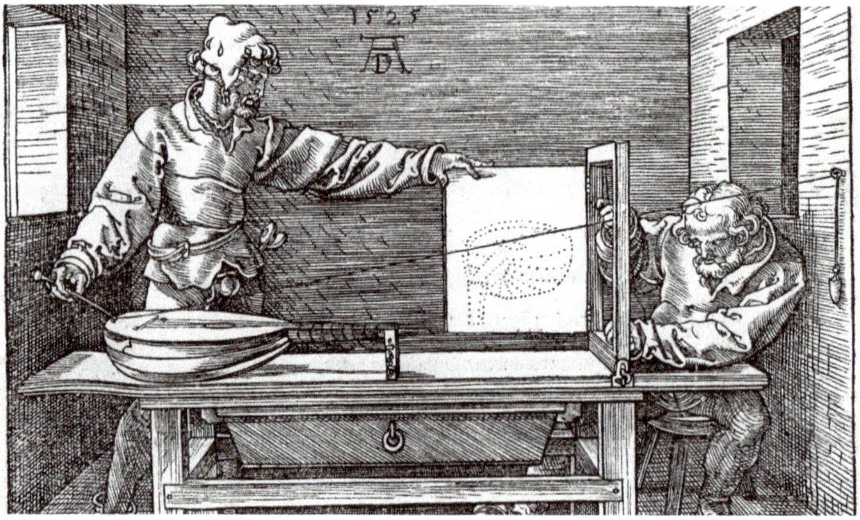

2 In Canaletto's painting *The stairs of the Santa Maria della Salute* one can easily locate the central vanishing point by finding the intersection of the base lines of the quay below the stairs and those of the houses on the opposite side. All the other lines start from this point and add the perspective illusion to the picture.

3

In Rafael's painting *The marriage of the Virgin* we can see the shrinking effect of the floor pattern and of the length of the persons and the pillars the more distant they are from the observer.

4 The Fresco in the office of the Secretary of State in the Vatican City shows the arches that are semi-circles but appear in perspective as semi-ellipses, of which the axes shrink the more they recede.

2.2.3 Contradictory figures

In the perception of perspective drawings the visual interpretation of the rules applied is paramount. After all, the representation is two-dimensional while the interpretation is three-dimensional. This interpretation is a particular faculty of our brain, combined with experience and optical features of sight, as is shown from the difficulties we have when analogously trying to process the two-dimensional representation of four-dimensional objects. This means that we can also think of representation rules for one more dimension, but because of a lack of appropriate experience, our imagination and intuition fail in the interpretation. In the following examples we show how the technique of perspective may cause false and contradictory impressions to arise. Also these aspects of perspective have been given form by artists in an ironical way, as is shown in the following examples:

1 In the litho *Up and down* by M.C. Escher (1898–1972) the men on the inside descend the staircase and return to their point of departure, which is contradictory. The men on the outside ascend the staircase, with the same result.

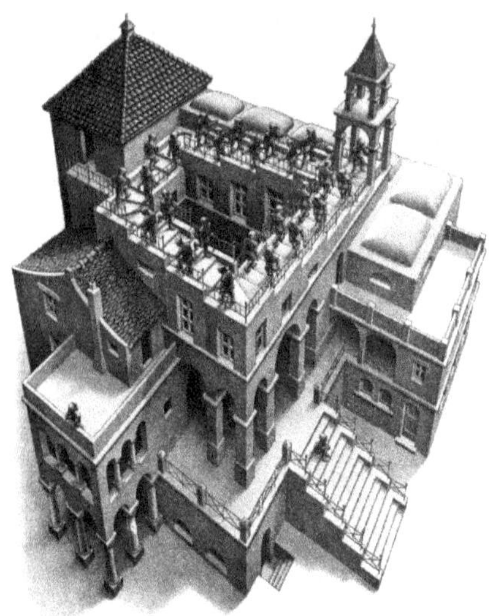

2 In the picture below from 1754 by William Hogarth, there are several deceptions related to the relative distance of details in the scene. The lady in the window effortlessly lights the pipe of the man further away on the hill. The line of the angler in front appears to hang in front of the nearby sign. This sign seems to hang behind the trees on the hill but at the same time extends halfway across the pond behind the house. The cattle and the trees seem to get bigger as they move further away from the observer, etc.

2.2.4 Planar decorations by means of symmetry groups

A *symmetry group* is a set of geometric transformations that are compositions of reflections. In the plane, compositions of reflections in respect to lines generate, for instance, translations (compositions of two reflections with parallel axes), rotations around a point (compositions of two reflections with intersecting axes) and point reflections (in fact rotations of 180°, or a composition of two reflections with perpendicular axes). These transformations preserve lengths and angles. Therefore they map figures onto congruent figures and are called *isometries*. The overall picture of a figure, together with all its mappings by such a symmetry group, is a symmetric figure representing this group. Many planar decorations by means of tiles, arabesques or other motives are based on symmetry groups. In three-dimensional space, reflections in respect to planes generate the different isometries, among which are translations, rotations around a line and torsions. These symmetry groups can, for instance, be seen in the forms of crystals.

1 Friezes

Friezes are ornamental borders with a repetitive feature. They are generally created by repetitive translation of a motif in a fixed direction, or by compositions of a translation with a reflection in respect to an axis of the same or the perpendicular direction of the translation. Also point reflections can be combined. Sometimes the motif itself can have one or more symmetries, in which case it is possible to reduce it to a more elementary form. On the basis of the transformations selected from all the possible ones, all friezes can be classified into no more than seven distinct types. We illustrate these types starting from a simple (not self-symmetrical) motive, namely **L**.

type 1 translation

type 2 reflection in respect to a horizontal line and translation

type 3 reflection in respect to a vertical line and translation

type 4 point reflection and translation

type 5 reflection in respect to a horizontal as well as a vertical line and translation

type 6 glide reflection, i.e. the composition of translation and a reflection in respect to a horizontal line

type 7 reflection in respect of a vertical line and glide reflection

In the following ornamental borders we recognize the types:

type 1

'Dragon and Phoenix', carpet from Asia Minor

motif 

type 2

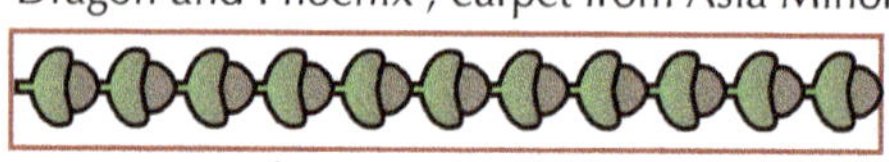

Window, Bourges cathedral

motif 

type 3

Decoration of a shrine , (French Renaissance)

motif 

type 4

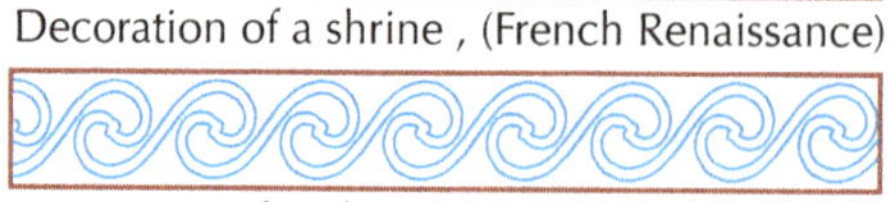

Decorative border of a parchment, Greece

motif 

type 5

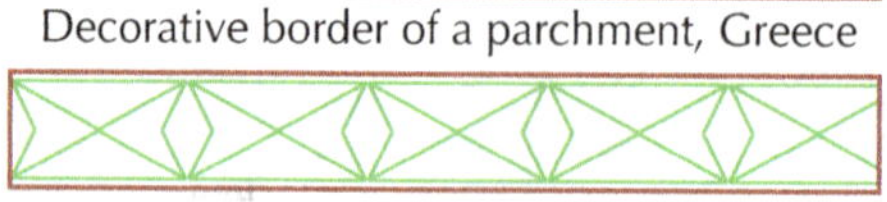

Mosaic in Pompei

motif 

type 6

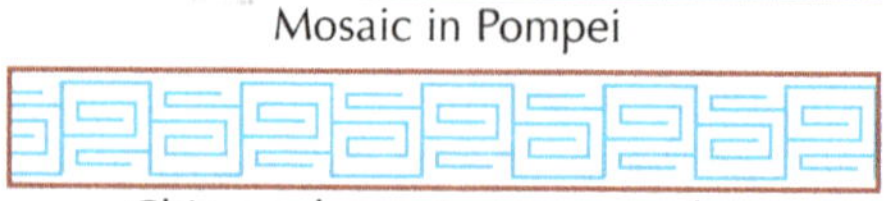

Chinese decoration on porcelain

motif 

type 7

Damask from the Italian Renaissance

motif 

2 Tilings

Also tessellations of the plane (tilings) can be described using these isometries. Depending on restrictions imposed on the shape and the diversity of the tiles, all their possible variations can also be classified. For instance, if all tiles have to be the same regular polygon then only three types can exist, namely those using equilateral triangles, those using squares and those using regular hexagons. The reason for this is that the sum of the angles in the common vertices has to equal 360°. An angle of a regular pentagon equals 108°, and this is not a divisor of 360°. So it is not possible to tile a floor with only regular pentagons, but in combination with other forms it would be possible. It can be proved that due to the possible symmetry groups in the plane, precisely seventeen symmetry patterns can be distinguished. Because rotation symmetries only allow angles of 180°, 120°, 90° and 60°, the patterns are thus all of the axial, the triangular, the square or the hexagonal type.

We illustrate the seventeen possible situations with the following figures:

While analysing wall paintings of Egyptian graves in 1923, the Swiss mathematician Andreas Speizer discovered that the Egyptians succeeded in producing all these 17 possible cases in their decorations. Also in Islamic art, the abstract art par excellence, he discovered these structures.

Examples

1 The following Egyptian decoration of a grave in Thebes is generated by a group of 8 elements, namely four rotations and four reflections in respect to lines. These are the symmetries of a square (quadratic lattice). The centre of the rotations we recognize in the midpoint of a curl. By means of translations in two perpendicular directions the motif is reproduced in a quadratic lattice.

2 In the window of a mosque in Cairo depicted on the left, several decorative techniques have been used. The ingenious inner plaiting is based on one of the symmetries that can be recognized in a regular hexagon (hexagonal lattice). In a hexagonal lattice there are rhombic and triangular cells. Depending on the symmetries occurring in these partial motifs one can distinguish five different types of hexagonal lattices. The plaiting in the picture represents one of these types. The frame cuts out a part of an endless picture that also extends by means of translations in two directions.

3 In the following picture by M.C. Escher the abstract lizards are arranged according to a variation of the symmetry patterns of the quadratic lattice. The points where two white and two black paws meet are centres of a quaternary rotation group with angles of respectively 0°, 90°, 180° and 270°. The points where two snouts of the same colour meet are centres of point reflections. No reflections in respect to a line occur. The same pattern can be recognized in a swastika figure.

4 The tilings of the walls in the Alhambra in Granada offer a symphony of symmetry groups, in which different patterns are combined. These are, for instance, the forms that so strongly inspired Escher's graphic work.

2.2.5 Limit tendencies and fractals

By repetition of the same pattern on an ever-decreasing scale, a self-similar figure comes into being that possesses a limit feature and that often adopts a fractal form. In the figure on the left, called *Circle Limit* by M.C. Escher, both aspects can be found.

2.2.6 Computer Art

Throughout the history of art, artists never hesitated to use everything they had at their disposal as resources for their design. Beside the brush and the chisel, an enormous variety of instruments were taken into artistic hands, often giving birth to new styles. Funk art, pop art, performance art, fluxus, minimalism, conceptual art, body-art, installation art, film and video art, all of these have acquired legitimacy in the artistic universe. So, why can't the computer be considered an artistic instrument? At first it might seem reprehensible to link a deterministic apparatus to imagination and creativity. However, it is the programmer who stipulates which algorithms are introduced to achieve an intended result. In this sense also the computer is but a medium. It is unbelievable what fascinating plastic forms are produced, for instance, by the calculus of variations or by fractals. Below, an example of both:

Summary

Also in the other visual arts besides architecture, mathematical backgrounds or principles can often be traced that give a more profound structure to the artwork and so contribute to greater awareness of the aesthetic experience. In some styles of especially modern art like Cubism (Picasso, Braque,...), Futurism (Marinetti, Cara,...), Suprematism (Malevitch, Rodtschenko,...), Constructivism (Tatlin, Lissitsky; ...), 'de Stijl' (Mondriaan, van Doesburg,...), Art Deco (Mackintosh, Hoffmann,...), Op-art (Vasarelly, Riley,...) and Minimalism (Judd, André,...), geometrical forms are legion. However, it was not our intention to give prominence to these items in this section, because of their too obvious references to mathematics, as can be seen in the figures below.

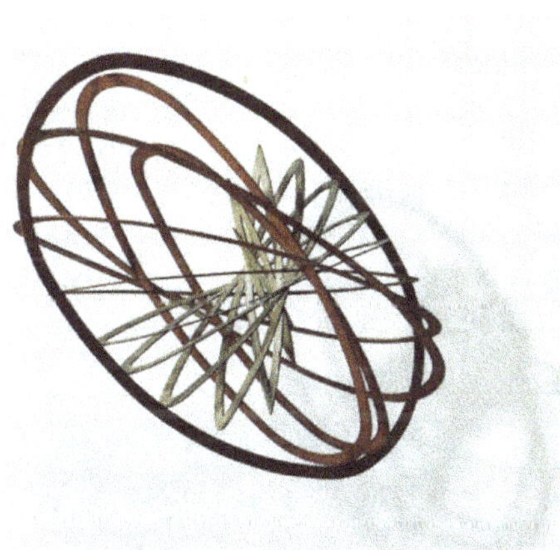

Three-dimensional oval-shaped spatial construction by Alexander Rodtschenko, exponent of Russian avant-garde.

Composition in red, yellow and blue by Piet Mondriaan, member of 'De Stijl', a group of artists around Theo van Doesburg that aimed for a formal modernism.

Of course, some structure always exists in most artworks, which is not necessarily expressed by basic geometrical forms or numerical ratios. Although in describing the more general structures a rational and therefore mathematical path can be followed, we decline to include controversial examples in this field.

2.3 Mathematics and music

Already in antiquity a strong link was made between music and mathematics. The discovery of the harmonic proportions based on integers, such as the octave, the fifth, the third and the fourth, excited the Pythagoreans so much that they started to believe in a celestial music of the harmony of spheres. This numerical vision of the cosmos influenced astronomy for a long time, as it lasted throughout the Middle Ages until Kepler.

Although music is probably even more than any other form of art associated with mood and emotions, there is nevertheless no unambiguous relation between the sense of sound that the ear receives and the impressions it evokes through our imagination. However, the Fechner law, which is valid for all senses, might be relevant because it states that if the stimulus increases in a geometric progression then the intensity of the sensation increases in an arithmetic progression. In other words, there is a linear relation between the intensity of the sensation and the logarithm of the stimulus. This, however, only concerns the intensity and not the kind of emotion that is experienced. Although practising and listening to music are activities that at an early stage were still carried out spontaneously and intuitively, gradually a music science came into being that became very abstract in both its theory and its symbolism, just as mathematics did.

Pythagoras at the monochord. He discovered the relation between the length of a string and the harmonic tone produced.

Pythagoras shows the figure of the Tetractys built by the numbers 1, 2, 3 and 4 (adding up to ten) that constitute the basis of the harmonic proportions in music.

Music is in the first place sound and therefore a physical phenomenon with its own patterns and laws. Just like other physical phenomena, sound can be described successfully with mathematical means. In the same way as white light is a composition of several monochromatic colours, also each sound is composed of several simple tones. Light and sound are described by wave theory. A sound wave, consisting of vibrations of the air, is periodic and characterized by its wavelength (length of one period) and its amplitude (peak of the wave). The number of wave periods per second is the frequency. The form of the wave relates to the specific timbre ('tone colour') of the source of sound and results from the simultaneous sounding of several related tones.

Example

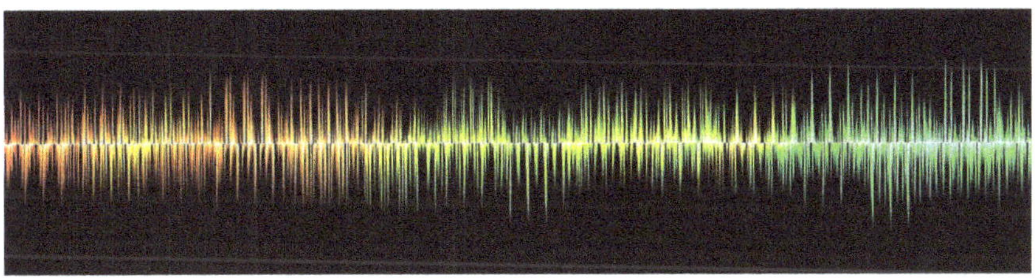

graphic representation of a sound wave

The tone scales are therefore based on physical laws that are expressed in mathematical terms. Musical sounds develop along several dimensions, such as time, chords, melodies, rhythm, cadence, tempo, loudness, timbre and periodicity. This is why a musical composition does not only take shape in the different notes on the staff but it has an infinite number of possible combinations and variations at its disposal. By introducing changing elements and all kinds of transformations, such as repetitions, crescendos, diminuendos and transpositions, the complex structure of a piece of music eventually comes into being. Among the dimensions mentioned above, especially time plays an important part. The melomaniac's delight consists precisely in analysing the local and overall structure of a composition and in recognising the techniques employed as they pass through time. The simultaneous and harmonious sound of instruments and voices has irrefutable rules that when violated result in painful cacophony. Their is no need to have received a theoretical musical training to notice this (although this might help), a trained ear is enough. The fact that in certain styles one can learn to feel how a tune or melody will continue proves that one can intuitively analyse the forms that are applied. There is no doubt that the structure that is heard can be recovered from the symbolic writing of the score and that reciprocally each composition is written by applying standard patterns. We will try to illustrate this with concrete examples.

2.3.1 Harmonic proportions and the different scales

A sound in music is a vibration of air that can be represented as a periodical wave function.

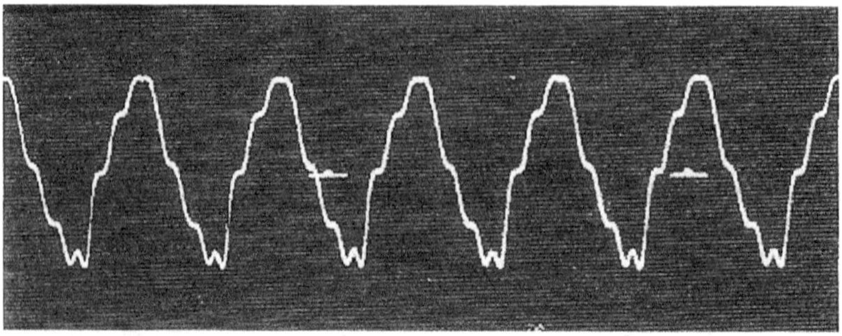

Graphic representation of the 258 Hz C tone on the clarinet. The marks indicate time spans of 0,01 second. The periodicity is very clearly shown in this picture.

By the technique of so-called Fourier analysis this wave function can be expanded as a sum of several regularly shaped sine functions. This way an arbitrary sound can be considered as a consonance of pure tones produced by an appropriate set of tuning forks. A pure tone is a definite number of vibrations per second of the air. This number is the *frequency* ν of the tone and is expressed in hertz (Hz). The average human ear is able to hear frequencies between 16 Hz and 20000 Hz. In this continuum of audible tones music uses only distinguished discrete values between about 30 Hz and 4000 Hz that can be arranged in several tone scales. When analysing the quantitative proportions of the frequencies of these distinct tones, the so-called *consonant intervals* that give a natural impression of harmony are essential for setting up the different tone scales. It is these proportions that were already discovered by Pythagoras in his monochord experiments, in which he studied the relation between the length of a vibrating string and the tone it produces. The observation that the most pleasing chords were reached precisely by simple proportions of integer numbers, such as 1/2 (octave), 2/3 (perfect fifth) and 3/4 (perfect fourth), not only permanently influenced western music theory but it also nourished speculative views of the cosmos.

1 Fundamental and harmonious tones

If A is a tone with frequency ν, then every tone of which the frequency is a multiple of ν, in other words of the form $n\nu$ where $n \in \mathbb{N}$, is called an *overtone* of A, which itself is called the *fundamental* or *root* tone of this sequence. As a consequence, the frequencies of two overtones of the same fundamental tone are related by a factor that can always be expressed by a rational number (fraction).

Example

If A has frequency ν then the third overtone of A has frequency 3ν and the fourth overtone of A has frequency 4ν. It applies that $(3/4)(4\nu) = 3\nu$. So, the frequency of the third overtone of A equals 3/4 times the frequency of the fourth overtone of A. This is the relation between a tone and its perfect fourth.

The sequence of a fundamental tone and its overtones is called the *harmonic sequence* or *chord progression* of this fundamental tone. The extent to which the consonance between two tones from the harmonic sequence is experienced as being pleasing depends on the arithmetic factor that expresses the relationship between the frequencies. The main harmonic intervals have the following proportions:

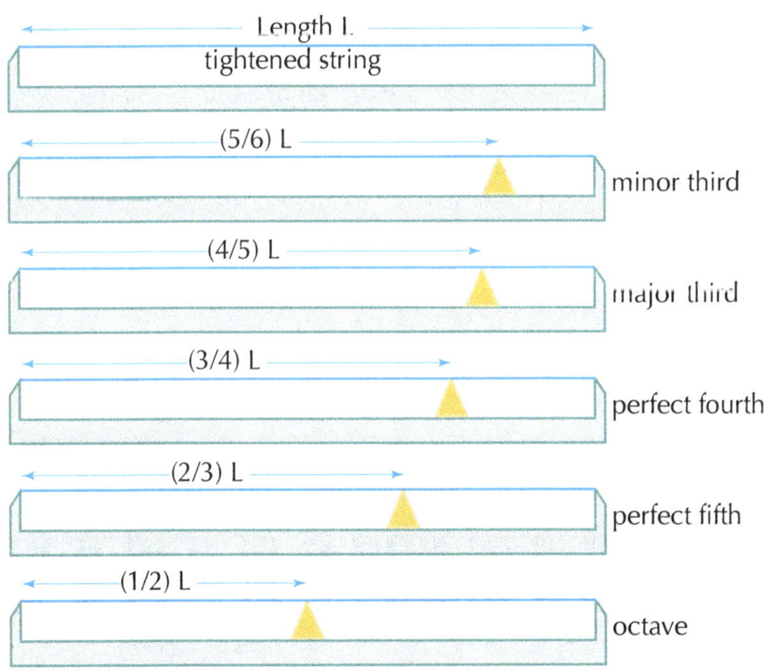

Each tone produced by a musical instrument or by the human voice is in fact a consonance of a fundamental tone with some of its overtones. This specific mix of overtones gives the typical timbre of the instrument or the voice.

2 Octave sequence

Let ν be the frequency of a certain tone, for instance A, then the tone with frequency 2 ν is called the (first) *octave* above A and the tone with frequency (1/2) ν the (first) octave below A. This relationship can be generalized for every tone of which the frequency equals $(2^z)\nu$, where z is a positive or a negative integer. In this way the tone with frequency $(2^3)\nu = 8\nu$ is the third octave above A and the tone with frequency $(2^{-4})\nu = (1/16)\nu$ is the fourth octave below A. The successive octaves make up the *octave sequence* of the tone A.

If one lets a tone sound together with one of its octaves then they are perceived as just one tone. Their relationship is the most harmonious that exists. The interval between the frequencies of a tone and its first octave is therefore taken as a unit for the distribution of the tones in a tone scale. The transition from a tone to one of its octaves is called *transposition*. Octave transposition can be interpreted as an equivalence relation in the set of tones and arranges them in equivalence classes that have a representative within each octave. Each tone scale consists of a number of tones with discrete frequency values in the continuum of the octave interval. Tones that differ by an octave are given the same name because of the equivalence relation between them. A modern piano usually contains seven successive full octaves that can be recognized by the periodical position of the white and the black keys. The low tones are on the left and the high ones on the right. The octave got its name because it spans eight white keys on the keyboard, first and last key included. Octo is Latin for eight.

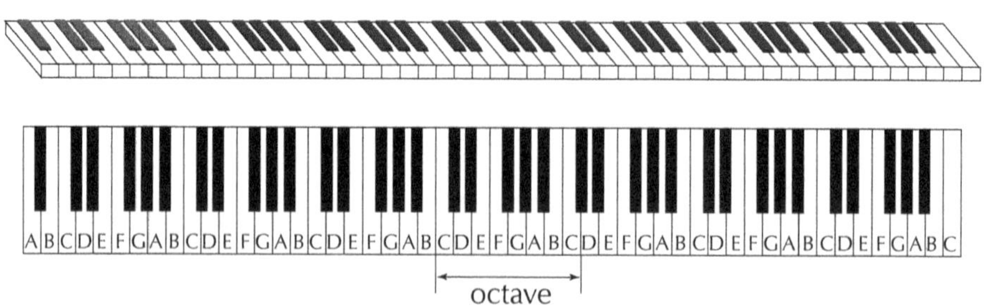

A complete piano keyboard. The double-headed arrow indicates an octave interval, here between two C notes..

3 The twelve-tone equally tempered scale

As from the beginning of the nineteenth century, pianos are usually tuned according to the so-called twelve-tone equally tempered scale, whereby an octave is divided into twelve equal intervals so that the frequencies corresponding to the dividing points set up a geometrical sequence with factor $\sqrt[12]{2} = 2^{(1/12)}$. Starting from a tone with frequency ν, this makes a sequence of 13 dividing points (making 12 intervals) with the following frequencies:

$$\nu,\ 2^{\frac{1}{12}}\nu,\ 2^{\frac{2}{12}}\nu,\ 2^{\frac{3}{12}}\nu,\ 2^{\frac{4}{12}}\nu,\ 2^{\frac{5}{12}}\nu,\ 2^{\frac{6}{12}}\nu,\ 2^{\frac{7}{12}}\nu,\ 2^{\frac{8}{12}}\nu,\ 2^{\frac{9}{12}}\nu,\ 2^{\frac{10}{12}}\nu,\ 2^{\frac{11}{12}}\nu,\ 2\nu$$

According to Fechner's law we experience the stimuli of the successive frequencies of this geometric sequence as impressions of increasing intensity. This, however, happens according to the successive terms of an arithmetic sequence, obtained by taking the logarithm of the terms in the geometric sequence. For convenience, we opt for the logarithmic function with base 2. This function maps the geometric sequence with initial term ν and factor $2^{(1/12)}$ onto an arithmetic sequence with initial term $\log_2 \nu = a$ and increment of 1/12, namely:

$$a, \frac{1}{12}+a, \frac{2}{12}+a, \frac{3}{12}+a, \frac{4}{12}+a, \frac{5}{12}+a, \frac{6}{12}+a$$
$$\frac{7}{12}+a, \frac{8}{12}+a, \frac{9}{12}+a, \frac{10}{12}+a, \frac{11}{12}+a, 1+a$$

This way the proportions of the frequencies of the corresponding tones are transformed into differences of their logarithms. The difference between two consecutive dividing points of the sequence is called a *semitone*. So, an octave contains twelve semitones. These can be recognized in the seven white and the five black keys of a piano, using a system of arrangement and names that will be discussed further on.

By subtracting $a = \log_2 \nu$ of all the terms of the arithmetic sequence above we obtain an arithmetic sequence that divides the interval [0, 1[into twelve equal parts, namely:

$$0, \frac{1}{12}, \frac{2}{12}, \frac{3}{12}, \frac{4}{12}, \frac{5}{12}, \frac{6}{12}, \frac{7}{12}, \frac{8}{12}, \frac{9}{12}, \frac{10}{12}, \frac{11}{12}, 1$$

We could have mapped the geometric sequence directly onto this last arithmetic sequence by means of a mapping *f* given by the formula:

$$f(x) = \log_2 x - \log_2 \nu = \log_2 (x/\nu).$$

On a five-lined staff, eleven different notes can be marked by writing them on a line, between two lines or directly above or under the staff. According to international agreement, the fifth note in this arrangement corresponds to a tone of 440 Hz and is indicated by the letter A or by la.

The tones that correspond to the seven white keys of a piano respectively from the left to the right are called A (la), B (ti), C (do), D (re), E (mi), F (fa) and G (so). Because the black keys also represent semitones, the difference in tone distance is not the same between each pair of consecutive white keys. For instance, the distance between A and B amounts to a whole tone (or two semitones) but between B and C it is only one semitone. Higher and lower notes can be noted on the staff by adding ledger lines above or below the staff. By means of the clef, the actual range of the staff is determined respective to a keynote. This is where the names of compositions come from, e.g. 'piano concerto in A-major', 'symphony in G-minor' etc. Two consecutive notes on the staff do not necessarily represent the same tone distance; this distance is determined by the system of the twelve-tone equally tempered scale (notice the position of the piano keys and their respective names).

D	E	F	G	A	B	C	D	E	F	G
re	mi	fa	so	la	ti	do	re	mi	fa	so

4 The sequence of fifths

After the octave relationship with its factor 2^z ($z \in \mathbb{Z}$), the relation of fifths with factor 3^z ($z \in \mathbb{Z}$) between a fundamental tone and its overtones, produces the second most harmonious impression.

Example

If A is a tone with frequency ν then we call the tones with frequencies

$\ldots,\quad 3^{-2}\nu,\quad 3^{-1}\nu,\quad \nu,\quad 3\nu,\quad 3^2\nu,\quad \ldots$ the *sequence of fifths* of A.

The image of a term $3^z\,\nu$ of this sequence by the mapping f of the previous subsection is:

$$f(3^z\,\nu) = \log_2(3^z\,\nu) - \log_2\nu = (z\log_2 3 + \log_2\nu) - \log_2\nu = z\log_2 3 \qquad (*)$$

Through the octave relation the perfect fifth and the perfect fourth are related to the sequence of fifths, since:

$$\frac{2}{3}\nu = 2\cdot\frac{1}{3}\nu \quad \text{and} \quad \frac{3}{4}\nu = \frac{1}{4}\cdot 3\,\nu$$

A fifth got its name from the fact that it spans five white keys on the piano, first and last included (*quintus* is Latin for fifth).

5 Tone scales

Music is composed with a limited set of tones that harmoniously fit together, meaning that they are chosen from the same set of overtones. Mainly octave and fifth relations are used. By combining these two, the fourths evidently become involved. The choice of the number of tones in a tone scale is the result of an historical process, but it can more generally be based on the theory of quotient groups, in which prime numbers play the main part as generating elements. The sub-group of octaves, generated by factor 2, determines the transposition classes. Theoretically tone scales with only one or two tones would be possible, but because they are so 'monotonous' and have restricted possibilities for variation, we won't discuss them. We will keep to the traditional constructions of the pentatonic (five tones), the heptatonic (seven notes) and the chromatic (twelve tones) scales as successive extensions. This happens by stacking successive perfect fifths within an octave.

6 The pentatonic scale

The pentatonic scale is obtained by taking four consecutive perfect fifths, starting from a certain tone, let's say F. Then, we transpose them within the interval of one and the same octave and put them in their natural order again. Observing the keys of a piano, we produce the selection of tones F, C, G, D and A.

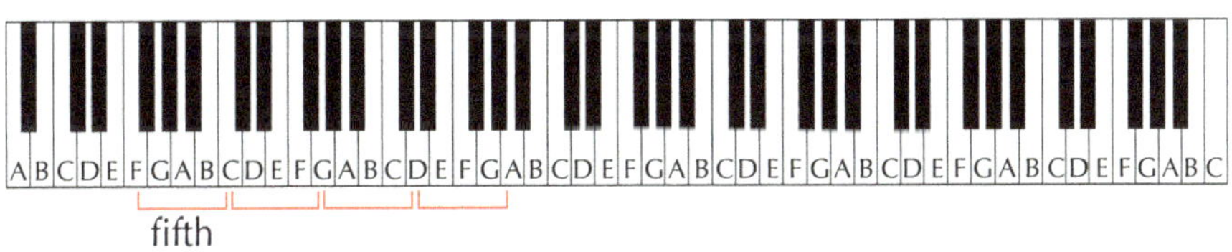

Because two consecutive fifths span five keys twice, this exceeds the octave starting from F that only contains eight white keys. So the tone G, which is reached after two fifths, must be drawn back by one octave so as to stay in the original octave that started from the chosen F. We have to do this once more with the tone A. This procedure could be represented schematically as:

$$F \rightarrow C \rightarrow G \downarrow G \rightarrow D \rightarrow A \downarrow A$$

By putting the tones in their natural order within the original octave starting from F we get the scale F, G, A, C, D.

Observe that these are exactly the five white keys that precede the black ones. By augmenting all tones in the scale by a semitone we are able to play pentatonic music only on the black keys of the piano. A lot of ethnic music has pentatonic themes and melodies.

Old Chinese music is also mainly written using this scale. Although some compositions in classical music restrict themselves to this scale, most of them use a broader scale, namely the heptatonic scale.

7 The heptatonic scale

The heptatonic or diatonic scale can be obtained in an analogous way to the pentatonic, but by stacking six fifths instead of four starting from the tone F. This leads to the selection of the tones: F, C, G, D, A, E, B.

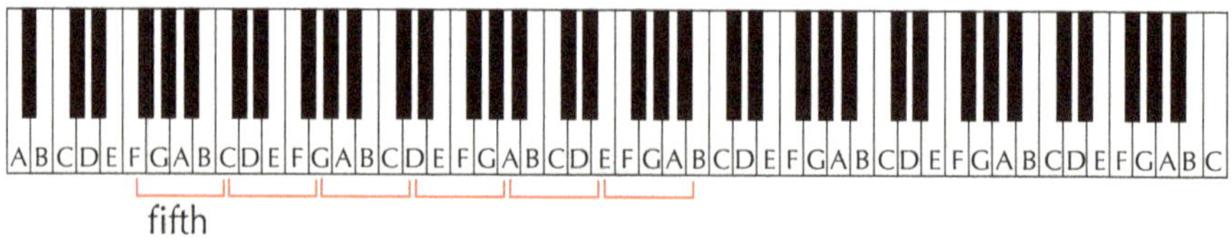

These have to be set in order within the same octave starting from F by the same procedure described above for the pentatonic scale:

$$F \rightarrow C \rightarrow G \downarrow G \rightarrow D \rightarrow A \downarrow A \rightarrow E \rightarrow B \downarrow B$$

When putting them in their natural order this leads to the tone scale: F, G, A, B, C, D, E, corresponding precisely to the seven successive white keys on the piano. This is also called the *'Pythagorean scale'*.

This explanation is a lot simpler than the historical construction because we considered the piano keyboard as already existing. Our way of working was in fact a reversal of the chronology of the construction of the tone scale. Tone scales have not been constructed starting from a piano that appeared out of the blue, but the other way round: the piano was arranged precisely according to an intended tone scale. The stacking of fifths, however, is the basic principle for the construction of a tone scale. With no piano available, one has to calculate frequencies and logarithmic values. We can illustrate this for the heptatonic scale.

We start from the internationally agreed frequency of $\nu = 440$ Hz that we mentioned before, and that is mapped onto 0 by the function f from subsection 3 and indicated by the letter A or by la. The function f maps in general the perfect fifth with index z starting from ν, namely $(3^z)\nu$, onto $z \cdot \log_2 3$, where $z \in \mathbb{Z}$ (see subsection 4)

We now select the fifths with successive indices:

$$-4, -3, -2, -1, 0, 1, 2$$

The function f maps the corresponding frequencies onto numbers for which we have to find the equivalent representatives p_z in the octave interval $[0, 1[$. This we perform by transposing them over an integer number of octave intervals, which here is simply 1; in other words we just have to add an appropriate integer. Like this:

$$(-4)\cdot \log_2 3 = -6.33985\ldots \rightarrow p_{-4} = -6.33985\ldots + 7 = 0.66014\ldots \in [0, 1[$$
$$(-3)\cdot \log_2 3 = -4.75488\ldots \rightarrow p_{-3} = -4.75488\ldots + 5 = 0.24511\ldots \in [0, 1[$$
$$(-2)\cdot \log_2 3 = -3.16992\ldots \rightarrow p_{-2} = -3.16992\ldots + 4 = 0.83007\ldots \in [0, 1[$$
$$(-1)\cdot \log_2 3 = -1.58496\ldots \rightarrow p_{-1} = -1.58496\ldots + 2 = 0.41503\ldots \in [0, 1[$$
$$0\cdot \log_2 3 = 0 \rightarrow p_0 = 0 \in [0, 1[$$
$$1\cdot \log_2 3 = 1.58496\ldots \rightarrow p_1 = 1.58496\ldots - 1 = 0.58496\ldots \in [0, 1[$$
$$2\cdot \log_2 3 = 3.16992\ldots \rightarrow p_2 = 3.16992\ldots - 3 = 0.16992\ldots \in [0, 1[$$

Next we have to put these representatives p_z in the interval [0, 1[in their natural order and give them a name:

p_0	<	p_2	<	p_{-3}	<	p_{-1}	<	p_1	<	p_{-4}	<	p_{-2}
(la)		(ti)		(do)		(re)		(mi)		(fa)		(so)
(A)		(B)		(C)		(D)		(E)		(F)		(G)

The names have been established by tradition.

Notice that the distance between the numbers p_1 and p_0 equals 0.58496... and that this distance precisely indicates a perfect fifth, corresponding to the fact that there are exactly five positions in the ordered sequence contained in the interval $[p_0, p_1]$.

8 The chromatic scale. Sharps and flats.

In order to bring even more refinement to compositions, the heptatonic scale was later extended to the chromatic scale, with twelve notes. To do this, the sequence of fifths is extended at both the upper and the lower end. Naming the new tones starts from those already existing followed by 'sharp' if the new tone is seven-fifths higher than than the affected tone, or by 'flat' if it is seven-fifths lower. These notes are indicated by adding a *'#'* (sharp) or a *'b'* (flat).

Example

If we also add the fifths with indices –6, –5 and 3, 4, 5 to the sequence of fifths with indices –4, –3, –2, –1, 0, 1, 2, then we respectively get the tones E^b, B^b and $F^\#$, $C^\#$, $G^\#$.

In order to also enter these new tones in the interval [0, 1[and to set them in order, we have to apply the mapping *f* of the previous paragraph and perform the appropriate transpositions.

$$(-6) \cdot \log_2 3 = -9.50977\ldots \rightarrow p_{-6} = -9.50977\ldots + 10 = 0.49022\ldots \in [0, 1[$$
$$(-5) \cdot \log_2 3 = -7.92481\ldots \rightarrow p_{-5} = -7.92481\ldots + 8 = 0.07518\ldots \in [0, 1[$$
$$3 \cdot \log_2 3 = 4.75488\ldots \rightarrow p_3 = 4.75488\ldots - 4 = 0.75488\ldots \in [0, 1[$$
$$4 \cdot \log_2 3 = 6.33985\ldots \rightarrow p_4 = 6.33985\ldots - 6 = 0.33985\ldots \in [0, 1[$$
$$5 \cdot \log_2 3 = 7.92481\ldots \rightarrow p_5 = 7.92481\ldots - 7 = 0.92481\ldots \in [0, 1[.$$

Together with the earlier values and after setting all of them in order we get:

$$p_0 < p_{-5} < p_2 < p_{-3} < p_4 < p_{-1} < p_{-6} < p_1 < p_{-4} < p_3 < p_{-2} < p_5$$

(A) (B^b) (B) (C) ($C^\#$) (D) (E^b) (E) (F) ($F^\#$) (G) ($G^\#$)

Notice that a flat tone is precisely a semitone below the tone of the same name and a sharp tone precisely a semitone above. On a piano keyboard these added tones are the black keys. If one goes on extending the sequence of fifths in one or other direction, the result is double sharps and double flats.

9 Natural tone scales

A natural tone scale is a selection of seven successive tones from the sequence of fifths. One of these tones is chosen as the keynote. The other notes are representatives from the transposition classes of the other six successive fifths, chosen so that they fit within a one-octave interval above the keynote. So, after setting these representatives in order, the keynote will always be the first tone of the scale.

Example 1

From the sequence of fifths we select the seven successive tones with respective indices: 0, 1, 2, 3, 4, 5 and 6. From these we choose the one with index 1 (the second of these tones) as the keynote. Since the function *f* maps this tone onto $p_1 = 0.58496\ldots$ (see subsection 7) we have to select representatives P_i within the interval $I = [0.58496\ldots, 1.58496\ldots[$ of the other tones p_i. Recovering the values of p_i from the previous pages, this results in:

$$p_0 = 0 \rightarrow P_0 = 0 + 1 = 1 \in I$$
$$p_1 = 0.58496\ldots \rightarrow P_1 = 0.58496\ldots \in I$$
$$p_2 = 0.16992\ldots \rightarrow P_2 = 0.16992\ldots + 1 = 1.16992\ldots \in I$$
$$p_3 = 0.75488\ldots \rightarrow P_3 = 0.75488\ldots \in I$$
$$p_4 = 0.33985\ldots \rightarrow P_4 = 0.33985\ldots + 1 = 1.33985\ldots \in I$$
$$p_5 = 0.92481\ldots \rightarrow P_5 = 0.92481\ldots \in I$$
$$p_6 = 0.50977\ldots \rightarrow P_6 = 0.50977\ldots + 1 = 1.50977\ldots \in I$$

The result of arranging these P_i in their natural order, is the scale:

P_1 <	P_3 <	P_5 <	P_0 <	P_2 <	P_4 <	P_6
E	F#	G#	A	B	C#	D#

This is called *the major scale of keynote* E, briefly E *major.*

When selecting the tones with indices –4, –3, –2, –1, 0, 1 and 2 from the sequence of fifths, as in subsection 7, and selecting the second tone as the keynote, i. e. the one with index –3 (C) for which $p_{-3} = 0.24511...$, and then placing their representatives within the interval [0.24511... , 1.24511...[in their natural order, we get the scale C, D, E, F, G, A, B. This is called *the major scale of keynote* C, or C *major.*

Example 2

From the sequence of fifths we select the seven successive tones with respective indices: –6, –5, –4, –3, –2, –1 and 0.

This time we choose the fifth tone among them as the keynote, namely the one with index –2 (G).

Since $p_{-2} = 0.83007...$ (see subsection 7), we have to arrange the representatives P_i of the other selected tones in the interval [0.83007... , 1.83007...[. This leads to:

P_{-2} <	P_0 <	P_{-5} <	P_{-3} <	P_{-1} <	P_{-6} <	P_{-4}
G	A	B^b	C	D	E^b	F

This is called *the minor scale of keynote* G, briefly G *minor.*

When selecting the tones with indices –4, –3, –2, –1, 0, 1 and 2 from the sequence of fifths, as in subsection 7, and selecting the fifth tone as the keynote, i. e. the one with index 0 (A) and for which $p_0 = 0$, and then setting their representatives within the interval [0, 1[in their natural order, we again obtain the scale A, B, C, D, E, F ,G of subsection 7. This is *the minor scale of keynote* A, or A minor.

10 Deviations, false chords and fluctuations

Most scales are based on stacks of octaves and fifths and they show small discrepancies because the frequencies of the tones in the sequence of octaves do not periodically coincide with those of the sequence of fifths. These deviations obviously also appear in the arithmetic sequences obtained by the logarithmic mapping of these frequencies.

These small differences induce fluctuations caused by the interference of the respective sound waves. The related chords are experienced as sounding unpleasant and are designated as being false in contrast to harmonious.

Example 1

The stacking of seven octaves starting from a given tone produces a geometric sequence with factor 2/1. In the seventh term this shows an increase by the factor $(2/1)^7 = 128$.

The stacking of twelve fifths starting from the same tone produces a geometric sequence with factor 3/2. In the twelfth term this shows an increase by the factor $(3/2)^{12} = (1.5)^{12} = 129.746...$

The corresponding terms in these two sequences are actually close but do not really coincide.

Evidently this also implies a deviation in the terms of the arithmetic sequences obtained by taking the logarithm (with base 2) of the terms of the geometric sequences:

- the seventh term of the first sequence has increased by $\log_2(2^7) = 7$
- the twelfth term of the second sequence by $\log_2(1.5)^{12} = 12\log_2(1.5) = 12 \cdot 0.58496... = 7.01955...$

This difference $7.01955... - 7 = 0.01955...$ is called *the Pythagorean comma*.

This means that the sequence of octaves and of fifths cannot be combined in a perfect way. This observation led to alternative opinions with regard to the construction of tone scales, into which thirds were also introduced as components, besides octaves and fifths. These, however, also show slight deviations.

Example 2

Four fifths up followed by two octaves down makes a factor:

$$\left(\frac{2}{3}\right)^4 \cdot \left(\frac{2}{1}\right)^2 = \frac{64}{81}$$

This is called a wide major third, because it equals almost 64/80 = 4/5, the factor that corresponds to a major third.

The deviation of the logarithmic values is $\log_2(4/5) - \log_2(64/81) = 0.0179....$

This is less than the Pythagorean comma, but nevertheless still a deviation.

Example 3

Naturally, the twelve tone equally tempered scale, in which the irrational number $\sqrt[12]{2}$ $\left(\log_2 \sqrt[12]{2} = \frac{1}{12}\right)$ is used as the factor for the partition of the octave, can also not be combined perfectly with the harmonic proportions obtained by integers that appear in the scales built by octaves, fifths, major thirds, etc. Here again, the approximations of the rational proportions show small deviations. In the following table the factors obtained as the logarithms (with base 2) of the natural numbers are compared to the fractions with denominator 12 that are their best approximations:

$\log_2(1) = 0$	$0/12 = 0$
$\log_2(2) = 1$	$12/12 = 1$
$\log_2(3) = 1.5849...$	$19/12 = 1.5833...$
$\log_2(4) = 2$	$24/12 = 2$
$\log_2(5) = 2.3219...$	$28/12 = 2.3333...$
$\log_2(6) = 2.5849...$	$31/12 = 2.5833...$
$\log_2(7) = 2.8073...$	$34/12 = 2.8333...$
$\log_2(8) = 3$	$36/12 = 3$
$\log_2(9) = 3.1699...$	$38/12 = 3.1666...$

These small differences have an influence on the perception of the corresponding tones. Where there is a deviation between the frequencies of two tones sounding together, fluctuations occur because of the interference of their waves. These fluctuations make a chord sound unpleasantly out of tune. Therefore they are called 'imperfect' or 'tempered' third and fifth, as opposed to the 'perfect' third and fifth.

The perception of false chords can be reinforced by the specific mixture and the amplitude of the overtones in the sound of an instrument or a voice. In a piano these are moderate, but in a harpsichord or an organ they are rather considerable, which is why the tuning system is different for different instruments. For instance, an organ is tuned differently than a piano.

2.3.2 Geometric models for periodic tone scales

The order and the periodicity of the tones is specific for every tone scale. These characteristics can be represented in a geometric model that lends a visual structure to the musical compositions in the tone scale concerned, and so links the eye to the ear, thus making music a combined experience of two senses.

Example 1

The pentatonic system from 2.3.1, section 6 was constructed by the stacking of four fifths starting from the tone F (fa). Thus we obtained the selection of the five tones F(fa), C(do), G(so), D(re) and A(la). We can represent this construction with the following diagram:

The arrows indicate how many white keys we have to move to the right on a piano keyboard so as to reach the next tone. Each arrow thus defines a fifth.

Disregarding the multitude of octaves that is covered by this construction, we can represent this collection by the vertices of a regular pentagon on a circle. The equal distances of the points that represent the successive tones symbolise the fifths and the circuit of the circle specifies the periodicity.

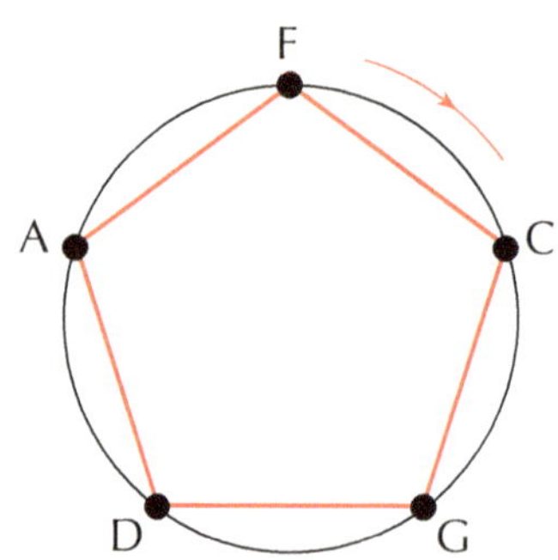

When we represent the tones within one octave and set them in order, we get respectively F(fa), G(so), A(la), C(do) and D(re). If we let this octave correspond to the circumference, then in our representation we have to take into account the unequal distances between the ordered tones.

We measure the tone distances in semitones as we can read them from the piano keyboard (see 2.3.1, section 6).

This makes:

If we divide the circle into twelve equal parts like on a clock face, then the points representing the ordered tones are respectively at positions 0, 2, 4, 7 and 9 o'clock. This results in an irregular pentagon of which the vertices are at distances (measured by arcs on the circle) that represent multiples of the semitone. The perimeter of the circle corresponds to one octave.

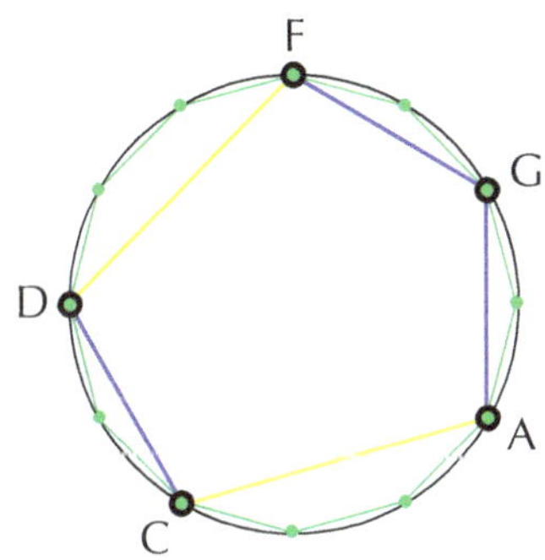

Thus, this pentatonic model is suitable for tracking the melody of a children's song as a fanciful polygon that comes into being by connecting the vertices of the pentagon respective to the consecutive notes of the melody. This could be represented dynamically on a computer screen.

In this way, the visual structure can underline the auditive structure and add some dizziness to the euphoric sensation of acoustic pleasure.

Children's song

Example 2

The pentatonic scale is essentially one-dimensional, as can be seen in the model of the circular line. The heptatonic scale, being an extension of the pentatonic, has a second dimension. This we also want to integrate into its geometric model.

By stacking six fifths starting from the tone F(fa), like we did in 2.3.1, section.7, we obtained the following seven tones: F(fa), C(do), G(so), D(re), A(la), E(mi) and B(ti). We can represent this construction geometrically with the following two-dimensional diagram, an extension of the one for the pentatonic scale, which we recognize in the horizontal dimension of its upper row:

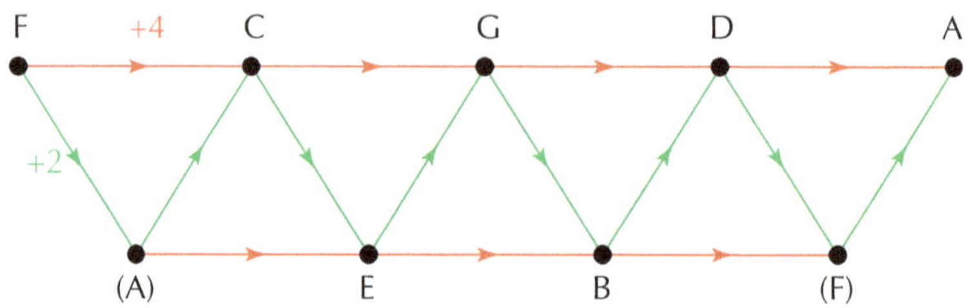

The arrows indicate how many white keys we have to move to the right on a piano keyboard so as to reach the next tone. The composition of two '+2 ' arrows hence makes one '+4' arrow, which defines a fifth.

If we coil, i.e. join the two ends of the strip in the diagram above, so that the F on the top left coincides with the (F) on the bottom right and the (A) on the bottom left coincides with the A on the top right, then we get a Möbius band. The diatonic scale is arranged periodically on the single edge of this one-faced Möbius band. In this model the edge is a single circuit which can be moved along continuously.

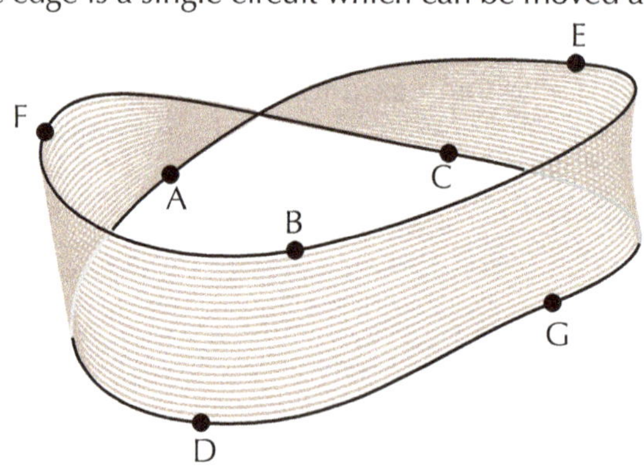

One can also make a model with the tones put in their natural order within one octave. If we implement this model on a computer, and combine it with a piece of heptatonic music (Haydn), then we can visually trace the melody as a swirling line connecting the positions of the consecutive notes of the melody on the edge of the Möbius band.

London Trio II

Franz Jozeph Haydn

Example 3

In the twentieth century the heptatonic scale was further extended and refined to the chromatic scale by adding sharps and/or flats.

In 2.3.1 subsection 8, we constructed a chromatic scale by selecting the tones with indices –6, -5, -4, -3, -2, -1, 0, 1, 2, 3, 4 and 5 from the sequence of fifths. This resulted in the scale:

E^b (mi^b), B^b (ti^b), F (fa), C (do), G (so), D (re), A(la), E (mi), B (ti), $F^\#$ (fa$^\#$), $C^\#$ (do$^\#$), $G^\#$ (so$^\#$)

On the piano keyboard the sharps and flats correspond to the black keys. After setting the tones in order within one octave, a flat appeared to be a semitone (one key) below (left of) the tone with the same name, and a sharp a semitone above (right of) the one with the same name.

We can represent the construction of the chromatic scale with the following diagram:

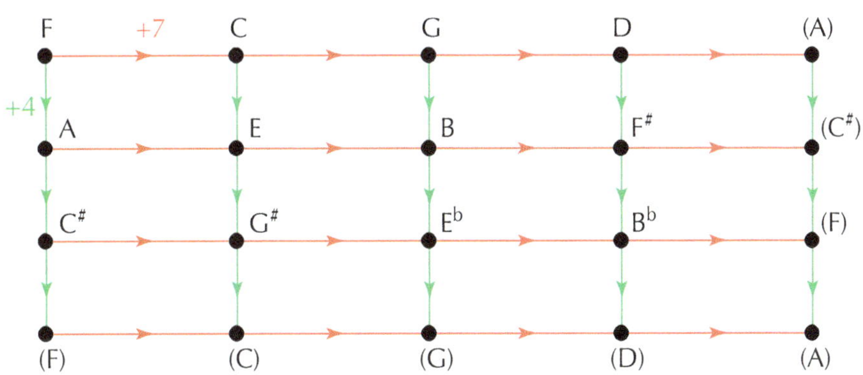

The arrows indicate how many keys (both white and black) we have to move to the right on a piano keyboard as to reach the next tone. Mind that the distances here are indicated as multiples of a semitone.

If we coil this diagram, i.e. join the top and bottom edges, so that the notes on the top row coincide respectively with the notes of the same name on the bottom row, then we get a cylinder on which the first three rows are represented as parallel lines at equal distances.

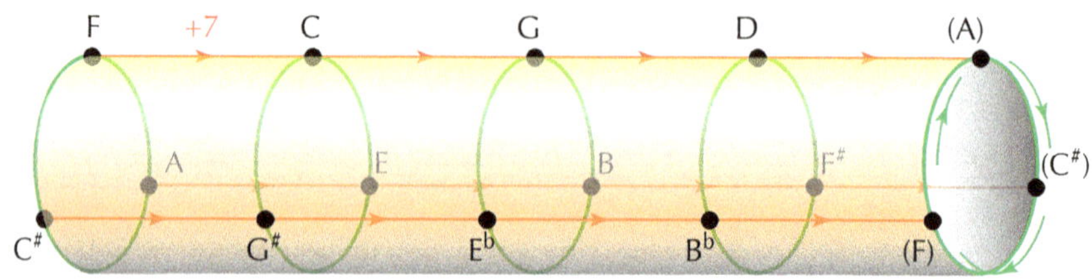

If we coil this cylinder into a hoop, after twisting it a bit so that the (A) on the right side coincides with the A on the left side, then the (C#) will fit on the C# and the (F) on the F. In this way we get the figure of a torus (doughnut) on which the twelve notes are represented on a closed curve that winds around the torus but that is neither a parallel nor a meridian circle.

By implementing such a model on a computer, we can visualize the melody of a piece of chromatic music (Proost) as the dynamic web of line segments connecting the consecutive notes of the melody.

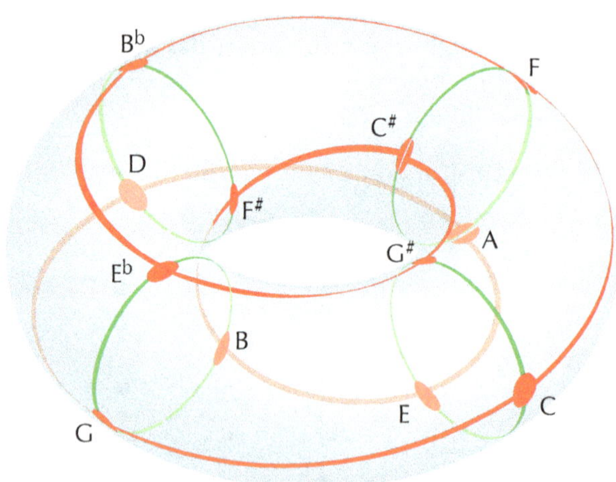

Sò III

Dirk Proost

Summary

The composition of music according to tone scales, based on the proportions of the frequencies of soundwaves, shows the undeniable close relationship between music and mathematics. Through a symbolic language, this underlying system helps us to understand the patterns and structures heard in these compositions. Moreover, these patterns can be made explicit for the other senses such as sight, not only by the abstract models we have presented, but also by choreographic figures in dance, for instance, which is an art form inseparably related to music. In this section, though, it was also not our intention 'to pull out all possible stops'.

2.4 Mathematics and literature

At first sight it might seem far-fetched to refer to mathematical aspects also in literature. When reading an epic, a novel or a poem we do not pay attention to the work's form and structure, but we are primarily carried away by the story, the adventures of the characters, the language and the style of the author. We are not always aware to what extent aspects of form also contribute to the expressiveness and the beauty of the composition. In classical epic works like those of Homer, Vergilius and Shakespeare, the strict patterns of metrical feet not only lend a solemn rhythm to the sentences, but they also lift the plot's banality to a serious dramatic level. Also in poetry that directs attention to the musical aspects of language, certain forms are subject to metre, stanza and rhyme schemes. In Dante's *Divine Comedy*, symbolic structures are the foundations of the composition of the work; the circles of hell, purgatory and paradise are arranged in three times thirty-three cantos. This lends logical coherence and clarity to its development.

Dante Alighieri (1265–1321), Florentine poet. In his famous *La Divina Commedia*, a tribute to his beloved Beatrice, he describes the environment of the medieval human being with his descriptions of hell, purgatory and paradise. The work is constructed very systematically and symbolically. After an introductory canto, each section is divided into thirty-three cantos, the number of which refers to the age at which Jesus Christ died. Hell is described in nine circles, Mount Purgatory in seven rounds and Paradise in nine circles again. The total of one hundred cantos makes the perfect divine number.

Besides its purpose of providing reading pleasure and information, there is also another view of language that deals with the intrinsic study of linguistics. Just remind yourself of the difficulties that have to be overcome in the deciphering of extinct languages, such as Egyptian hieroglyphic writing or Babylonian cuneiform, the decoding of encrypted messages in times of war, the setting up of transformation rules for creative linguistic usage and the characteristics of an author's typical style, which makes it as recognizable as the composition style of a musician.

Besides, the dramatic development of a play or a novel can be classified according to its structure, by recognizing simularities to models and regardless of the apparent difference in the stories. The combined action of protagonists, antagonists, fate and 'dei ex machina' is not only to be found in classical drama, but also in Ian Fleming's popular James Bond novels and Karl May's cowboy and indian stories.

We'll illustrate some of these aspects of literature with some examples.

Marcel Proust (1871–1922), French author. His famous series of novels *A la recherche du temps perdu* (In Search of Lost Time), consisting of seven volumes, is with no doubt one of the most important and original works of the twentieth century. Proust's style is characterized by the 'monologue intérieur' that meanders and extends in page-long sentences. The central role of memory and emotional recollection is forever portrayed in the scene where a madeleine biscuit is dipped into a cup of tea. The taste launches him into a search for the lost paths in the labyrinth of his youth.

2.4.1 Feet, verses and poetic forms

In poetry, the theory of the structure of verses is called *metrics*. The (traditional) poets organize the syllables according to the laws of prosody (the theory of the quantity of syllables) so that the stressed and unstressed syllables (rises and falls) make a phrase with a musical rhythm. This they manage by arranging them according to a variation of poetical feet. A foot consists of a specific succession of rises, symbolized by a dash —, and falls, symbolized by a pit ∪.

Frequently used feet are the following::

name	scheme	example
amphibrach	∪—∪	amicus
amphimacer	—∪—	carmina
anapest	∪∪—	affectus
dactyl	—∪∪	vincula
iamb	∪ —	vita
spondee	— —	amor
trochee	— ∪	primus

The successive feet in a verse are separated by a slash /. The so-called *caesura sign* (; or :) is used to mark a pause in a longer verse. The technique of marking feet and caesuras is called scansion. Some of you may remember this 'enjoyment' from your Latin and Greek lessons.

Example:

— ∪ ∪ /— ∪ ∪ /— : — / — — / — ∪ ∪ / — —
Ar ma vir um que ca no Tro jae qui pri mus ab o ris
— ∪ ∪ / — —/— ∪ ∪/— : ∪∪/— ∪ ∪ / —∪
It al i am fa to pro fu gus Lavi ni a que ve nit

(Opening lines of Vergil's Aeneid)

Observe that both of the preceding verses contain six feet. They are called hexameters. Most classical epics, like Homer's Odyssey and Vergil's Aeneid, are written in hexameters. Other commonly used poetical forms are:

- — the alexandrine: a hexameter consisting of six iambs
- — the pentameter: consists of five feet
- — the tetrameter: consists of four feet

The symbolic notation of verses strongly reminds us of Morse code messages (just replace ∪ by a dot •). In such an interpretation the feet correspond to the letters of the alphabet. For instance, the dactyl corresponds to the letter d and the iamb to the letter a. Mathematical combinatorics (the theory of calculating the number of possible combinations) provides the means to calculate the maximum number of signs for each length of the string of symbols. In this way, the number of feet could be considerably increased. But maybe that would be too much of a good thing. Naturally there also exists non-metric poetry, where those rules are consciously broken. This poetical form is called *antimetric*.

Verses are combined into stanzas that make up a poem. Also here strict numerical patterns are respected, so that the rhythm is once more reinforced by the specific rhyme schemes. Some well-known variations of stanzas, poems and rhyme schemes are the following:

stanzas:

- — distich: consists of two verses
- — terza rima: consists of three verses
- — quatrain: consists of four verses
- — cinquain: consists of five verses

poems:

- — ballad: consists of three stanzas of eight or ten verses, followed by a refrain of four or five verses
- — elegy: consists only of disticha
- — sonnet: consists of four stanzas, namely two quatrains (the octave) and two terza rima (the sextet)
- — limerick: consists of a cinquain

rhyme schemes:

—for the distich:	AA
—for an elegy:	AA BB CC ...
—for (successive) terza rima:	ABA BCB CDC ...
—for a quatrain:	ABBA AABB ABAB AABA ...
—for a limerick:	AABBA

Examples

Shall I compare thee to a summer's day?	A	Sonnet
Thou art more lovely and more temperate:	B	
Rough winds do shake the darling buds of May,	A	
And summer's lease hath all too short a date:	B	
Sometimes too hot the eye of heaven shines,	C	
And often is his gold complexion dimm'd:	D	
And very fair from fair sometimes declines,	C	
By chance, or nature's changing course untrimm'd;	D	
But thy eternal summer shall not fade,	E	
Nor lose possession of that fair thou ow'st,	F	
Nor shall death brag thou wander'st in his shade,	E	
When in eternal lines to time thou grow'st;	F	
So long as men can breathe, or eyes can see,	G	
So long lives this, and this gives life to thee.	G	

William Shakespeare (1564–1616)

There was an Old Man with a beard	A	Limerick
Who said, 'It is just as I feared!	A	
Two Owls and a Hen,	B	
Four Larks and a Wren,	B	
Have all built their nests in my beard!	A	

Edward Lear (1812–1888)

It is clear that already because of the form of traditional poetry, numbers and structures are allotted a role that contributes to the emotional tone of the poem: narrative in the ballad, lamenting in the elegy, ironic and cheerful in the limerick, etc.

2.4.2 Linguistic theory and style analysis

In a study comparing an unknown or extinct language with a language that is still living or at least known, there is always an intention to start from certain correspondences: which letters of the alphabet appear the most, which concepts are frequent in the relevant context, which names of kings or other rulers date from the time concerned? It is always the case that advanced numerical analysis and a technique of logical association have to be called upon in order to achieve any initial success. These are then extrapolated over more extensive passages, so as to lead to a full transcription of the text. The history of the unveiling of Egyptian hieroglyphics by Jean François Champollion and Thomas Young, as well as Babylonian cuneiform by Henry C. Rawlinson and George Grotefend, offer striking examples.

Knowledge of the old Egyptian hieratic script (used by the priests) and the demotic script (used by the people) as well as hieroglyphics got lost over the course of time. It was a brilliant idea of Champollion's to search for references in texts written in the still known Coptic language, which was the final stage of Egyptian that adopted a Greek-based alphabet. After discovering the cartouches of Ptolemy and Cleopatra on the 'Rosetta Stone', some characters could be linked to the corresponding phonetic sounds. Based on this, the name 'Ramses' was also later unveiled, and finally through these keys the whole Egyptian alphabet was revealed. Rawlingson followed an analogous method by distinguishing three categories in the cuneiform, in one of which he saw similarities to an old Persian language.

Jean François Champollion
23 December 1790 – 4 March 1832

The Rosetta Stone

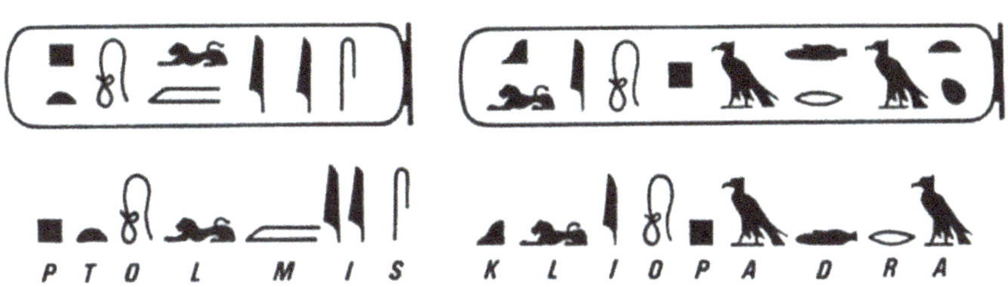

Cartouches of Ptolemy and Cleopatra

Noam Chomsky

Besides words, a language also consists of a grammar that determines the rules for the correct manipulation of the words in acceptable sentences. Because it is impossible to learn all potential sentences of a language by experience and save them in the memory, grammatical rules must be set up so that, starting from certain patterns, one can deal with the language in a creative way and so produce new sentences that are comprehensible to any person using the same language. Especially transformational generative grammar had this dynamic use of language as its objective. In his *Syntactic Structures* of 1957, Noam Chomsky showed new ways of how to perform this, using a very mathematical approach with substitutions, transformations, algorithms, formulas, tree diagrams and combinatorics. With these the three main problems of selection, arrangement and formation are solved and the necessary models established. The approach is strongly deductive and scientific in the sense that the suggested models serve as working hypotheses that can be tested. This theoretical approach to linguistics transcends the particularity of a given language (surface structure) and seeks the intrinsic universality of all languages (deep structure), thus making the genesis of language, quoting Chomsky, a *competence*.

Each author has his own typical style that is connected to his personality, education and aesthetic standards. Quoting George-Louis Buffon: "Le style, c'est l'homme même" (the style is the man himself). How can one recognize somebody by their writings? Experience and familiarity are certainly factors that contribute to the feeling of 'déjà lu', but are probably too vague to be determining factors. An explicit analysis of the specific features would lead to better results. What features are appropriate? First of all, writers are classified according to well-determined historical trends, like romanticism, realism, magic-realism, symbolism, modernism, avant-gardism, surrealism, existentialism, etc. The main representatives of these trends have themselves formulated the substantial ingredients of their movement in contempory literary magazines, which can be easily checked in essayistic literary chronicles. In the second place, the preference for certain themes can be examined: social engagement, family chronicles, autobiographic experiences, adventure, fiction, etc. There are few authors that can not talk about their own field of interest, which, in spite of themselves always reappears in the fruits of their pen. In the third place comes linguistic usage: short sentences or longer ones, containing a string of subordinate clauses one inside the other; direct or indirect speech; first or third person account; descriptive or dialogic; few adjectives or abundant attributes; vocabulary preferences with recurring fetish words or a tendency to exhaustive excess. This last tendency goes so far in doctoral theses that computers are called in to produce numerical results from which some conclusions can then be drawn. Finally, there are the global aspects of ethics and aesthetics: optimist or pessimist, charming or shocking; moralizing or impartial; plain or embellished; etc. All of these criteria mainly determine qualitative categories that, due to their taxonomic structure, result in a subtle matrix by means of which an author can often be placed in a cell where he or she is the sole occupant. In this sense, style analysis also respects general mathematical methods of classification and arrangement.

2.4.3 Dramaturgy and Catastrophe theory

René Thom

The course of the dramatic development of a play or a novel can have certain characteristics of the same kind as the dynamic phenomena of various scientific disciplines, such as a stock exchange crash (economics), a heart attack or brain hemorrhage (biology), a landslide (geology), a change of aggregate state (chemistry) or the collapse of a bridge (physics). They all concern a process of apparently continuous change from one stable state to another, but by crossing a critical border somewhere, the whole system is suddenly in a totally different condition. Contrary to discontinuity it concerns an unexpectedly strong, but nevertheless continuous acceleration in the development. These phenomena are described and classified in a mathematical theory called *catastrophe theory*, developed in the 1960's by the French mathematician René Thom. Depending on the number of parameters in the dynamic process, all possible catastrophes in the three-dimensional world can be arranged in seven classes (one for each day of the week). According to E.C. Zeeman, the so-called cusp model type which has two parameters, and was already used, for instance, for explaining the functioning of the heart and the brains even before the introduction of chaos theory, would also be applicable to the dramatic development of certain literary novels. In numerous lectures he reported on examples and analyses in this context.

In order to make clear what a catastrophe means in the context of this theory, we can take a look at a much simplified physical model that Zeeman thought of to illustrate it. It consists of a disc that rotates smoothly around an eccentric pivot *A* so that a rubber band, which is attached to point *B* on the disc and to a fixed point *C* away from the disc, will gradually stretch during this rotation. When *B* is pulled, the rubber band will at first just slow down the rotation of the disc, but will keep it in a stable position.

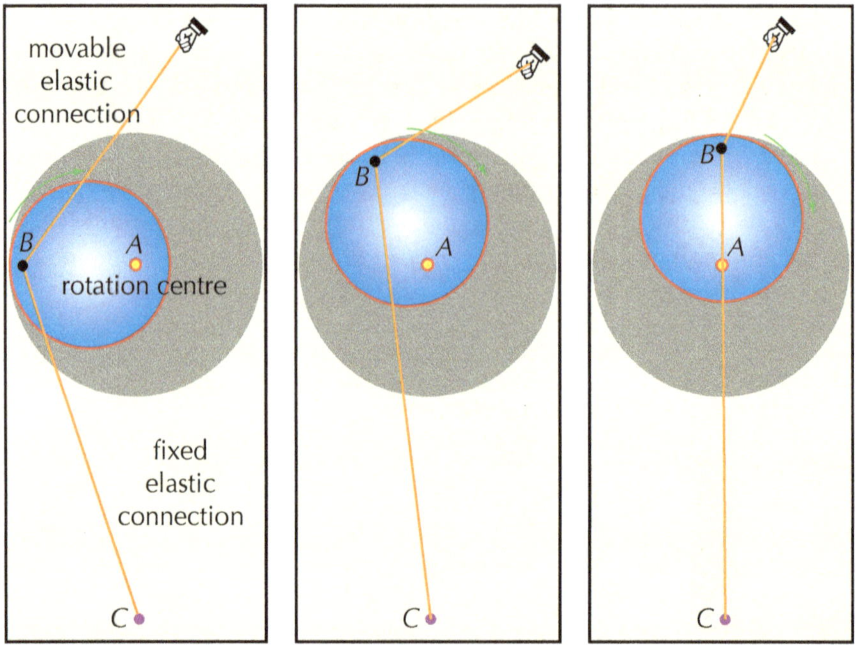

But at the very moment the disc comes into the position where the points *B*, *A* and *C* line up, and the tension in the rubber band reaches its maximum, the situation becomes unstable. The disc suddenly accelerates, caused by the tension of the rubber band, so that point *B* abruptly moves to another stable position lower down, but at a far faster speed than it moved before reaching the critical point.

Besides its typical equation in two variables, the cusp model that Zeeman used for the description of the dramatic development that occurs in several literary works, also has a geometric representation that looks like a folded sheet, as shown in the figure below.

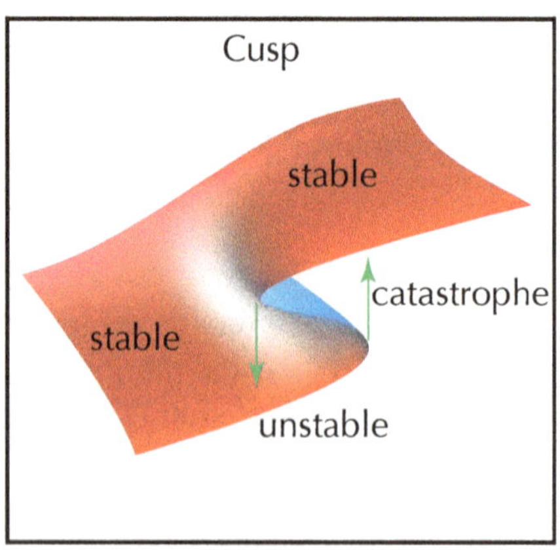

The red area indicates the stable zone for a point moving on the surface, the blue area is the unstable zone. The edge between both zones determines the critical transition from the one stable state to the other. The fold indicates that there an unusual acceleration of the movement will take place.

In a drama, such as the one of Oedipus (Sophocles), Othello (Shakespeare) or Madame Bovary (Flaubert), one can point out two simple parameters that determine the condition in which a person is situated, namely a parameter that indicates the changes in the external circumstances and a parameter that indicates the internal tensions. From the fluctuations of these two the whole drama evolves from an initially apparently stable state to a critical bifurcation point, after which everybody involved in the plot (system) is caught up in an acceleration of their own inescapable fate. This development is indeed legion in classical dramas and tragic novels.

A descriptive and explanatory model can indeed help to understand a story, but as a post hoc activity it is not relevant for the process itself. Only predictive power can give real scientific substance to such a model. For this shortcoming, catastrophe theory still meets with a lot of criticism and scepticism. It would indeed be useful if we could predict critical moments from certain symptoms in stock exchange fluctuations, but in order to do this, the equations and geometric models of catastrophe theory do not seem to offer sufficient clarity, at least not so far. When reading a story, this predictive power would more likely be unpleasant. So, in literature a post hoc model is quite satisfactory.

Summary

Also in literature, both in its form as well as in its content, numerical and structural patterns borrowed from the mathematical arsenal can be pointed out. The fact that these evidently do not make up the essence of what the author wants to express, does not mean that knowledge of these aspects would not enhance the pleasure of reading.

On the other hand, pure linguistics can to a great extent benefit from mathematical methods, just as is the case in other domains of science. Although no living language is logically constructed, nevertheless it is in permanent evolution, partially because of its users but also due to periodical interventions of linguists. So, there would be no objection to gradually giving language a more transparent logical structure that would simply make it more convenient for its users.

Conclusion

In this chapter we have given examples showing that even nature itself works by means of disciplined forces taking shape in mathematical patterns. In the phenomena of both the greatest and the smallest possible, models appear that can be described successfully by means of mathematics. Even closer to us, shells, flowers and butterflies show off their shapes and patterns with a clear mathematical signature. These scenes inspire the artist to use proportions and structures springing from the same source. The muses of the visual arts, as well as those of music and literature, love to potter around in the suggestive gardens of number and form. This can't have gone unnoticed by the attentive art lover.

> *"It is true that a mathematician who is not also something of a poet will never be a perfect mathematician."*
>
> **Karl Weierstrass**

Statue of James Joyce (1882–1941), Irish author who wrote the famous *Ulysses*, without a doubt the most remarkable book of the twentieth century. In an unequalled way an ironic parallel is drawn in this book between Ulysses' wanderings, returning home to his faithful Penelope after the Trojan War, and Leopold Bloom's adventurous walk through Dublin on June 16th, 1904, with no choice but to return to his yes-saying Molly. Apart from the tight structure of the book, full of references to the great works of world literature and allusions to Irish national history, there is the creation of an expressive and elaborate language, full of deeper meanings, which without expert guidance is not easily accessible to everybody. As Joyce himself declared to Eugene Jolas: "I can do everything with language". But also in his approach to language and word manipulations there is a system that can be revealed by mathematical analyses.

Epilogue

Although limited both in number and ability, it is through our senses that we get a grip on the world. They give us a whole range of kaleidoscopic impressions of reality. It is our mind that uses the dimensions of time and space to create order from this multiplicity of impressions. From this self-created order we put together the indispensable maps for navigating through our day-to-day worries. Just like our senses, our mind is a tool that has been finely honed by an immeasurably long evolutionary process. We still have no clear idea of what precisely happens in those billions of neural connections in the brain and neural system. For now we can only identify the activities of our mind through the manifestation of its language and logic. Both aspects culminate in the abstractions of mathematics. This equipment provided by nature is not just coincidental, for no single species has been as successful in surviving as mankind has. The fact that this mind and this world do not exist in a dualistic way, but make up part of one and the same history, should give us confidence that what we can find in the one must also be present in the other. So are we providing evidence of consistent thinking if we keep on being astonished at how understanding and success occur in all domains of science where mathematical methods have been introduced? The intentions of this book are precisely to provide examples of this obviousness, admittedly with some enthusiasm. Then how can it still be that so many people who are convinced that all children should exercise their senses, train their body and try to learn languages, unquestioningly accept that building a critical, logical mind should not necessarily be part of each pupil's standard curriculum? Why should this pure romantic antagonism between emotion and reason, between art and science, continue to be perpetuated? It would be a nice reward if this book could contribute to changing this attitude and mentality.

R. Verhulst, *Under the Spell of Mathematics*, https://doi.org/10.1007/978-3-031-63871-8

Reference list of illustrations

R. Verhulst, *Under the Spell of Mathematics*, https://doi.org/10.1007/978-3-031-63871-8

p. 113 Roulette: https://www.google.be/search?as_st=y&tbm=isch&hl=nl&as_q=roulette&as_epq=&as_oq=&as_eq=&cr=&as_sitesearch=&safe=images&tbs=sur:fc#imgrc=15lR9w8nilYAzM:
p. 116 crowd: Live, Les Mathématiques, 1963, p.130
p. 122 avalanche:
https://en.wikipedia.org/wiki/Avalanche#/media/File:Avalanche_on_Everest.JPG
p. 126 Königsberg 1: https://commons.wikimedia.org/wiki/File:Euler_rid6exp.png
Königsberg 2: https://commons.wikimedia.org/wiki/File:7_bridges.svg
p. 141 credit cards: https://commons.wikimedia.org/wiki/File:Credit-cards.jpg
p. 143 bar code: https://www.hetboekenschap.nl/product/barcode/
p. 144 Marsexplorer: https://en.wikipedia.org/wiki/Mars_Exploration_Rover#/media/File:NASA_Mars_Rover.jpg
p. 144 CD: https://commons.wikimedia.org/wiki/File:Cd-r-650mb-basf_hg.jpg
p. 147 Hamming medal: https://ethw.org/IEEE_Richard_W._Hamming_Medal
p. 148 Enigma: https://commons.wikimedia.org/wiki/File:Enigma_(crittografia)_-_Museo_scienza_e_tecnologia_Milano.jpg
p. 150 Julia fractal: https://commons.wikimedia.org/w/index.php?curid=31888310, By Binette228 - Own work, CC BY-SA 3.0,
p. 150 Mandelbrot fractal: https://commons.wikimedia.org/wiki/File:Demm_2000_Mandelbrot_set.jpg
p. 158 Julia set: https://pixabay.com/nl/mandelbrotverzameling-fractal-979322/
p. 158 https://en.wikipedia.org/wiki/Mandelbrot_set#/media/File:Mandel_zoom_14_satellite_julia_island.jpg
p. 159 Mandelbrot set: https://commons.wikimedia.org/wiki/File:Mandel_zoom_00_mandelbrot_set.jpg
p. 159 Mandelbrot set 2: https://pixabay.com/nl/fractal-mandelbrotverzameling-74055/
p. 159 Mandelbrot set 3: https://pixabay.com/nl/fractal-mandelbrot-julia-veel-2993372/
p. 159 Mandelbrot set 4: https://commons.wikimedia.org/wiki/File:Mandel_zoom_04_seehorse_tail.jpg
p. 159 Mandelbrot set 5: https://commons.wikimedia.org/wiki/File:Mandel_zoom_09_satellite_head_and_shoulder.jpg
p. 159 Mandelbrot set 6: https://commons.wikimedia.org/wiki/File:Mandel_zoom_06_double_hook.jpg
p. 162 Lorentz attractor: https://en.wikipedia.org/wiki/File:Lorenz_system_r28_s10_b2-6666.png
p. 163 Henri Poincaré: https://commons.wikimedia.org/wiki/File:PSM_V82_D416_Henri_Poincare.png
Omar Al-Khayyam: https://ro.m.wikipedia.org/wiki/Fişier:Omar_Khayyam2.JPG
p. 184 Oriental research centre: Live, Les Mathématiques, 1963, p.71
p. 187 cartoons computerparadise: https://gkvpn.nl/coppermine/displayimage.php?album=2&pid=96,
p. 187 http://www.pgm-stadskanaal.nl and Rik Verhulst 2008; with friendly permission of © Garant 2018. All Rights Reserved

dynamosquito/2431828694
p. 234 Copernicus: https://commons.wikimedia.org/wiki/File:Copernicus.jpg
p. 236 Simon Stevin: https://vls.wikipedia.org/wiki/Ofbeeldienge:Stevin.jpg
p. 241 Kepler: https://nl.m.wikipedia.org/wiki/Bestand:Johannes_Kepler_1610.jpg
p. 242 Galileo Galilei: https://commons.wikimedia.org/wiki/File:Justus_Sustermans_-_Portrait_of_Galileo_Galilei,_1636.jpg
p. 243 Tartaglia: https://it.m.wikipedia.org/wiki/File:Tartaglia-Opere-portrait.jpg
p. 245 Descartes : https://commons.wikimedia.org/wiki/File:PSM_V37_D740_Rene_Descartes.jpg
p. 246 Kristina of Sweden: https://nl.m.wikipedia.org/wiki/Bestand:Ebba_Sparre.jpg
p. 246 Blaise Pascal: https://nl.m.wikipedia.org/wiki/Bestand:Blaise_Pascal_Versailles.JPG
p. 248 Pierre de Fermat: https://fr.m.wikipedia.org/wiki/Fichier:Pierre_de_Fermat2.png
p. 249 Desargues: https://commons.wikimedia.org/wiki/File:G%C3%A9rard_Desargues.jpeg
p. 251 Cavalieri: https://nl.wikipedia.org/wiki/Bestand:Bonaventura_Cavalieri.jpg
p. 254 Newton: https://ha.wikipedia.org/wiki/File:SS-newton.jpg
p. 255 Leibniz: https://nl.wikipedia.org/wiki/Bestand:Gottfried_Wilhelm_Leibniz,_Bernhard_Christoph_Francke.jpg
p. 256 Leibniz' calculator: http://www.staff.science.uu.nl/~doorm101/rekenmachine/htm/deel3.htm
p. 259 Saccheri: https://commons.wikimedia.org/wiki/File:Saccheri_1733_-_Euclide_Ab_Omni_Naevo_Vindicatus.gif
p. 260 Nikolai Lobatschewski: http://learn-math.info/dutch/historyDetail.htm?id=Lobachevsky
p. 260 Janos Bolyai: https://commons.wikimedia.org/wiki/File:JanosBolyai.jpg
p. 260 Carl Friedrich Gauss: https://commons.wikimedia.org/wiki/File:Carl_Friedrich_Gauss.jpg
p. 260 Bernhard Riemann: https://www.flickr.com/photos/smithsonian/2551069295
p. 262 Rowan Hamilton: https://commons.wikimedia.org/wiki/File:William_Rowan_Hamilton_painting.jpg
p. 262 Rowan-Hamilton-Gedenktafel: https://en.wikiquote.org/wiki/William_Rowan_Hamilton
p. 263 Arthur Cayley: https://commons.wikimedia.org/wiki/File:Arthur_Cayley_Engraving.jpg
p. 265 Augustin Louis Cauchy: https://commons.wikimedia.org/wiki/File:Cauchy_Augustin_Louis_dibner_coll_SIL14-C2-03a.jpg
p. 265 Karl Weierstrass: https://www.flickr.com/photos/smithsonian/2552869153
p. 265 Julius Wilhelm Richard Dedekind: https://nl.m.wikipedia.org/wiki/Bestand:Richard_Dedekind_1900s.jpg
p. 268 Bertrand Russell: https://commons.wikimedia.org/wiki/File:Bertrand_Russell,_Bestanddeelnr_909-1508.jpg
p. 271 Eniac - 50 : https://maths.ucd.ie/~plynch/eniac/
p. 274 The violin maker of Venice, Altamira, letzte Seite
p. 276 eclipse: https://en.wikipedia.org/wiki/File:Solar_eclipse_1999_4_NR.jpg

p. 276 shells : https://pxhere.com/nl/photo/819242
https://pixabay.com/nl/zee-shell-clam-oceaan-zee-schelpen-1162757/
https://nl.m.wikipedia.org/wiki/Bestand:Nautilus_species_shells.png
https://www.flickr.com/photos/44603071@N00/3041743367
https://pxhere.com/en/photo/573468
https://www.maxpixel.net/Shell-Creature-Spiral-331922
p. 277 Pantheon : https://vls.wikipedia.org/wiki/Ofbeeldienge:Pantheon_Rome_(1).jpg
p. 277 cross section: https://www.pinterest.com/pin/910572236864546 22/
p. 278 horse head: https://en.wikipedia.org/wiki/File:Barnard_33.jpg
p. 279 Hubble telescope: https://www.af.mil/News/Article-Display/Article/120334/46th-test-wing-integral-part-of-hubble-repair/
p. 281 Georges Lemaître: https://www.flickr.com/photos/tonynetone/14578237295
p. 282 trajectories of projectiles: Les Mathématiques Live, p. 109
p. 283 F.W. Bessel: https://de.wikipedia.org/wiki/Friedrich_Wilhelm_Bessel#/media/File:Friedrich_Wilhelm_Bessel.jpeg
p. 284 Einstein: https://en.wikipedia.org/wiki/File:Einstein_1921_by_F_Schmutzer_-_restoration.jpg
p. 284 deflection of light 29 may 1919: Le livre de la nature. Dr. F. Kahn, p. 141
p. 285 Paul A. M. Dirac: https://commons.wikimedia.org/wiki/File:Diracb.jpg
p. 286 particle accelerator, CERN: https://he.wikipedia.org/wiki/%D7%A7%D7%95%D7%91%D7%A5:CERN_LHC_Tunnel1.jpg
p. 286 bubble bath: https://thenewstack.io/discovering-higgs-boson-cern-integrates-openstack-kubernetes/
p. 287 L. Fibonacci: https://commons.wikimedia.org/wiki/File:Fibonacci.jpg
p. 287 pine cone: https://pixabay.com/nl/pijnappel-dennenappels-ruwe-2818203/
p. 287 pineapple: https://pxhere.com/nl/photo/599758
p. 287 sunflower: https://pixabay.com/nl/zonnebloem-bloem-1537724/
p. 287 Nautilus: https://en.wikipedia.org/wiki/File:NautilusCutawayLogarithmicSpiral.jpg
p. 290 Phyllotaxis: https://www.flickr.com/photos/arenamontanus/3540498156
p. 291 Venus of Milo: https://commons.wikimedia.org/wiki/File:The_Louvre,_the_Venus_de_Milo,_Paris,_France-LCCN2001698517.jpg
p. 291 da Vinci, head : Les Mathématiques Live pag. 94
p. 293 embryo: https://commons.wikimedia.org/wiki/File:Human_Embryo.JPG
p. 294 seastar: https://commons.wikimedia.org/wiki/File:Sea_star_bg_01.jpg
p. 296 symmetry: tiger: https://www.pexels.com/photo/close-up-photography-of-tiger-792381/
p. 296 symmetry: butterfly : https://pixabay.com/nl/vlinder-png-3192713/
p. 296 symmetry: human body: https://www.pexels.com/de-de/foto/der-akt-der-madchen-portrat-schwester-207890/
p. 296 symmetry: flower: https://commons.wikimedia.org/wiki/File:Iris_versicolor_3.jpg
p. 297 Inventing the Circle, Johan Gielis, Geniaal
p. 297 honeycomb: https://pxhere.com/nl/photo/1183648
p. 297 cristal : https://pixabay.com/nl/blauw-kleur-kristallen-kleurrijke-881074/

p. 298 clouds: https://pixabay.com/nl/wolk-blauw-wolken-white-zomer-2354837/
p. 301 Romanesco: https://commons.wikimedia.org/wiki/File:Chou_Romanesco_au_marché.jpg
p. 302 clay house: https://inventaris.onroerenderfgoed.be/erfgoedobjecten/2538
p. 303 Cheops: eigenes Foto, Annemie Segers
p. 303 Babel https://nl.wikipedia.org/wiki/Bestand:WLANL_-_Quistnix!_-_Museum_Boijmans_van_Beuningen_-_Toren_van_Babel,_Bruegel.jpg
p. 303 Zikkurat: https://en.wikipedia.org/wiki/File:SumerianZiggurat.jpg
p. 304 Parthenon: https://www.flickr.com/photos/fruey/1372547275
p. 304 Epidauros https://commons.wikimedia.org/wiki/File:The_great_theater_of_Epidaurus,_designed_by_Polykleitos_the_Younger_in_the_4th_century_BC,_Sanctuary_of_Asklepeios_at_Epidaurus,_Greece_(14058124143).jpg
p. 304 Epidauros, cross cut: https://www.flickr.com/photos/126377022@N07/14741564716
p. 305 Titus arch: https://www.flickr.com/photos/rogersg/11796861345
p. 305 Hagia Sophia: https://en.wikipedia.org/wiki/File:Hagia_Sophia_Mars_2013.jpg
p. 305 Hagia Sophia, Querschnitt: https://tr.wikipedia.org/wiki/Dosya:Hagia-Sophia-Laengsschnitt.jpg
p. 306 Triforium, Kathedrale von Rochester: Ornamenten, Alexander Spelt, Gaade, Amerongen, p. 193.
p. 306 Patrixbourne Church: https://commons.wikimedia.org/wiki/File:St_Mary,_Patrixbourne,_Kent_-_Window_-_geograph.org.uk_-_328860.jpg
p. 307 Notre-Dame, Paris: https://commons.wikimedia.org/wiki/File:Notre_Dame_de_Paris_DSC_0846w.jpg
p. 307 pillar: Dirk Van Hemeldonck, own photograph
p. 307 Chambord: https://fr.wikipedia.org/wiki/Fichier:Escalier_Chambord.jpg
p. 308 Eiffel tower: https://www.pexels.com/photo/worms-view-of-eiffel-tower-during-daytime-149419/
p. 308 Hotel Tassel: https://commons.wikimedia.org/wiki/File:Tassel_House_stairway.JPG
p. 308 opera house in Sidney: https://commons.wikimedia.org/wiki/File:Sydney_opera_house_side_view.jpg
p. 309 Guggenheim Museum New York: https://pixabay.com/nl/guggenheimmuseum-new-york-2707258/
p. 309 house of Le Corbusier: http://www.phys.tue.nl/TULO/guldensnede/architectuur.html
p. 309 Seagram Building New York: https://nl.wikipedia.org/wiki/Bestand:Seagrambuilding.JPG
p. 309 Seagram Building New York: https://www.flickr.com/photos/43391993@N03/4114889237
p. 310 modulor of Le Corbusier http://www.neermanfernand.com/corbu.html
p. 311 Taj Mahal: https://en.wikipedia.org/wiki/File:Taj_Mahal_(Edited).jpeg
p. 312 Cordoba Mesquita pillars: https://pixabay.com/nl/mezquita-kathedraal-van-c%C3%B3rdoba-2842852/

p. 312 frieze; Symmetry, Lauwerier, Aramith
p. 313 cathedral in Chartres, northern rosetter: https://www.flickr.com/photos/eusebius/3318918018
p. 313 castle of Chenonceau: https://fr.wikipedia.org/wiki/Fichier:Chateau_de_Chenonceau.JPG
p. 314 interior of the Pantheon: https://nl.m.wikipedia.org/wiki/Bestand:Giovanni_Paolo_Panini_-_Interior_of_the_Pantheon,_Rome_-_Google_Art_Project.jpg
p. 316 Lyon station Satolas: https://www.flickr.com/photos/2613-say-yeah/3941996959/
p. 316 Cristo Obrero church in Atlántida Uruguay: https://commons.wikimedia.org/wiki/File:Iglesia_Atlántida_Dieste_5.jpg
p. 317 Sagrada Familia, Barcelona: https://commons.wikimedia.org/wiki/File:Sagrada_Familia_nave_roof_detail.jpg
p. 318 Olympia stadion Munich: https://pixabay.com/nl/olympisch-stadion-stadion-2612223/
p. 318 experiment railway station Stuttgart: https://www.researchgate.net/figure/Experiment-by-Frei-Otto-and-team-for-the-new-high-speed-Stuttgart-train-station-Image_fig20_314116498
p. 319 Boticelli, the birth of Venus: https://nl.wikipedia.org/wiki/Bestand:Botticelli_geboorte_venus.jpg
p. 320 La primavera, Boticelli: https://pt.wikipedia.org/wiki/Ficheiro:Primavera_(Botticelli).jpg
p. 320 the man of Vitruvius, da Vinci: https://commons.wikimedia.org/wiki/File:Leonardo_da_Vinci-_Vitruvian_Man.JPG
p. 320 Mona Lisa of da Vinci: https://upload.wikimedia.org/wikipedia/commons/6/6a/Mona_Lisa.jpg
p. 321 the last supper, La Ultima Cena, Da Vinci: https://pixabay.com/nl/het-laatste-avondmaal-1921290/
p. 321 Luca Pacioli: https://commons.wikimedia.org/wiki/File:Jacopo_de%27_Barbari_-_Portrait_of_Fra_Luca_Pacioli_and_an_Unknown_Young_Man_-_WGA1269.jpg
p. 322 railway: https://www.pinterest.com/pin/485262928571768531/
p. 323 Albrecht Dürer, the artist and the femal model: https://www.rijksmuseum.nl/nl/collectie/RP-P-OB-1492
p. 324 Albrecht Dürer, the artist and the lute: https://commons.wikimedia.org/wiki/File:Duerer_Underweysung_der_Messung_fig_001_page_181.jpg
p. 324 Canale Grande e Santa Maria della Salute, Canaletto: https://nl.m.wikipedia.org/wiki/Bestand:Canaletto_Entrance_to_the_Grand_Canal_Venice.jpg
p. 325 spozal, Raffael: https://www.artsalonholland.nl/renaissance/rafael-de-verloving-van-maria-en-jozef
p. 325 secretary in the Vatican: *Standaard Encyclopedie*, Band 3, p.452
p. 326 Klimmen en dalen, Escher: https://www.alletop10lijstjes.nl/top-10-beroemde-werken-m-c-escher/
p. 327 print by William Hogarth: https://commons.wikimedia.org/wiki/File:The_

importance_of_knowing_perspective_-_Satire_on_False_Perspective,_by_William_Hogarth_(1753).jpg
p. 331 imitation of the wall decoration in the Mexuar hall, Alhambra: inside cover of *Islam kunst en architectuur*, Könemann
p. 332 circle limit, Escher: https://static.kunstelo.nl/ckv2/bevo/vorm/rotatie1.htm
p. 332 computerl andscape: http://www.skytopia.com/gallery/mountains/big/mountmaroon_mouldy.jpg
p. 332 fractal: https://pxhere.com/nl/photo/400565
p. 333 Rodtschenko, three dimensional construction: https://www.vice.com/en_au/article/mg9zd4/original-creators-constructivist-aleksandr-rodchenko
p. 333 Mondriaan: https://nl.m.wikipedia.org/wiki/Bestand:Piet_Mondriaan,_1921_-_Composition_en_rouge,_jaune,_bleu_et_noir.jpg
p. 334 Pythagoras at the strings: https://commons.wikimedia.org/wiki/File:Gaffurio_Pythagoras.png
p. 334 Pythagoras: http://www.travelingtemplar.com/2018/03/the-great-pythagoras.html
p. 335 graphic representation of a sound wave: https://pixabay.com/nl/geluid-golf-stem-luisteren-856770/
p. 336 graphic representation of a clarinet tone: http://www.nieuwarchief.nl/serie5/pdf/naw5-2001-02-2-136.pdf
p. 354 Dante Aleghieri: https://commons.wikimedia.org/wiki/File:Dante_Alighieri%27s_portrait_by_Sandro_Botticelli.jpg
p. 354 La Divina Commedia, Dante: https://nl.m.wikipedia.org/wiki/Bestand:Firenze,_divina_commedia,_xiv_sec.,_cod._tempi_1,_01.JPG
p. 355 Marcel Proust: https://nl.wikipedia.org/wiki/Bestand:Jacques-Emile_Blanche_Portrait_de_Marcel_Proust_1892.jpg
p. 359 Jean-François Champollion: https://en.wikipedia.org/wiki/File:Leon_Cogniet_-_Jean-Francois_Champollion.jpg
p. 359 the Rosetta stone: https://nl.wikipedia.org/wiki/Bestand:Rosetta_Stone_BW.Jpeg
p. 360 Noam Chomsky: https://commons.wikimedia.org/wiki/File:Noam_Chomsky,_2004.jpg
p. 361 René Thom: https://www.ihes.fr/en/oeuvres-mathematiques-volume-1-by-rene-thom-just-published-by-the-french-mathematical-society/
p. 364 statue of James Joyce: https://www.flickr.com/photos/rosenkranz/294078948

Bibliography

This list is intended for the reader who wants to gain more in-depth knowledge of certain topics of this book. Because the works mentioned also include a bibliography, the list is intentionally kept short.

Chapter 1

BIRKHOFF G. Lattice Theory. American Mathematical Society, 1948
CALVIN W. The Cerebral Symphony. Bantam Books, 1987
CHANG C.C., KEISLER H.J. Model Theory. North-Holland, 1973
COHN P.M. Algebra Volume 1. John Wiley & Sons, 1974
COMBÈS M. Fondements des Mathématiques. Presses Universitaires de France, 1971
DENNETT D. Consciousness Explained. Little, Brown and Co. 1991
GOODSTEIN R.L. Boolean Algebra. Pergamom Press, 1963
GOODSTEIN R.L. Development of Mathematical Logic. Springer Verlag, 1971
HALMOS P.R. Lectures on Boolean Algebras. Van Nostrand.
HILTON P., PEDERSEN J. Fear No More. Addison Wesley, 1982
HILTON P., HOLTON D., PEDERSEN J. Mathematical Reflections. Springer, 1998
HUME D. An Enquiry Concerning Human Understanding. Collier and Son, 1910
HUNTER G. Metalogic. University of California Press, 1973
KLEEN S.C. Introduction to Mathematical Logic. John Wiley, 1967
LAKATOS I. Proofs and Refutations. Cambridge University Press, 1976
LANG S. Algebra. Addison-Wesley, 1980
LOCKE J. An Essay Concerning Human Understanding. Mc Clintock's Essays, 1995
NAGEL E., NEWMAN J.R. Gödel's Proof. University Press, New York, 1958
PAPY G. Groups. MacMillan, London, 1964
PINKER S. How the Mind Works. Penguin, 1998
SEARLE J. Minds, Brains and Science. Harvard University Press, 1984
TARSKI A. Introduction à la logique. Gauthier-Villars en Nauwelaerts, 1960
VAN DALEN D. Logic and Structure. Springer Verlag, 1997
VAN DALEN D. Mystic, Geometer and Intuitionist. The life of L.E.J. Brouwer. Oxford University Press, 1999
VERHULST R. The birth of bidecimal numbers. VWNL Cahier K.U.Leuven, 1982
WILDER R. An introduction to the Foundations of Mathematics. John Wiley, 1952
WITTGENSTEIN L. Tractatus logico-philosophicus. Atheneum-Polak en Van Gennep

Chapter 2

ADLER I. Probability and Statistics for Everyman. The John Day Company, New York, 1963
BERGÉ C. Théorie des graphes et ses applications. Dunod, 1963
CHOQUET G. L'enseignement de la géometrie. Hermann, 1964
COHEN P. Set Theory and the Continuum Hypothesis. W.A. Benjamin, 1966
DIEUDONNÉ J. Algèbre lineaire et géometrie élémentaire. Herman1968
DIEUDONNÉ J. Fondements de l'Analyse Moderne. Gauthier-Villars, 1965

R. Verhulst, *Under the Spell of Mathematics*, https://doi.org/10.1007/978-3-031-63871-8

DIEUDONNÉ J. Pour l'honneur de l'esprit humain. Hachette, 1987
FLEGG G. Numbers. Their history and meaning. Barnes & Noble Books, 1993
HALMOS P.R. Naive Set Theory. D. Van Nostrand Company, Princeton, 1964
HARDY G.H. An introduction to the Theory of Numbers. Oxford University Press,1960
HILTON P., HOLTON D., PEDERSEN J. Mathematical Vistas. Springer, 2002
LAUWERIER H. Fractals Endlessly Repeated. Princeton Science Library, 1992
MANDELBROTT B.B. Fractals: Form, Chance and Dimension. W.H. Freedman, 1977
MANDELBROTT B.B. The fractal geometry of nature. Freedman, 1983
MARTIN G.E. The Foundations of Geometry and the Non-Euclidean Plane. Springer-Verlag, 1975
McCORD J.R., MORONEY R.M. Introduction to Probability Theory. The MacMillan Company, New York, 1964
PAPY G. Nombres Réels et Vectoriel Plan. Didier, Bruxelles, 1968
PAPY G. Mathématique Moderne. Marcel Didier, 1966
PEITGEN H.O., RICHTER P.H. The Beauty of Fractals. Springer, 1986
PRIGOGINE I., STENGERS I. Order out of Chaos. Mountain Man Graphics, Australia, 1984
SCHUSTER H.G. Deterministic Chaos, an introduction. Physik-Verlag, 1984
SINGH S. The Code Book. Simon's Shop, 2000
STEWART I. From here tot Infinity. Oxford University Press, 1996
STEWART I. Does God play dice?
VAN DALEN D., DOETS H.C. and DE SWARTH.C.M. Sets, Naive, Axiomatic and Applied. Pergamom Press, Oxford, 1978
VERHULST R. Nomograms for the calculation of Roots. Belgische Sub-Commissie ICMI, 1982

Chapter 3

AMIR D. ACZEL Fermat's Last Theorem. Four Walls Eight Windows, New York, 1996
BELL E.T. Men of Mathematics. Simon and Schuster, 1965
Bibliothèque pour la Science. Les Mathématiciens. Pour la Science Diffusion Belin, 1996
BOCHNER S. The Role of Mathematics in the Rise of Science. Princeton University Press, 1965
CAJORI F. A History of Mathematics. Chelsea Reprint New York, 1980
DAVIS P.J., HERSH R. Descartes Dream. Viking Press, 1988
DEVREESE J.T., VANDEN BERGHE G. Magic is no Magic, The Wonderful World of Simon Stevin.
EVES H. An Introduction to the History of Mathematics. Holt, Rinehart and Winston, 1964
HILTON P., PEDERSEN J. Build Your Own Polyhedra Addison Wesley, 1988
IFRAH G. From One to Zero. Penguin Group
KLINE M. Mathematical Thought from Ancient to Modern Times. Oxford University Press, 1972
KUHN T. The structure of Scientific Revolutions. University of Chicago Press, 1962
MANKIEWICZ R. The Story of Mathematics.
MASINI G. Il Romanzo dei Numeri. Firenze, 1973
MORRISON P., E. Charles Babbage and his Calculating Engines. Dover Publications, 1961

ODIFREDDI P. The mathematical Century: the 30 Greatest Problems of the last 100 years.
POPPER K. Conjectures and Refutations: The Growth of Scientific Knowledge. Harper Torch Books, 1963
REGIS E. Who Got Einstein's Office? Addison Wesley, 1987
SINGER C. A Short History of Scientific Ideas to 1900. Oxford University Press, 1962
STRAY G. The Mayan and Other Ancient Calendars. Wooden Books, 2007
STRUIK D.J. A Concise History of Mathematics. Dover Publications, New York, 1987
VAN DALEN D. Brouwer's Cambridge Lectures on Intuitionism. Cambridge University Press, 1981
VERHULST R. The Universe of the Coloured Arrows: Papygrams. Bulletin of the Belgian Mathematical Association, 1991

Chapter 4

ASHTON A. Harmonograph: A Visual Guide to the Mathematics of Music. Wooden Books, 2005
BENSON B.J. Music, A Mathematical Offering. Cambridge University Press, 2006
CHADWICK J. The Decipherment of Linear B. Aula Het Sectrum, 1962
COOKE D. The Language of Music. Oxford University Press, 1959
DIEUDONNÉ J. Mathematics-The music of Reason. Springer Verlag, 1992
DAWKINS R. The Selfish Gene. Oxford University Press. 1976
DYSON F. Imagined Worlds. Harvard University Press, 1997
ERNST B. The Magic Mirror of M.C. Escher. Taschen
ESCHER M.C. The Graphic Work. Taschen
GIELIS J. Inventing the Circle. Geniaal Press, 2003
GIELIS J., HAESEN S., VERSTRAELEN L. Universal Shapes: from the super eggs of Piet Hein to the cosmic egg of George Lemaître. Kragujevac Journal of Mathematics. Vol 28.
GRAHAM G. Philosophy of The Arts. Routledge Taylor and Francis. 1997
GREEN B. The Elegant Universe. Nova, 1999
GUTH A.H. The Inflationary Universe. London Vintage Publishers, 1998
HICKS J. Piet Hein Bestrides Art and Science. Life, October 1966
HOFSTADTER D. Gödel, Escher, Bach. Vintage Books, 1980
LAUWERIER H. Symmetry. Aramith, 1988
MEYER L.B. Emotion and Meaning in Music. University of Chicago Press, 1956
MONOD J. Chance and Necessity. Random House Vintage Books, 1971
SUTTON D. Islamic Design. Wooden Books, 2007
PEAT D. Superstrings and the search for the Theory of Everything. Cardinal, 1991
VERSTRAELEN L. The Geometry of Eye and Brain. Soochow J. Math. 30, 2004
WITTEN E. The end of Science. Helix Books, 1992
WADE D. Symmetry: The ordering Principle. Wooden Books, 2006
WALSER H. , PEDERSEN J. and HILTON P. The Golden section. The Mathematical Association of America, 2001

Index

R. Verhulst, *Under the Spell of Mathematics*, https://doi.org/10.1007/978-3-031-63871-8

www.ingramcontent.com/pod-product-compliance
Lightning Source LLC
Chambersburg PA
CBHW080620060425
24660CB00010B/41
9783031638701